WITHDRAWN

D1120242

Apoptosis and Cancer

Apoptosis and Cancer

Editor *Seamus J. Martin, San Diego, Calif.*

53 figures and 5 tables, 1997

Basel · Freiburg · Paris · London · New York ·
New Delhi · Bangkok · Singapore · Tokyo · Sydney

Seamus J. Martin, Ph.D.
The National University of Ireland at Maynooth, Republic of Ireland, and La Jolla Institute for Allergy and Immunology, San Diego, Calif., USA

Library of Congress Cataloging-in-Publication Data
Apoptosis and cancer / edited by Seamus J. Martin
Includes bibliographical references and index.
1. Carcinogenesis. 2. Apoptosis. 3. Cancer cells. 4. Cancer – Growth – Regulation.
I. Martin, Seamus J., 1966–
[DNLM: 1. Neoplasms – physiopathology. 2. Apoptosis – physiology. QZ 200 A644 1997]
RC268.5.A652 1997 616.99'407 – DC21 DNLM/DLC
ISBN 3–8055–6579–8 (hard cover: alk. paper)

Drug Dosage. The authors and the publisher have exerted every effort to ensure that drug selection and dosage set forth in this text are in accord with current recommendations and practice at the time of publication. However, in view of ongoing research, changes in government regulations, and the constant flow of information relating to drug therapy and drug reactions, the reader is urged to check the package insert for each drug for any change in indications and dosage and for added warnings and precautions. This is particularly important when the recommended agent is a new and/or infrequently employed drug.

Printed in Switzerland on acid-free paper by Reinhardt Druck, Basel
ISBN 3–8055–6579–8

Contents

Preface

Understanding how apoptosis is regulated, both intra– as well as extracellularly, is undoubtedly one of the hottest areas of current research. It is now well accepted that apoptotic cell death plays a pivotal role in many fundamental biological processes. Cells die for many reasons; during the sculpting of developing tissues, as part of the ongoing renewal of dividing cell populations, because they have suffered damage, or have become infected. Apoptosis represents a cellular disposal mechanism where the dying cell is dismantled from within while simultaneously signalling its imminent demise to surrounding cells capable of aiding its disposal. Cells do not always die on cue however, and when this happens they can accumulate and cause several problems. Cells that live longer than their originally intended life span may not be life threatening per se, however, if such cells are dividing (or subsequently) acquire the capacity to do so they may present serious problems for the organism of which they are a part.

This book attempts to address the role that apoptosis plays in malignant transformation, and discusses how insights into how apoptosis is regulated at a cellular and molecular level may be exploited in the treatment of various malignancies. The first two chapters of the book are designed to bring the reader unfamiliar with the basics of apoptosis, or how apoptosis may contribute to malignancy, up to date on the broad concepts involved. In subsequent chapters we have attempted to cover most aspects of apoptosis research as they apply to the understanding and treatment of cancer. Inevitably, some very recent developments in this area may not have been covered in as much detail as we would have liked, but that is the nature of any fast moving field.

The molecular face of apoptosis is finally emerging. The past two years have seen the caspase family of proteases come to the fore as the likely effectors of death at the molecular level. Although the details are still far from clear, it seems that activation of a subset of these proteases may directly lead to the phenotypic changes characteristic of apoptosis, probably via cleavage of specific structural and regulatory proteins. As the key players of the cell death machinery fully emerge, they will undoubtedly provide attractive new targets for drug design strategies aimed at eliminating malignant cells.

Acknowledgment

I sincerely thank all of the authors for contributing to this volume and for responding to my nagging, cajoling, and pleading as deadlines passed. I thank the production staff at Landes Bioscience for their help and patience, and Doug Green and Gustavo Amarante-Mendes for many discussions. Finally, to Geraldine, Mia and Madeleine, a huge thank you for putting up with the absences, the late nights and for the smiles.

Chapter 1

Apoptosis and Cancer, edited by Seamus J. Martin.

Apoptosis and Cancer: An Overview

Seamus J. Martin

Molecular Cell Biology Laboratory, Department of Biology,
National University of Ireland, Maynooth, County Kildare, Ireland

Introduction

Until recently, cancer was thought of, in almost exclusive terms, as a disease of cell proliferation. However, developments within the cell death field over the past five years have provided a new perspective on how cell populations are normally maintained at equilibrium and have revealed how defects in cell death regulation can also contribute to the development of malignancy.

The present volume is a synthesis of many of the recent developments in this area, written by experts in their respective fields. The layout proceeds from a basic introduction to apoptosis (chapter 2) for those unfamiliar with the field, on to a discussion of the role that apoptosis and necrosis play in tumor development and progression, and the implications of this for patient prognosis and therapy (chapter 3). Chapter 3 also provides many general insights into tumor biology and introduces the topic of how apoptosis may be exploited in cancer therapy.

A theme that crops up throughout this volume—the involvement of cell cycle–related proteins, such as the cyclin–dependent kinases (CDKs), CDK inhibitors, Rb, E2–F and p53, in apoptosis—is initially introduced in chapter 4, which provides a comprehensive overview of this topic.

This is followed by a series of focused discussions on the involvement of the Bcl–2 family, Abl, p53, and Rb in the regulation of cell death generally and in the development of malignancy (chapters 5–8). Overviews of apoptosis in the context of specific neoplasms are presented in chapters 9–12. These chapters discuss the role of apoptosis and apoptosis–regulating proteins in the normal prostate (chapter 9), the developing kidney (chapter 10), the hematopoietic system (chapter 11), and in neuroblastomas (chapter 12), and detail how derangements of apoptosis in these situations can contribute to tumor development. Several of the latter chapters also highlight the contribution of apoptosis repressor proteins such as Bcl–2 and Bcr–Abl in particular malignancies.

Finally, the TNF ligand/receptor system and prospects for the exploitation of TNF and CD95–based death pathways for cancer therapy are discussed in the context of recent developments in these areas (chapter 13).

Most of the chapters are self–contained; however, an effort has been made to minimize redundancy between chapters. To enable the reader to dip into the book a little more easily, some of the basic concepts concerning the role of apoptosis in the development of cancer are summarized in this chapter.

Cell Division and Cell Death: A Delicate Balance

Why do we fail to grow throughout our lives? Presumably, at some point during our development into mature adults the cells within our tissues either receive signals to stop dividing, or the rate of cell death within dividing tissues competes on more or less equal terms with the rate of cell division (Fig. 1.1*a*). This all seems obvious enough, so it is all the more remarkable that the role of cell death in the maintenance of cell populations at equilibrium was practically ignored until recent years.

Clearly, the size of a dividing cell population can be affected by changes in the rate of cell death as well as in the rate of cell division. Too much death and involution will occur, too little, and accumulation of unwanted cells will result (Fig. 1.1*b*,*c*). The rate of cell death is dictated by many factors, both extra– as well as intracellular, which interact with the complex array of gene products (of which we will hear much more throughout the present volume) that either induce or facilitate apoptosis, or conversely, block or delay it. These inducers or repressors are thus perfect targets for mutagenic events that result in a transformed phenotype.

But is all of this merely hypothetical or do derangements involving genes that are involved in the regulation of cell death actually contribute to the transformed state? As Arends and Toft discuss in chapter 3, it is well known that net growth rates of tumors generally do not correlate well with the rate of cell division within the tumor—as assessed by the abundance of mitotic figures—suggesting that rates of cell death and/or other factors play a significant role. Throughout this volume, evidence will be discussed from a variety of in vitro and in vivo model systems to argue that defects in genes that regulate cell death do contribute in a very significant way to the development of many tumor types.

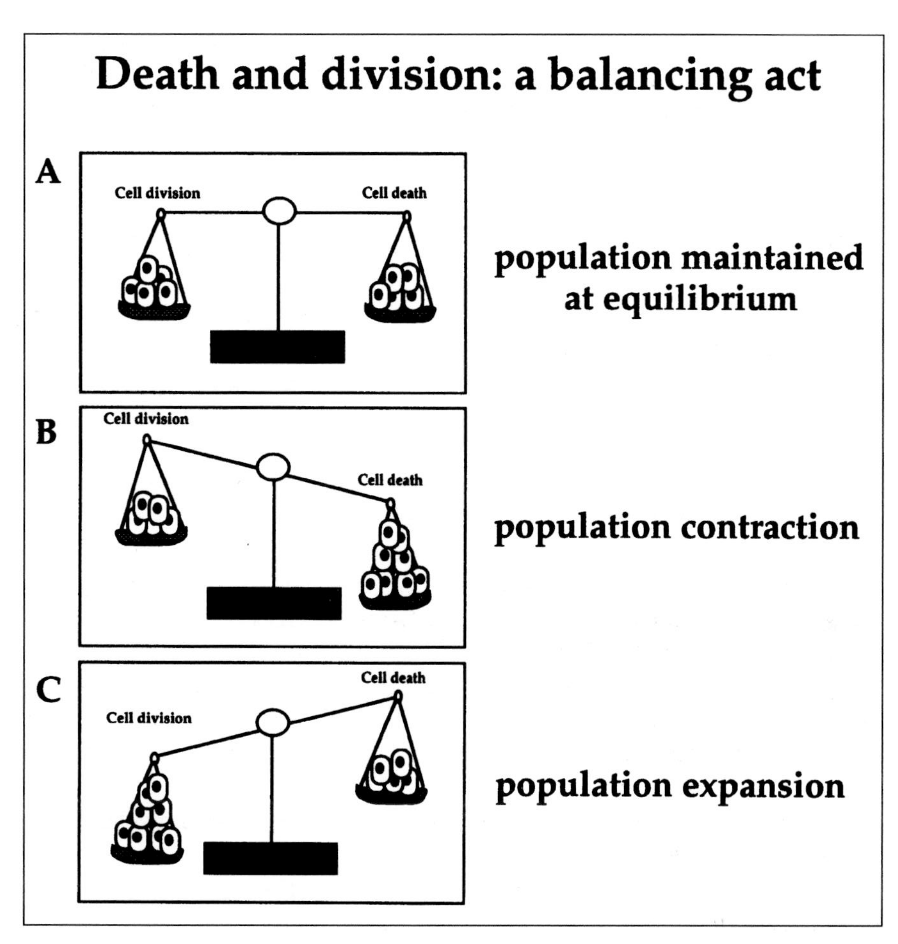

Fig. 1.1. (a) In the adult the rates of cell division and cell death are approximately equal. Perturbations to either side of this equation will result in a net population *(b)* contraction or *(c)* contraction.

Before some examples are presented, it is useful to point out that somatic mutations that result in overexpression of proteins that can block or repress apoptosis would not in themselves be expected to be transforming events since a cell possessing this type of mutation would simply live longer. However, extending the lifespan of a cell in this way increases the probability of it acquiring a second mutation. In the case where a secondary mutation affects a proliferation–related gene, such mutations would be expected to have cotransformating properties, and indeed this has been found to be the case. Alternatively, if a mutation in a cell death regulator enhances the normal lifespan of a cell type that is normally rapidly dividing, a marked population expansion would be expected to result.

When Controls on Death Fail

As outlined above, there are two main types of cell death regulatory molecules: inducers and repressors. Cell death inducers facilitate apoptosis and include, amongst others, the transcription factor Myc [1, 2], the DNA repair–associated protein p53 [3–5], certain cyclin–dependent kinases [6, 7], and a subset of the Bcl–2 family [8–11] (discussed in detail in chapter 5). Cell death repressors block or delay apoptosis and include the archetypal death repressor protein Bcl–2 [12–14], the Bcr–Abl chimeric protein found in chronic myeloid leukemia [15–17] (discussed in chapters 6 and 11), the inhibitor of apoptosis proteins (iaps) [18–20] that were originally discovered in baculovirus [21], the cowpox virus–derived serpin CrmA [22], baculovirus–derived p35 [23], adenovirus–derived E1B [24], as well as many others.

Clearly, two types of situation can result in too little cell death. A cell death repressor protein may become overexpressed or its function enhanced such that the cell acquires resistance to conditions that would have normally killed it. Alternatively, loss–of–function mutations in cell death inducer proteins may result in death signals failing to be registered within the cell, again with the result that the cell survives under conditions where it would have otherwise died.

Examples of both types of mutation are given below and this theme will dominate much of the discussion in subsequent chapters.

Overexpression of an Apoptosis–Repressor: Bcl–2

Bcl–2 was originally discovered at the breakpoint between chromosomes 14 and 18 that is characteristic of many B cell lymphomas [25]. Bcl–2 is typically expressed at constitutively low levels in many tissues. The t14;18 translocation results in a massive upregulaton in Bcl–2 expression in B cells since this translocation places the *bcl–2* gene under the control of the IgH promoter–enhancer at 14q32. Initially, *bcl–2* was thought to be a proliferation–related oncogene; however, it was subsequently established that Bcl–2 was in fact a potent repressor of many forms of apoptosis when overexpressed [12, 13].

Mice transgenic for a *bcl–2–Ig* minigene which mimics the t(14;18) rearrangement seen in follicular lymphoma initially develop a polyclonal lymphoid hyperplasia as a result of accumulation of B cells that would normally have been eliminated in these animals [26]. However, after a latency period of approximately 15 months these mice then develop clonal malignant B cell lymphomas [27], which in 50% of cases was due to secondary rearrangements of *c–myc*. Cells with elevated levels of Bcl–2 were therefore available for a second mutagenic event, involving a proliferation–associated oncogene that facilitated progression of the hyperplasia to a much more aggressive malignant phenotype. More direct evidence that *bcl–2* and *c–myc* can cooperate in lymphomagenesis was provided by experiments where mice transgenic for *c–myc* were crossed with *bcl–2* transgenic mice [28]. The resultant double transgenics developed tumors much more rapidly than either of the parental strains alone. This appears to be directly related to the ability of Bcl–2 to block the apoptosis that would normally have occurred as consequence of inappropriate Myc–induced entry into the cell cycle [29, 30].

In man, a similar derangement of Bcl–2 expression contributes to the development of follicular (non–Hodgkin's) B cell lymphomas [25] (also see chapters 5, 10 and 11 for a detailed discussion of the role of the Bcl–2 family in malignancy). This type of lymphoma is characterized by elevated levels of Bcl–2 protein, resulting in failure of B lymphocyte death, with the result that these cells accumulate. In this type of malignancy there appears to be no change in proliferation rates. Fortunately, this type of lymphoma usually responds well to chemotherapy. Elevated levels of Bcl–2 expression have also been found to correlate with the progression of prostatic cancer from a treatable androgen–dependent state to a much more aggressive androgen–independent one (see chapter 9) [31]. Bcl–2 protein levels are also elevated in many other malignancies, possibly due to loss of p53 function in these cases (p53 is a transcriptional repressor of the *bcl–2* gene [32]). Rough estimates indicate that 50% of human cancers may possess abnormally high levels of Bcl–2, implying an important role for this protein in either the development and/or the maintenance of malignancy.

Inactivation of Apoptosis–Inducing Proteins: p53

p53 is a multifunctional tumor suppressor protein that is charged with the task of monitoring the integrity of the genome [33]. Upon detection of DNA damage—by direct binding to damaged DNA—p53 arrests the cell cycle in G_1 and can directly recruit DNA repair–associated proteins by binding to these proteins [34, 35]. In addition, p53 can also act as a direct transcriptional activator of the genes encoding DNA–repair associated proteins [36]. When repair of the damage is not feasible, p53 can also activate the apoptosis machinery within the cell, once again via transcriptional–dependent as well as independent means [3,

37, 38]. For example, p53 can act as a positive transcriptional regulator of the apoptosis–inducing protein Bax [39] and can also negatively regulate the cell death repressor protein, Bcl–2, [32] (as mentioned above). However, although the ability of p53 to interfere with the Bcl–2:Bax ratio by transcriptional regulation is likely to be a significant component of p53–associated apoptosis, p53–dependent modes of apoptosis that are transcriptional–independent have also been documented [38]. In any event, cells with inactivating p53 mutations can continue to cycle under conditions where repair or apoptosis would normally have been invoked. It is no wonder then that p53 is thought to be the most widely targeted gene for mutation, with over 50% of human malignancies exhibiting a defective form of this protein [40] (see chapter 7 for a detailed review of p53 function and role in malignancy). Studies using "knockout" mice to assess the impact of wild type p53 inactivation have confirmed that this type of mutation can result in tumor development [41]. In addition, thymocytes from these animals were abnormally resistant to undergoing apoptosis when treated with DNA damaging agents in vitro [4, 5].

Implications for Cancer Therapy

Having established the significance of mutations in genes encoding apoptosis repressor and inducer proteins for the malignant phenotype, researchers are now trying to improve on existing therapeutic strategies by targeting these defects.

It is well established that most chemo– and radiotherapeutic agents eliminate cells by triggering apoptosis [42–44]. When this is not the case it usually indicates that the cells within the tumor have acquired resistance to the triggering of apoptosis. In these cases, necrosis is typically observed. Although it may seem paradoxical at first sight, the occurrence of necrosis instead of apoptosis within a tumor is usually a bad prognostic indicator since it is likely that the tumor has acquired a block in the cell death pathway and will be more difficult to treat (i.e., it will require higher doses of therapy or a more prolonged period of therapy which the patient may not be able to tolerate) [45]. Oncologists have known this for some time, but the underlying reason for this was unclear until the importance of apoptosis in the whole equation was finally appreciated.

The fact that many genes involved in the regulation of cell death are targeted for mutation during the development of malignancy makes these molecules doubly attractive as potential targets in cancer therapy. Lowering of the levels of apoptosis–repressor proteins within solid tumors may result in spontaneous death of the tumor cell due to the harsh environment that typically exists within this type of tumor. In addition, driving down the levels of apoptosis–repressor proteins would be expected to make the tumor less resistant to existing chemotherapeutics. Tumors that have suffered inactivating mutations in apoptosis inducers such as p53 are a somewhat more difficult proposition and may require restoration of p53 function using gene therapy, for example.

In either case, drugs that can manipulate apoptosis–sensitivity within malignant cells (such as Bcl–2 antagonists) are expected to provide a valuable new weapon in the armory of clinical oncologists and will hopefully significantly expand the growing list of treatable malignancies.

References

1 Evan GI, Wyllie AH, Gilbert CS et al. Induction of apoptosis in fibroblasts by c–myc protein. Cell 1992; 69:119–128.

2 Shi Y, Glynn JM, Guilbert LJ et al. Role for c–myc in activation–induced apoptotic cell death in T cell hybridomas. Science 1992; 257:212–214.

3 Yonich–Rouach E, Resnitzky D, Lotem J et al. Wild–type p53 induces apoptosis of myeloid leukemic cells that is inhibitable by interleukin–6. Nature 1991; 352:345–347.

4 Lowe SW, Schmitt EM, Smith SW et al. p53 is required for radiation–induced apoptosis in mouse thymocytes. Nature 1993; 362:847–849.

5 Clarke AR, Purdie CA, Harrison DJ et al. Thymocyte apoptosis induced by p53–dependent and independent pathways. Nature 1993; 362:849–852.

6 Shi L, Nishioka WK, Th'ng J et al. Premature p34cdc2 activation required for apoptosis. Science 1994; 263:1143–1145.

7 Fotedar R, Flatt J, Gupta S et al. Activation–induced T cell death is cell cycle dependent and regulated by cyclin B. Mol Cell Biol 1995; 15:932–942.

8 Oltvai ZN, Milliman CL, Korsmeyer SJ. Bcl–2 heterodimerizes in vivo with a conserved homolog, Bax, that accelerates programmed cell death. Cell 1993; 74:609–619.

9 Boise LH, Gonzalez–Garcia M, Postema CE et al. Bcl–x, a bcl–2–related gene that functions as a dominant regulator of apoptotic cell death. Cell 1993; 74:597–608.

10 Yang E, Zha J, Jockel J et al. Bad, a heterodimeric partner for Bcl–XL and Bcl–2, displaces Bax and promotes cell death. Cell 1995; 80:285–291.

11 Chittenden T, Harrington EA, O'Connor R et al. Induction of apoptosis by the Bcl–2 homolog Bak. Nature 1995; 374:733–736.

12 Vaux DL, Cory S, Adams JM. Bcl–2 gene promotes hemopoietic cell survival and cooperates with c–myc to immortalize pre–B cells. Nature 1988; 335:440–442.

13 Hockenbery D, Nunez G, Milliman C et al. Bcl–2 is an inner mitochondrial membrane protein that blocks programmed cell death. Nature 1990; 348:334–338.

14 Miyashita T, Reed JC. Bcl–2 oncoprotein blocks chemotherapy–induced apoptosis in a human leukemia cell line. Blood 1993; 81:151–157.

15 Evans CA, Owen–Lynch PJ, Whetton AD, Dive C. Activation of the abelson tyrosine kinase activity is associated with suppression of apoptosis in hemopoietic cells. Cancer Res 1993; 53:1735–1738.

16 McGahon A, Bissonnette R, Schmitt M et al. Bcr–Abl maintains resistance of chronic myelogenous leukemia cells to apoptotic cell death. Blood 1994; 83:1179–1187.

17 McGahon AJ, Nishioka WK, Martin SJ et al. Regulation of the Fas apoptotic cell death pathway by Abl. J Biol Chem 1995; 270:22625–22631.

18 Liston P, Roy N, Tamai K et al. Suppression of apoptosis in mammalian cells by NAIP and a related family of IAP genes. Nature 1996; 379:349–353.

19 Duckett CS, Nava VE, Gedrich RW et al. A conserved family of cellular genes related to the baculovirus iap gene and encoding apoptosis inhibitors. EMBO J 1996; 15:2685–2694.

20 Rothe M, Pan MG, Henzel WJ et al. The TNFR2–TRAF signaling complex contains two novel proteins related to baculoviral inhibitor of apoptosis proteins. Cell 1995; 83:1243–1252.

21 Crook NE, Clem RJ, Miller LK. An apoptosis–inhibiting baculovirus gene with a zinc finger–like motif. J Virol 1993; 67:2168–2174.

22 Tewari M, Dixit VM. Fas– and tumor necrosis factor–induced apoptosis is inhibited by the poxvirus crmA gene product. J Biol Chem 1995; 270:3255–3260.

23 Beidler DR, Tewari M, Friesen PD, Poirier G, Dixit VM. The baculovirus p35 protein inhibits Fas– and tumor necrosis factor–induced apoptosis. J Biol Chem 1995; 270:16526–16528.

24 White E, Sabbatini P, Debbas M et al. The 19–kilodalton adenovirus E1B transforming protein inhibits programmed cell death and prevents cytolysis by tumor necrosis factor alpha. Mol Cell Biol 1992; 12:2570–2580.

25 Tsujimoto Y, Jaffe E, Cossman J, Croce CM. Involvement of the bcl–2 gene in human follicular lymphoma. Science 1985; 228:1440–1443.

26 McDonnell TJ, Deane N, Platt FM et al. Bcl–2–immunoglobulin transgenic mice demonstrate extended B cell survival and follicular lymphoproliferation. Cell 1989; 57:79–89.

27 McDonnell TJ, Korsmeyer SJ. Progression from lymphoid hyperplasia to high grade malignant lymphoma in mice transgenic for the t(14:18). Nature 1991; 349:254–256.

28 Strasser A, Harris AW, Bath ML, Cory S. Novel primitive lymphoid tumors induced in transgenic mice by cooperation between c–myc and bcl–2. Nature 1990; 348:331–334.

29 Bissonnette RP, Echeverri F, Mahboubi A, Green DR. Apoptotic cell death induced by c–myc is inhibited by *bcl–2*. Nature 1992; 359:552–554.

30 Fanidi A, Harrington EA, Evan GI. Cooperative interaction between *c–myc* and *bcl–2* proto–oncogenes. Nature 1992; 359:554–556.

31 McDonnell TJ, Troncoso P, Brisbay SM et al. Expression of the protooncogene bcl–2 in the prostate and its association with emergence of androgen–independant prostate cancer. Cancer Res 1992; 52:6940–6944.

32 Miyashita T, Harigai M, Hanada M, Reed JC. Identification of a p53–dependent negative response element in the bcl–2 gene. Cancer Res 1994; 54:3131–3135.

33 Lane D. Cancer. p53, guardian of the genome. Nature 1992; 358:15–16.

34 Wang XW, Yeh H, Schaeffer L et al. p53 modulation of TFIIH–associated nucleotide excision repair activity. Nat Genet 1995; 10:188–195.

35 Leveillard T, Andera L, Bissonnette N et al. Functional interactions between p53 and the TFIIH complex are affected by tumor–associated mutations. EMBO J 1996; 15:1615–1624.

36 Shivakumar CV, Brown DR, Deb S, Deb SP. Wild–type human p53 transactivates the human proliferating cell nuclear antigen promoter. Mol Cell Biol 1995; 15:6785–6793.

37 Yonish–Rouach E, Resnitzky D, Lotem J et al. Wild–type p53 induces apoptosis of myeloid leukemic cells that is inhibited by interleukin–6. Nature 1991; 352:345–347.

38 Caelles C, Helmberg A, Karin M. p53–dependent apoptosis in the absence of transcriptional activation of p53–target genes. Nature 1994; 370:220–223.

39 Miyashita T, Reed JC. Tumor suppressor p53 is a direct transcriptional activator of the human bax gene. Cell 1995; 80:293–299.

40 Rotter V, Foord O, Navot N. In search of the functions of the normal p53 protein. Trends in Cell Biol 1993; 3:46–49.

41 Donehower LA, Harvey M, Slagle BL et al. Mice deficient for p53 are developmentally normal but susceptible to spontaneous tumors. Nature 1992; 356:215–221.

42 Searle J, Lawson TA, Abbot PJ. An electron–microscope study of the mode of cell death induced by cancer–chemotherapeutic agents in populations of proliferating normal and neoplastic cells. J Pathol 1975; 116:129–139.

43 Dive C, Hickman JA. Drug–target interactions: only the first step in the commitment to a programmed cell death? Br J Cancer 1991; 64:192–196.

44 Lennon SV, Martin SJ, Cotter TG. Dose–dependent induction of apoptosis in human tumor cell lines by widely diverging stimuli. Cell Prolif 1991; 24:203–214.

45 Arends MJ, McGregor AH, Wyllie AH. Apoptosis is inversely related to necrosis and determines net growth in tumors bearing constitutively expressed myc, ras and HPV oncogenes. J Pathol 1994; 144:1045–1057.

Chapter 2

Apoptosis and Cancer, edited by Seamus J. Martin.

Apoptosis: An Introduction

Seamus J. Martin

Molecular Cell Biology Laboratory, Department of Biology,
National University of Ireland, Maynooth, County Kildare, Ireland

Introduction

This chapter introduces some basic concepts about apoptosis and broadly outlines our current understanding about how this cell death program works at a molecular level. This section has been written primarily for those new to the apoptosis field and thus may be bypassed if the reader is already familiar with the basics of apoptosis and current development in this area.

Death Gets a Breath of Life

By now, it is something of a cliché to say that the apoptosis field is currently one of the hottest among the biomedical sciences, but it's true. This remarkable wave of interest in how cells die goes some way towards compensating for the previous disinterest shown in this topic by the majority of scientists. In retrospect, their disinterest is perfectly understandable—after all, we are used to studying processes geared towards sustaining life. Generations of cell biologists experimented with ways of keeping cells alive in culture—devising complicated mixtures of salts, amino acids and serum supplements. Killing cells was easy, that usually meant that your culture conditions were less than optimal. Who wanted to study dying cells, except for pathologists perhaps? Prior to the recent upsurgence of interest in apoptosis, the prevailing attitude towards cell death was that it was a disorganized and chaotic process, largely involving events beyond the cells control, which resulted in a nasty mess that was invariably called necrosis. The possibility that cell death could be a precisely controlled process, just as cell division was thought to be, was discounted by the majority of scientists. However, a few dedicated pioneers had long since recognized that certain instances of cell death were special. As early as the late 1940s Glücksmann had recognized that cell death played an integral part of embryogenesis and understood that cell death in this context was both normal as well as desirable [1]. Several earlier studies, in the late 1800s, had also described cell deaths that were different to the typical pattern that had been classified as necrosis, but the general significance of these findings was not appreciated (see ref. 2 for an overview of this topic).

The studies of Kerr and colleagues in the mid 1960s and the early 1970s culminated in what has become a landmark paper [3] in which the term apoptosis was proposed for those cell deaths that shared a distinctive set of morphological characteristics. In their seminal work [3], Kerr, Wyllie and Currie painted a remarkably comprehensive picture of apoptosis and emphasized the widespread biological significance of this "new" mode of cell death. Although biologists still did not rush to study this phenomenon, Wyllie and Kerr, working independently, continued to add to their initial observations [4–7], until finally, in the mid 1980s others began to take interest. It was largely immunologists, intent on understanding cytotoxic lymphocyte–mediated killing [8–10] and thymic selection [11–13], who were the ones to finally appreciate the wider significance of Kerr et al [3] and helped bring this topic to the attention of the wider scientific community.

The Process

Apoptotic Cells: Fast Food for Phagocytes

In essence, apoptosis is a process whereby a cell, upon receipt of an appropriate stimulus (see next section), engages a set of molecules that cooperate to dismantle the cell from within. This process culminates in the separation of the cell body into many intact pieces—called *apoptotic bodies*—that continue to maintain membrane integrity for several hours after they have been formed [3]. Alterations to the composition of the plasma membrane of the apoptotic cell attracts the attention of nearby phagocytes [14–19]. These phagocytes then eat the dying cell—and this is a critical point—before it has had a chance to burst open and release its contents into the extracellular space (Fig. 2.1).

Thus, apoptosis can be viewed as a kind of packaging process that facilitates the disposal of unwanted cells by breaking them up into "bite–sized" pieces, complete with an attached label which reads "I'm dying, come and get me!" The membrane changes that trigger macrophages and other cells with phagocytic capability to eat apoptotic cells remain incompletely understood; however, two

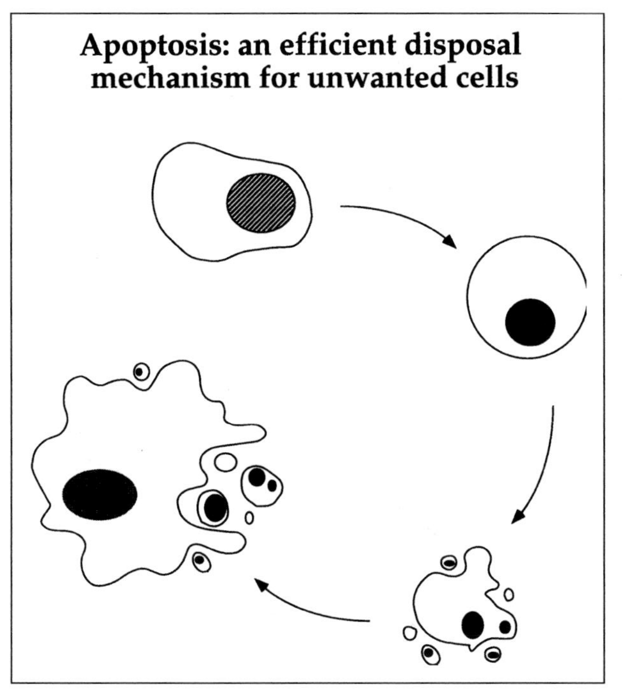

Fig. 2.1. Schematic representation of apoptosis. During apoptosis, cells initially round up and detach from their neighbors if they are part of a tissue, this is followed by condensation of the chromatin, dissolution of the nuclear envelope and separation of the nucleus into distinct fragments, followed by collapse of the cell into several small intact vesicles (apoptotic bodies). Apoptotic cells and apoptotic bodies are recognized by phagocytes as being in the throes of death and are eaten before their cytoplasmic contents escape.

main changes have been consistently observed. Savill, Haslett and colleagues have identified a mechanism that utilizes thrombospondin, in concert with CD36 and the vitronectin receptor, to recognize a thrombospondin–binding moiety that appears on apoptotic cells from many lineages [15, 16], as illustrated in Figure 2.2. The nature of the thrombospondin–binding moiety is, thus far, unknown. Another subset of phagocytes appear to recognize apoptotic cells due to the appearance of phosphatidylserine—a lipid normally confined to the inner leaflet of the plasma membrane—on the external leaflet of the plasma membrane of these cells [17–19], (see Figs. 2.2 and 2.3).

Part of the cellular dismantling process involves the destruction of the normal nuclear architecture and cleavage of the chromatin [3, 4,6, 9, 20]. It is these changes that result in the striking phenotypic alterations that were originally recognized by Kerr and colleagues [3] to be characteristic of apoptotic cells (Fig. 2.4). Collapse of the nucleus is thought to be due to destabilization of the

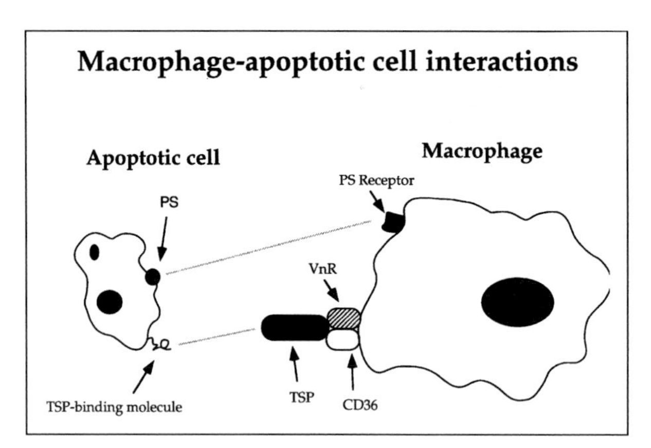

Fig. 2.2. This scheme depicts two of the membrane alterations that are thought to stimulate recognition of apoptotic cells by phagocytes. PS, phosphatidylserine; TSP, thrombospondin; VnR, vitronectin receptor.

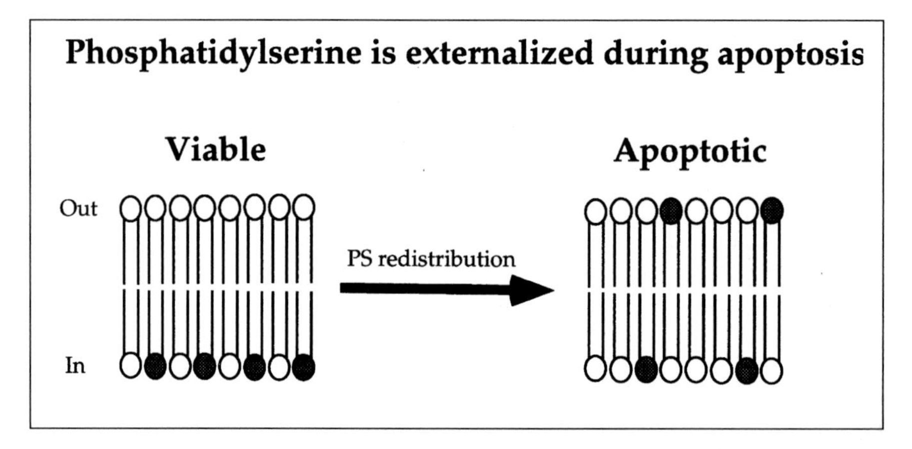

Fig. 2.3. Phosphatidylserine (PS) is normally almost totally confined to the inner leaflet of the plasma membrane on a healthy cell. During apoptosis, PS is selectively exported to the external leaflet of the plasma membrane, possibly due to activation of an inside–outside PS translocase or an aminophospholipid scramblase. This redistribution is rapid, occurs well before any increase in plasma membrane permeability and can be conveniently measured by means of a PS–binding protein—annexin V.

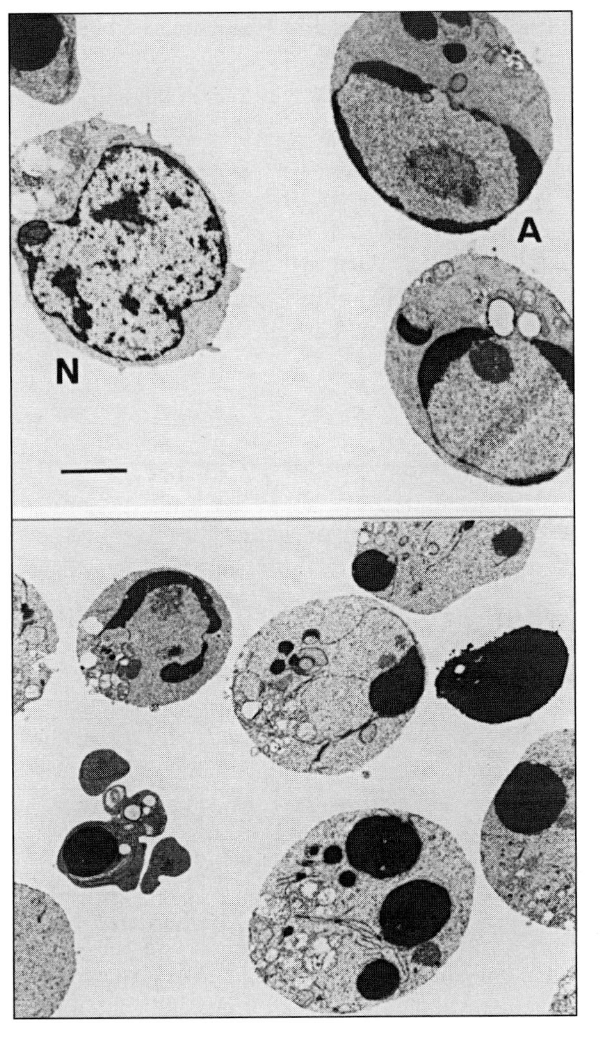

Fig. 2.4. Morphological features of apoptosis. Electron micrographs of normal (N) versus apoptotic cells (A) illustrating the nuclear condensation and fragmentation that are highly characteristic of this mode of cell death. The lower panel contains cells that are in various stages of apoptosis. Micrographs were kindly provided by Dr. Ruth M. Kluck.

nuclear envelope as a consequence of lamin proteolysis [the protease(s) that may mediate this event are discussed in the section on "The Machinery of Death" and will crop up throughout the present volume]. This results in the loss of matrix attachment regions—points at which the chromatin is attached to the nuclear envelope—causing the chromatin to compact [21]. Cleavage of the nuclear DNA occurs predominately at sites between nucleosomes, resulting in a fragmentation pattern of integer multiples of ~200 bp [4]. Larger fragments of 50–250 Kbp have also been widely reported and typically appear before the smaller fragments [20]. The nature of the enzyme(s) that cleave DNA during apoptosis are the subject of ongoing debate and no consensus has emerged as yet.

How Do Apoptosis and Necrosis Differ?

Broadly speaking, apoptosis is the mode of cell death that is observed where death is a normal part of the lifecycle of the organism and for this reason has been called "the physiological mode of cell death." This is not strictly true, since apoptosis is frequently observed when cells are dying due to a pathological process—during a viral infection for example [22, 23]. However, as a general rule it is true to say that apoptosis is typically observed during normal or physiological processes as well as during mildly pathological ones. Because apoptotic cells are eaten before they can release their contents into the extracellular space, damage to neighboring cells is largely curtailed—this is a critical difference between apoptosis and necrosis [3, 7, 24].

Necrotic cell death is generally the result of a gross departure from physiological conditions where the cell suffers a major insult [24–26]. Damage is generally so severe that the cell loses its ability to maintain membrane integrity within a matter of minutes, with the result that it undergoes rapid swelling (due to the influx of water) and bursts open, thus releasing its contents. Cells contain many enzymes (lipases, proteases, nucleases) and byproducts of metabolism (such as reactive oxygen species) that can cause severe injury, and even death, if these are simply released into the extracellular space. Thus, a necrotic cell often ends up damaging many neighboring cells as a consequence of its own demise. This sets up a chain reaction of more cell deaths, followed by still more deaths as the wave of necrosis radiates out from the initial site of damage. To make matters worse, all of these dead and dying cells soon attract the attention of neutrophils and inflammatory macrophages, that rush in expecting to find a bacterial infection or some other invasion. Confronted by this unsightly mess they then start to add to the general mayhem by releasing their own toxic enzymes (normally used for killing bacteria and the like), thereby adding to the carnage. This scenario reveals why necrosis was for a very long time considered to be the predominant mode of cell death—it was difficult to miss! By contrast, apoptosis can occur at high frequency within tissues and go practically unnoticed due to the efficiency of clearance of the dead cell corpses.

In summary, having a more controlled mode of cell death that prevents the escape of cellular contents, the body avoids not only the injury and death of cells surrounding the cell destined to die, but also the further cell death and wasted effort that results from mounting an inflammatory reaction.

The Stimuli

Some Cells Choose to Die, Others are Persuaded

So why do cells die? For a multitude of reasons. Because they are no longer required, have become damaged or infected, or because they need to be replaced because they have reached the end of their useful life. Some cells (such as neurons) can live throughout the whole life of an organism, while others (such as neutrophils) only live for a few hours before they are replaced. At present, we cannot fully explain why this is. We can draw some conclusions from the fact that cycling cells tend to have much shorter lives than noncycling cells, but we still do not know, in molecular terms, the reason for this.

Cells die during many physiological processes. It is a recurring theme during developmental processes that complex structures are shaped from much cruder masses of tissue by a kind of sculpting process (for example, see chapter 10 for an overview of development in the kidney). In this context, the sculpting is achieved by hormonal cues that provoke whole tracts of cells in strategic positions to die in a coordinated manner. The classical example of this is the development of the hand from a crude limb bud due to death of the cells in the interdigital spaces. In a less obvious but nonetheless very similar manner, the mature immune system arises by massively overproducing lymphocytes and then eliminating those that would interfere with the proper functioning of this system in the adult.

In the context of the immune system, it is relevant to mention that many immune reactions are simply sophisticated ways of killing cells that are not wanted—because they are foreign, are infected, are dysfunctional or are excess to requirements. Current evidence suggests that numerous cell deaths within the immune system are initiated due to encounter of susceptible CD95– (also called Fas or APO–1) bearing cells encountering other cells expressing CD95 ligand [27]. In some cases, it appears that both CD95 as well as its ligand can even be coexpressed on the same cell, provoking death of the cell even where there may be no other CD95 ligand–bearing cells in the vicinity [27, 28]. CD95–mediated cell deaths have been implicated in cytotoxic T cell as well as natural killer cell–mediated apoptosis, elimination of autoreactive mature T cells, elimination of T and B cells during the resolution of immune responses, the maintenance of immune privilege in delicate tissues such as the eye, as well as in numerous other circumstances (ref. 27 provides a comprehensive overview of this topic).

Repair or Die

It is also part of the normal scheme of things that cells will suffer injury during the course of their life. Accumulating evidence suggests that there are an array of cellular "sensors" that can register damage within the cell and can either recruit the necessary repair enzymes, or, if the injury is severe, can initiate apoptosis (Fig. 2.5). Few of these intracellular sensors have been identified thus far; however, it is likely that there are molecules that play an important role in detecting damage within the cytoplasm and at the plasma membrane, just as p53 appears to do within the nucleus (see chapters 4, 6, 8 and 11). The importance of damage–sensory and repair molecules for normal cellular function is highlighted by the fact that these molecules are frequent targets for mutation during the progression to malignancy (see chapters 7 and 8).

The Machinery of Death

Peeking into the Black Box

So how, in molecular terms, does apoptosis work? Less than three years ago it is probably true to say that we really didn't have a clue. Some of the pieces of the puzzle have since fallen into place and a mechanism is now beginning to emerge. There are still many missing components to find however. The current evidence suggests that the core of the cell death machinery is comprised of a family of proteases called caspases [29–31] (previously called the ICE/CED–3 family proteases). Caspases were originally discovered due to their homology with a gene (*ced–3*) that was found to be required for all of the developmental–related programmed cell deaths in *C. elegans* [32]. A number of caspases (**c**ysteine **asp**artate–specific prote**ases**) have now been identified [31], (see Fig. 2.6) but it is still unclear whether they all participate in apoptosis or whether different subsets of these proteases participate, depending upon the specific pro–apoptotic stimulus in question. Current evidence suggests that the subset con-

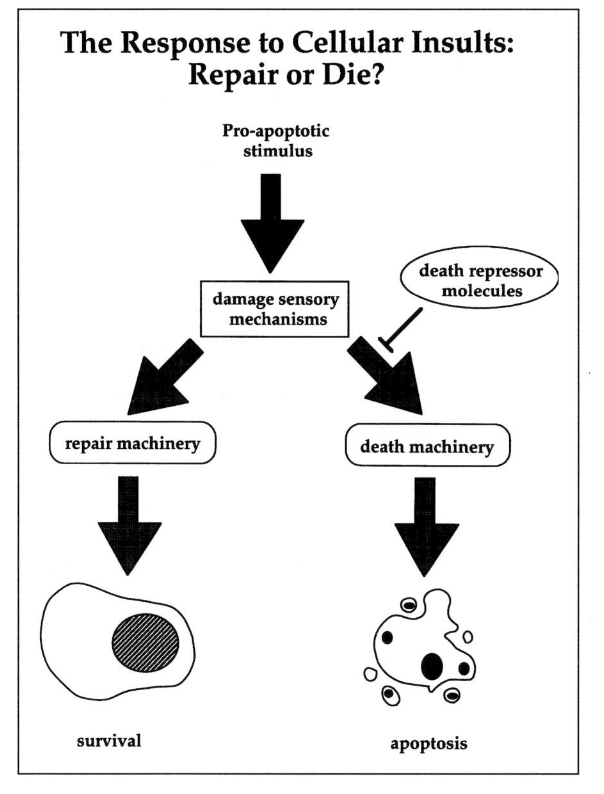

Fig. 2.5. Upon receipt of cellular damage, cells respond by either repairing the damage or activating their cell death machinery. The likelihood of either response can be modulated by the level of death repressor proteins within the cell.

taining the caspases most closely related to Ced–3 (caspases–3, 6, 7, 8 and 10) are likely to be the ones that are centrally involved in apoptosis.

What is clear, is that these proteases are normally present in healthy cells in inactive configurations (zymogens) that require proteolytic processing to achieve their active forms. Once active, they are responsible for cleaving other caspases—thus amplifying and propagating the protease cascade—as well as many substrate proteins within the cell (Fig. 2.7) [29, 30, 33–35]. It is probably the latter cleavage events that result in the profound structural changes typical of apoptosis. Precisely which of the caspases are responsible for the critical cleavage events have yet to be determined. Caspases cleave proteins that contain certain key amino acid sequences (e.g., YVAD, DEVD, DEAD). Each caspase has its own preference for a particular amino acid sequence, but only some of these have

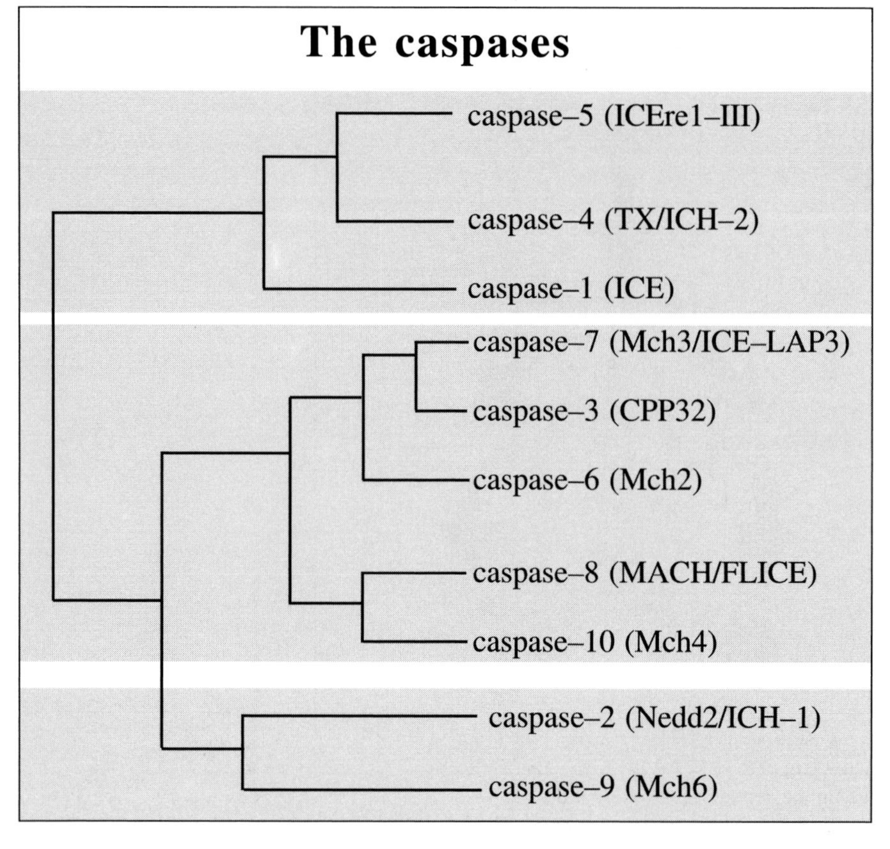

Fig. 2.6. The human caspase family of cysteine proteases, indicating the phylogenetic relationships between members of this family. Previous names (due to lack of space no more than two prior names are listed for each protease) are given in brackets. The middle shaded area contains the caspases most closely related to CED–3. This figure was adapted from Alnmeri et al, 1996, with modifications [31].

been identified to date. The consistent feature of each cleavage site is an Asp residue in the P1 position and a small amino acid in the P2 position.

An ever–growing number of proteins have been found to undergo proteolysis at caspase consensus sites during apoptosis. The consequences of these cleavage events are still the subject of speculation at present but it is likely that several of these directly result in some of the gross phenotypic changes (such as plasma membrane blebbing and nuclear envelope collapse) that take place during apoptosis. As a cautionary note, however, it is very likely that many of these cleav-

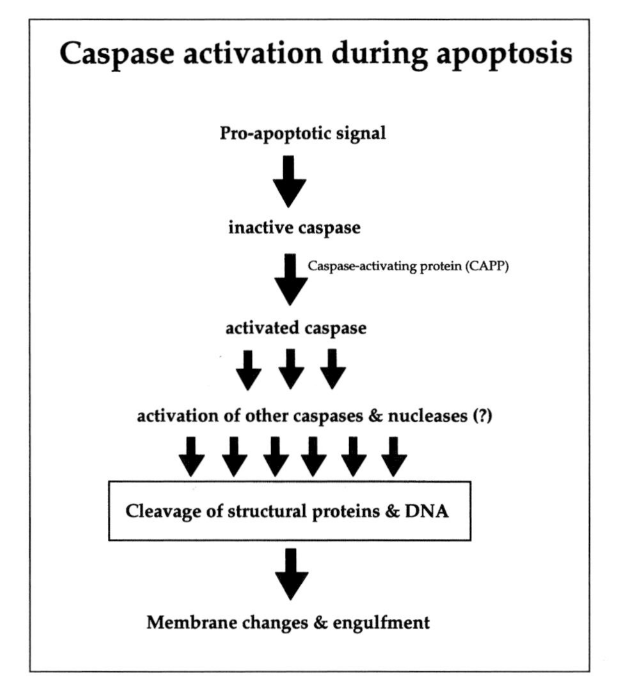

Fig. 2.7. Simplified scheme for caspase participation in apoptosis. In this scheme, the most proximal caspase activated in the chain undergoes processing via a putative caspase–activating protein (CAAP). The CAAP could be another caspase that is constitutively active within the cell (at a low level) and whose activity is regulated by means of a specific inhibitor or by phosphorylation/dephosphorylation. Alternatively, the CAAP could be the proximal caspase itself that can mediate autoactivation (in cis or trans) possibly due to receptor–mediated aggregation (as appears to be the case for caspase–8). Once the most proximal caspase in the chain becomes fully active, the death signal is depicted as propagating in a linear fashion with amplification of the response as more caspases are recruited (due to the proximal caspase processing more distal ones in the pathway). Alternatively, several caspases could become activated in parallel with an explosive activation of several caspases occurring simultaneously. During all of this, the caspases are also cleaving substrate proteins that are required for cell viability, DNA repair, maintenance of cell structure and regulation of membrane asymmetry. The consequence of these cleavage events is the cellular collapse that we recognize as apoptosis.

age events are purely coincidental. The challenge will be to identify the significant proteolytic events and to assess their role in the programmed cell death process.

Current studies in this area are aimed at understanding how the sensors of cellular damage, such as p53, activate the caspases when repair is no longer an option. Recent studies on the CD95 (Fas/Apo–1) molecule suggest that the pathway from so–called "death–receptor" molecules to the death machinery may be a surprisingly short one [36, 37].

CD95–Mediated Caspase Activation: A Short Trip

The CD95 molecule—a cell surface glycoprotein that is expressed on a variety of cells, particularly within the immune system—is capable of transducing a death signal into cells bearing this receptor upon trimerization via its specific ligand [38, 39]. CD95 ligand is much more restricted in its distribution, but can be upregulated on lymphocytes under appropriate conditions and is constitutively present on certain tissues [27]. The nature of the CD95–associated death signal has been the subject of intense investigation and was somewhat of an enigma since the cytoplasmic tail of the CD95 molecule had no obvious signaling motifs. By mutation analysis, a region within the cytoplasmic tail of CD95 (called the death domain) was identified that was necessary for the propagation of the death signal, but it transpired that this domain was simply a protein:protein interaction domain that enabled CD95 to recruit another death–domain containing protein called FADD (or MORT1) to the complex [40, 41]. Overexpression of FADD/MORT1 was found to be sufficient to trigger apoptosis but this property was found to lie within a region outside of the FADD death domain and was duly called the death–effector domain (DED) [42]. That is where things stood until recently when it was discovered that a member of the caspase family (caspase–8 or MACH/FLICE), contains two regions that share homology with the death–effector domain of FADD/MORT1 that enabled recruitment of this protease to the complex [36, 37]. Upon recruitment into the complex, caspase–8 then presumably undergoes activation, possibly due to adjacent caspases–8 molecules acting in trans. In the cellular context, this whole process takes place within seconds of ligation of the CD95 molecule [43].

The step distal to caspase–8 activation in the CD95–initiated death cascade has yet to be elucidated, but is likely to involve caspase–8–mediated activation of other caspase family members, thereby propagating the death signal.

Mitochondria Re–enter the Picture

Recent developments have implicated mitochondria as potential key participants in the propagation of death signals [44]. Although mitochondria do not undergo any obvious morphologic changes during apoptosis [3], a cessation of oxidative phosphorylation seems to occur due to loss of the pH and ionic gradient that exists between the two sides of the inner mitochondrial membrane. This loss in transmembrane potential has been hypothesized to be due to the opening of megachannels in the mitochondria—a process which has been called "permeability transition," (see ref. 44 for a comprehensive overview of this topic). The nature of the megachannel pore is still unknown. However, it does appear that concomitant with the opening of megachannels in the mitochondria, one or more mitochondrial proteins are released into the cytoplasm where they can trigger activation of several of the caspases [45–48] (Martin et al, submitted). Surprising as it may seem, cytochrome c is one of the mitochondrial components that exhibit caspase–activating properties [45, 47], and there are likely to be others [46]. Even more surprisingly, the role of Bcl–2 and related death–repressor molecules may be to abrogate or delay the release of such caspases–activating proteins [46–48], possibly by disrupting the transport mechanism.

This area is sure to be one of intensive investigation in the near future as it seems that release of caspase–activating mitochondrial factor(s) may be a general mechanism for activation of the caspases during many instances of apoptosis [44]. Death receptor molecules such as CD95 that integrate with the caspases directly via their cytoplasmic tails may either bypass the requirement for a mitochondrial–derived factor or may channel subsequent signaling events through the mitochondria.

The Controls

To Die or Not To Die, That is the Question

Clearly, not all cells die in response to the same level of injury. Some cells appear to be more resistant to death than others. Some of this variation is undoubtedly related to the relative availability of proteins within a given cell type that can either facilitate or block entry into apoptosis.

There are several families of proteins that are known for their ability to modulate apoptosis in a profound way, perhaps the most well known of which is the Bcl–2 family (see chapter 5 for a comprehensive overview of this area). The Bcl–2 family is comprised of an ever–expanding array of molecules that can either repress or facilitate apoptosis—how they achieve this is still under investigation but recent advances in this area suggest that we may be getting close to understanding the mechanism (see chapter 5). Other apoptosis repressor proteins in man include the protein product of the *bcr–abl* translocation observed in chronic myeloid leukemia, and the more recently discovered **i**nhibitors of **a**poptosis **p**roteins (iaps) [49–51]. Apoptosis repressor proteins have also been discovered in many other organisms, particularly viruses.

Proteins that can promote apoptosis include members of the Bcl–2 family (such as Bax and Bik), the transcription factor Myc, the DNA–repair–associated protein p53, as well as several others.

It is still unclear how all of these modulators of apoptosis are hooked up to the cell death pathway but it seems likely that they either promote or retard activation of the caspases under appropriate conditions. It will be interesting to see just how.

The repressors and modulators of apoptosis mentioned above play a very significant role in the remainder of this volume as they are, in many cases, the proteins that become either mutated or dysregulated during malignancy. Thus, understanding the way these proteins modulate apoptosis will provide critical insights for the treatment of many cancers.

Conclusion

Apoptosis research has come a long way in a relatively short time. Many of the key participants in this process have been identified but how they interact with each other is still far from clear. One thing is clear, if the secrets of the cell death machinery continue to be revealed at the present rate, the prospects of exploiting this knowledge in disease situations where cell death controls have gone awry—such as cancer—are excellent.

References

1 Glücksmann A. Cell deaths in normal vertebrate ontogeny. Biol Rev 1951; 26:59–86.
2 Gupta S. Apoptosis/programmed cell death: a historical perspective. Adv Exp Med Biol 1996; 406: 1–9.
3 Kerr JFR, Wyllie AH, Currie AR. Apoptosis: a basic biological phenomenon with wide ranging implications in tissue kinetics. Brit J Can 1972; 26:239–257.
4 Wyllie AH. Glucocorticoid–induced thymocyte apoptosis is associated with endogenous endonuclease activation. Nature 1980; 284:555–556.
5 Kerr JFR, Harmon B, Searle J. An electron–microscope study of cell deletion in the anuran tadpole tail during spontaneous metamorphosis with special reference to apoptosis of striated muscle fibres. J Cell Sci 1974; 14:571–585.
6 Wyllie AH, Morris RG, Smith AL, Dunlop D. Chromatin cleavage in apoptosis: association with condensed chromatin morphology and dependence on macromolecular synthesis. J Pathol 1984; 142:67–77.
7 Searle J, Kerr JFR, Bishop CJ. Necrosis and apoptosis: distinct modes of cell death with fundamentally different significance. Pathol Ann 1982; 17:229–259.

8 Russell JH, Masakowski V, Rucinsky T, Phllips G. Mechanisms of immune lysis III. Characterization of the nature and kinetics of the cytotoxic T lymphocyte–induced nuclear lesion in the target. J Immunol 1982; 128:2087–.

9 Duke RC, Chervenak R, Cohen JJ. Endogenous endonuclease–induced DNA fragmentation; an early event in cell–mediated cytolysis. Proc Natl Acad Sci USA 1983; 80:6361–6365.

10 Ucker DS. Cytotoxic T lymphocytes and glucocorticoids activate an endogenous suicide process in target cells. Nature 1987; 327:62–64.

11 Smith CA, Williams GT, Kingston R, Jenkinson EJ, Owen JT. Antibodies to CD3/T cell receptor complex induce death by apoptosis in immature T cells in thymic cultures. Nature 1989; 337:181–183.

12 Shi Y, Sahai BM, Green DR. Cyclosporin A inhibits activation–induced cell death in T cell hybridomas and thymocytes. Nature 1989; 339:625–626.

13 Shi Y, Szalay MG, Paskar L, Boyer M, Singh B, Green DR. Activation–induced cell death in T cell hybridomas is due to apoptosis. J Immunol 1990; 144:3326–3333.

14 Duvall E, Wyllie AH, Morris RG. Macrophage recognition of cells undergoing programmed cell death. Immunology 1985; 56:351–358.

15 Savill JS, Dransfield I, Hogg N, Haslett C. Vitronectin receptor–mediated phagocytosis of cells undergoing apoptosis. Nature 1990; 343:170–173.

16 Savill JS, Fadok V, Henson P, Haslett C. Phagocyte recognition of cells undergoing apoptosis. Immunol Today 1993; 14:131–136.

17 Fadok VA,Voelker DR, Campbell PA et al. Exposure of phosphatidylserine on the surface of apoptotic lymphocytes triggers specific recognition and removal by macrophages. J Immunol 1992; 148:2207–2216.

18 Fadok VA, Savill JS, Haslett C et al. Different populations of macrophages use either the vitronectin receptor or the phosphatidylserine receptor to recognize and remove apoptotic cells. J Immunol 1992; 149:4029–4035.

19 Martin SJ, Reutelingsperger CPM, McGahon AJ et al. Redistribution of plasma membrane phospholipids is an early and widespread event during apoptosis: inhibition by overexpression of Bcl–2 and Abl. J Exp Med 1995; 182:1545–1556.

20 Brown DG, Sun X–M, Cohen GM. Dexamethasone–induced apoptosis involves cleavage of DNA to large fragments prior to internucleosomal fragmentation. J Biol Chem 1993; 268:3037–3039.

21 Neamati N, Fernandez A, Wright S et al. Degradation of lamin B1 precedes oligonucleosomal DNA fragmentation in apoptotic thymocytes and isolated thymocyte nuclei. J Immunol 1995; 154:3788–3795.

22 Terai C, Kornblut RS, Pauza CD et al. Apoptosis as a mechanism of cell death in cultured T lymphoblasts acutely infected with HIV–1. J Clin Invest 1991; 87:1710–1715.

23 Martin SJ, Matear PM, Vyakarnam A. HIV–1 infection of human CD4+ T cells in vitro: Differential induction of apoptosis in these cells. J Immunol 1994; 152:330–342.

24 Wyllie AH. Cell death: a new classification separating apoptosis from necrosis. In: Bowen ID, Lockshin RA, eds. Cell Death in Biology and Pathology. London and New York: Chapman and Hall, 1981:9–34.

25 Trump BF, Berezesky IK, Osornio–Vargas AR. Cell death and the disease process: the role of calcium. In: Bowen ID, Lockshin RA, eds. Cell Death in Biology and Pathology. London and New York: Chapman and Hall. 1981:209–242.

26 Lennon SV, Martin SJ, Cotter TG. Dose–dependent induction of apoptosis in human tumor cell lines by widely diverging stimuli. Cell Prolif 1991; 24:203–214.

27 Nagata S, Goldstein P. The Fas death factor. Science 1995; 267:1449–1455.

28 Brunner T, Mogil RJ, La Face D et al. Cell autonomous Fas (CD95)/Fas–ligand interaction involved in apoptosis of activated T cells. Nature 1995; 373:441–444.

29 Martin SJ, Green DR. Protease activation during apoptosis: death by a thousand cuts? Cell 1995; 82:349–352.

30 Henkart P. ICE family proteases: mediators of all apoptotic cell death. Immunity 1996; 4:195–201.

31 Alnemri ES, Livingston DJ, Nicholson DW et al. Human ICE?CED–3 protease nomenclature. Cell 1996; 87:171.

32 Ellis HM, Horvitz HR. Genetic control of programmed cell death in the nematode *C. elegans*. Cell 1986; 44:817–829.

33 Lazebnik YA, Kaufmann SH, Desnoyers S et al. Cleavage of poly(ADP–ribose) polymerase by a proteinase with properties like ICE. Nature 1994; 371:346–347.

34 Martin SJ, Amarante–Mendes GP, Shi L et al. The cytotoxic cell protease granzyme B initiates apoptosis in a cell–free system by proteolytic processing and activation of the ICE/CED–3 family protease, CPP32, via a novel two–step mechanism. EMBO J 1996; 15:2407–2416.

35 Casiano CA, Martin SJ, Green DR, Tan EM. Proteolysis of a subset of nuclear autoantigens during CD95 (Fas/APO–1)–mediated T cell apoptosis. J Exp Med 1996; 184:765–770.

36 Boldin MP, Goncharov TM, Goltsev YV, Wallach D. Involvement of MACH, a novel MORT1/FADD–interacting protease, in Fas/APO–1 and TNF receptor–induced cell death. Cell 1996; 85:803–815.

37 Muzio M, Chinnaiyan AM, Kischkel F et al. FLICE, a novel FADD–homologous ICE/CED–3–like protease, is recruited to the CD95 (Fas/APO–1) death–inducing signaling complex. Cell 1996; 85:817–827.

38 Yonehara S, Ishii A, Yonehara M. A cell–killing monoclonal antibody (anti–Fas) to a cell surface antigen co–downregulated with the receptor of tumor necrosis factor. J Exp Med 1989; 169:1747–1756.

39 Trauth BC, Klas C, Peters AM et al. Monoclonal antibody–mediated tumor regression by induction of apoptosis. Science 1989; 245:301–305.

40 Chinnaiyan AM, O'Rourke K, Tewari M, Dixit VM. FADD, a novel death domain–containing protein, interacts with the death domain of Fas and initiates apoptosis. Cell 1995; 81:505–512.

41 Boldin MP, Varfolomeev EE, Pancer Z et al. A novel protein that interacts with the death domain of Fas/APO1 contains a sequence motif related to the death domain. J Biol Chem 1995; 270:7795–7798.

42 Chinnaiyan AM, Tepper CG, Seldin MF et al. FADD/MORT1 is a common mediator of CD95 (Fas/Apo–1) and tumor necrosis factor receptor–induced apoptosis. J Biol Chem 1996; 271:4961–4965.

43 Kischkel FC, Hellbardt S, Behrmann I et al. Cytotoxicity–dependent APO–1 (Fas/CD95)–associated proteins form a death–inducing signaling complex (DISC) with the receptor. EMBO J 1995; 14:5579–5588.

44 Kroemer G, Zamzami N, Susin SA. Mitochondrial control of apoptosis. Immunol Today 1997; 18:44–51.

45 Liu X, Kim CN, Yang J, Jemmerson R, Wang X. Induction of apoptotic program in cell–free extracts: requirement for dATP and cytochrome c. Cell 1996; 86:147–157.

46 Susin SA, Zamzami N, Castedo M et al. Bcl–2 inhibits the release of an apoptogenic protease. J Exp Med 1996; 184:1331–1341.

47 Kluck RM, Bossy–Wetzel E, Green DR, Newmeyer DN. The release of cytochrome c from mitochondria: a primary site for Bcl–2 regulation of apoptosis. Science 1997; 275:1132–1136.

48 Yang J, Liu X, Bhalla K et al. Prevention of Apoptosis by Bcl–2: Release of Cytochrome c from Mitochondria blocked. Science 1997; 275:1129–1132.

49 Crook NE, Clem RJ, Miller LK. An apoptosis–inhibiting baculovirus gene with a zinc finger–like motif. J Virol 1993; 67:2168–2174.

50 Liston P, Roy N, Tamai K et al. Suppression of apoptosis in mammalian cells by NAIP and a related family of *IAP* genes. Nature 1996; 379:349–353.

51 Rothe M, Pan MG, Henzel WJ et al. The TNFR2–TRAF signaling complex contains two novel proteins related to baculoviral inhibitor of apoptosis proteins. Cell 1995; 83:1243–1252.

Chapter 3

Apoptosis and Cancer, edited by Seamus J. Martin.

Apoptosis and Necrosis in Tumors

Neil J. Toft and Mark J. Arends

Cancer Research Campaign Laboratories, Department of Pathology,
University Medical School, Edinburgh, U.K.

Introduction

The patterns of cell death within the microenvironments of growing tumors are important, as they express fundamental features of tumor biology and therefore may relate to prognosis and treatment. The two patterns of cell death in tumors are apoptosis and necrosis and they differ in morphology, incidence, mechanism and regulation. Apoptosis usually affects single cells surrounded by viable neighbors. Apoptosis is a genetically programmed mode of cell death that is regulated by many genes, including oncogenes and oncosuppressor genes, which may be mutated, deleted or abnormally expressed in neoplasms, thus altering tumor cell susceptibility to apoptosis[1]. In contrast, necrosis often occurs in confluent zones and results from severe nonphysiological perturbation of the cellular environment, such as hypoxia due to inadequate blood flow in tumors [2]. The topography of necrosis in tumors is a complex function, a major determinant of which is the rate of proliferation of tumor cells relative to the process of angiogenesis, itself induced by tumor–derived growth and angiogenic factors. Blood flow within tumors is notably heterogeneous and is not always directly proportional to the anatomic extent of their vasculature [3]. Local regions of rapid tumor expansion may produce zones of relative hypoxia resulting in necrosis. This has been observed at a strikingly constant distance from blood vessels or oxygen supply as within experimental tumor "spheroids" in vitro and "corded carcinomas" in vivo [4–8].

It is interesting to consider whether the two modes of cell death occur independently or not within tumors and from this a number of questions arise. Why does it matter whether tumor cells die by apoptosis or necrosis? Is the rate of tumor cell death important? Is there a relationship between cellular susceptibility to apoptosis and necrosis? Does the preferred mode of death influence either selection of neoplastic cells during tumor growth, prognosis or response to treatment?

Morphology of Necrosis and Apoptosis

Sheets of cells dying by necrosis, such as following ischemic infarction, do so in synchrony and structurally, the necrotic cells show critically damaged membranes and organelles. Mitochondria undergo "high amplitude swelling," with formation of calcium matrix densities. Cellular plasma membranes become defective and eventually are ruptured, the cytoplasm swells as fluid and ions enter and there is dispersal of cytoplasmic elements into the extracellular space, stimulating an acute inflammatory reaction [2, 8]. Nuclear chromatin initially becomes more coarsely staining with retention of distinct regions of heterochromatin showing clumping and euchromatin (Fig. 3.1), but eventually the chromatin undergoes karyolysis as it fades away as the DNA and proteins are degraded. The mechanisms are various and include four major components that may interact to produce a vicious circle: membrane damage allowing uncontrolled fluid and ionic fluxes, uncontrolled calcium influx with activation of proteases, nucleases and phospholipases, failure of ATP production and generation of reactive oxygen species. None of these depend upon continuing synthetic activity and there is no evidence that specific signaling pathways are involved.

The morphological changes of apoptosis can be divided into three overlapping phases [9–12]. In the first, there is reduction in nuclear size with condensation of chromatin into crescentic caps or toroids at the periphery of the nucleus (Fig. 3.2). There is a characteristic pattern of disintegration of the nucleolus with dissociation of the transcriptional complexes of the dense fibrillar component from the fibrillar center. Early on in apoptosis, cells detach themselves from their viable neighbors, and along with the loss of contact regions, there is also loss of specialized surface structures, such as microvilli, and the cell initially adopts a smooth contour. Time lapse studies of apoptosis reveal the sudden onset of cell shrinkage, with surface blebbing and bubbling, sometimes known as zeiosis, as cells enter phase 1 [13]. As the cell shrinks, cytoplasmic organelles become compacted and the smooth endoplasmic reticulum undergoes dilatation. These dilated cisternae fuse with the cell membrane, giving rise to a cratered appearance at the surface on scanning electron microscopy. Cytoskeletal filaments may be seen to aggregate in side–to–side bundles, parallel to the cell surface, and ribosomal particles are dispersed from the endoplasmic reticulum and may form semicrys

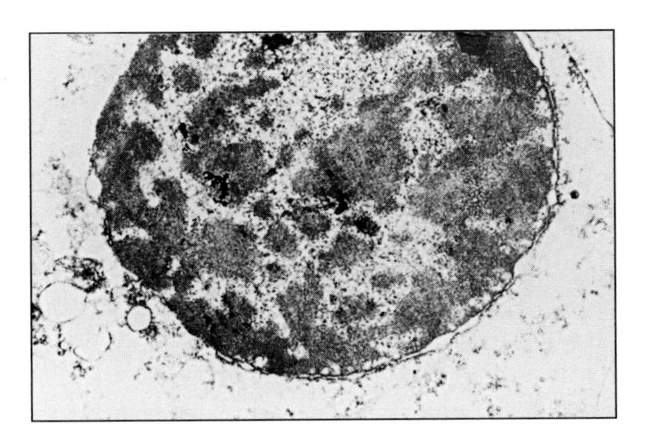

Fig. 3.1. Electron micrograph of necrotic cell showing coarse staining and clumping of the heterochromatin admixed with euchromatin with a leached out appearance of the cytoplasm due to membrane damage and loss of cytoplasmic contents with an influx of fluid and ions.

talline arrays, but otherwise the organelles remain intact. In contrast to necrosis, mitochondria do not show "high amplitude swelling," the cell membrane does not become permeable to vital dyes at this early stage, and apoptotic cells within tissues do not elicit an acute inflammatory reaction [2].

In phase 2 apoptosis, there is blebbing at the cell surface and crenation of the nuclear outline, leading to controlled fragmentation of both nucleus and cytoplasm which split up into particles of various sizes. Usually, the cell becomes a cluster of round, smooth, membrane–bounded "apoptotic bodies," some containing variably–sized, spherical, nuclear fragments of condensed chromatin (Fig. 3.2). These bodies are phagocytosed by neighboring viable tumor cells or macrophages [10].

In phase 3 apoptosis, the residual nuclear and cytoplasmic structures undergo progressive degradation. This is manifested as membrane rupture producing permeability to vital dyes and leaching out of the cytoplasmic contents at this late stage. Apoptotic nuclear masses lose their smooth, rounded appearance but are still recognizable as apoptotic as they consist of condensed chromatin forming unusual shapes with irregular outlines (Fig. 3.2). In tissues these changes, sometimes called "secondary necrosis," usually occur within the phagosome of the ingesting cell. Eventually membranes disappear, organelles become unrecognizable and the appearance is that of a residual lysosomal body. This is the appearance of the majority of apoptotic bodies seen by light microscopy and sometimes the smooth outline of the surrounding phagosome, but earlier phases can also be recognized by their rounded contours and deeply hyperchromatic,

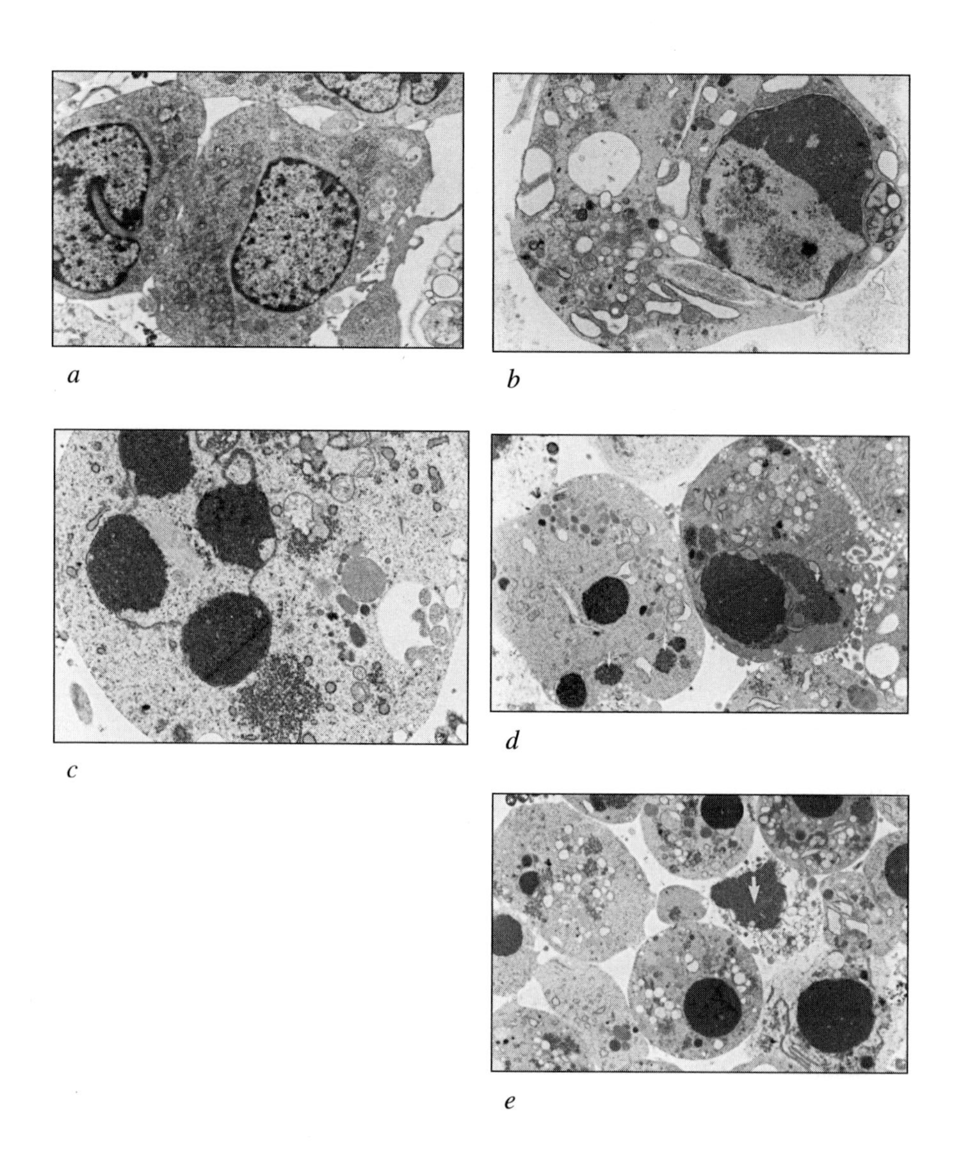

Fig. 3.2. Electron micrographs of (*a*) viable fibroblastic tumor cell showing a vesicular pattern of nuclear chromatin staining and the same tumor cell types at various stages of apoptosis showing: (*b*) crescentic caps of condensed chromatin at the nuclear periphery with nucleolar disintegration and cytoplasmic changes of dilatation of endoplasmic reticulum but normal–size mitochondria; (*c*) spheres of condensed chromatin at the nuclear periphery; (*d*) single or multiple spheres of condensed chromatin dispersed throughout the cell with crystalline–like arrays of ribosomes (small white arrows); and (*e*) several apoptotic cells and bodies including one displaying features of "secondary necrosis" including a triangular shaped mass of chromatin with an irregular outline (white arrow) and leached out appearance of the cytoplasm more akin to that seen in necrosis.

fragmented nuclei. Apoptotic cells remain recognizable within tissues for 2–9 hours, a time–course which coincides with that of complete degradation of other large biological structures within the phagosomes of macrophages. This relatively short period ensures that high rates of apoptosis produce only small increases in the proportion of apoptotic cells observed in tissue sections.

Cell Death and Tumor Growth

It is axiomatic that the biological aggressiveness of a tumor determines the prognosis for the patient. Yet, tumor aggression is a poorly defined property, which probably compounds many elements of tumor behavior. The growth rate of a tumor mass is often regarded as a key parameter that is of fundamental significance for tumor aggression, but for practical reasons, it is difficult to study directly in patients. Proliferation in tumors—both human and experimental—has been investigated extensively, but does not correlate well with overall growth. A major reason for this is that net tumor growth results from the balance of cell gain and cell loss. However, the contribution of cell loss to tumor growth and aggressiveness is less well understood. In the few instances where quantitative data are available tumor cell loss is considerable.

Cells may be lost from tumor cell populations by a variety of mechanisms. Cells may depart the primary tumor site by natural exfoliation or cell migration. They may leave the proliferating pool of cells by terminal differentiation, or they may die. Many descriptions of tumor cell death have often focused on necrosis [3, 4, 7, 8, 14], with less extensive study of apoptosis in tumors [9, 15]. Quantitative estimates of net tumor cell loss have been made [7, 8, 14]. It is possible to calculate the potential doubling time (Tp) from estimates of tumor cell production rates (mitotic indices, nucleotide incorporation or fraction of cycling cells),and the actual doubling time (Td) from measurements of tumor volume, and so derive the cell loss factor (CLF = 1 – Tp/Td). The CLF approaches zero with no cell loss, and unity if extensive loss, which makes the actual doubling time unmeasurably large. In deriving CLFs, errors can arise from several inevitable assumptions, but in almost every tumor studied the CLF is large. In rodent sarcomas and carcinomas CLFs of 0.65–0.78 have been recorded, and in human bronchial and colorectal carcinomas and a malignant melanoma CLFs of 0.73–0.96 have been recorded [7, 8, 14, 16]. The high level of tumor CLFs presumably explains the poor correlation between measures of cell proliferation, such as the thymidine labeling index, and the tumor volume doubling times for a whole range of human tumors [14] and points to the considerable importance of tumor cell loss in determining tumor growth rate. Large cell loss factors are likely to act as major regulators of the rate of tumor expansion and probably determine the overall growth pattern in terms of tumor enlargement, stasis or regression.

It has been argued that cell death is numerically the most significant form of cell loss from a growing tumor [9, 15, 17]. This may occur either by necrosis or apoptosis. However, despite the histologically conspicuous appearance of necrosis, it is by no means established that it can account for the high levels of cell loss observed in tumors. Apoptosis is less conspicuous histologically because it has a short time course, leaves no residual fingerprints and occurs in a scattered distribution within tumors. It is readily calculated that a surprisingly small proportion of apoptotic cells visualized in a tissue section can represent very substantial cell loss, and there are increasing quantitative reports on apoptosis in tumors [15, 18, 19].

Control of Cell Death in Tumors

The factors controlling cell death in a population of tumor cells are complex. Both the intrinsic genetic make up of the cell itself, and extracellular signals or insults from the surrounding microenvironment, influence the onset and mode of death. Necrosis tends to occur in contiguous sheets of cells and is related to severe, nonphysiological changes in the microenvironment, predominantly hypoxia, lack of nutrients and falling pH, as a result of progressively worsening ischemia in tumors [15, 20]. It is usually seen at a strikingly constant distance from blood vessels at about 1–2 mm. Tumors that are in their early stage of neovascularization may be initially well perfused, but as a result of both increased interstitial pressure from leaky vasculature and the relative absence of lymphatics within the tumor, compression of the main tumor vessels occurs and the central necrosis ensues [21]. Thus, the traditional view is that the single most important control over necrosis is the distance of the tumor cell from the tumor microvasculature, with the cell itself playing no part in its fate.

In contrast, apoptosis is a genetically programmed process that can be regulated by various oncogenes and tumor suppressor genes. Expression of *c–myc* and wild type *p53* enhance susceptibility to apoptosis, whereas *bcl–2*, *bcl–*X_L, *abl*, retinoblastoma gene product (*Rb*) and *ras*, protect against or suppress apoptosis [13, 22–31]. Cellular sensitivity or resistance to apoptosis in response to one type of stimulus is often associated with similar sensitivity or resistance to a variety of other agents [11, 33, 34]. We have argued elsewhere [12] that this is because some cell types are "primed" for apoptosis and that different oncogenes and tumor suppressor genes stimulate or suppress the "priming events" for apoptosis by modulating the expression of a number of effector proteins in apoptosis, for example ICE–like proteases, endonucleases or transglutaminases

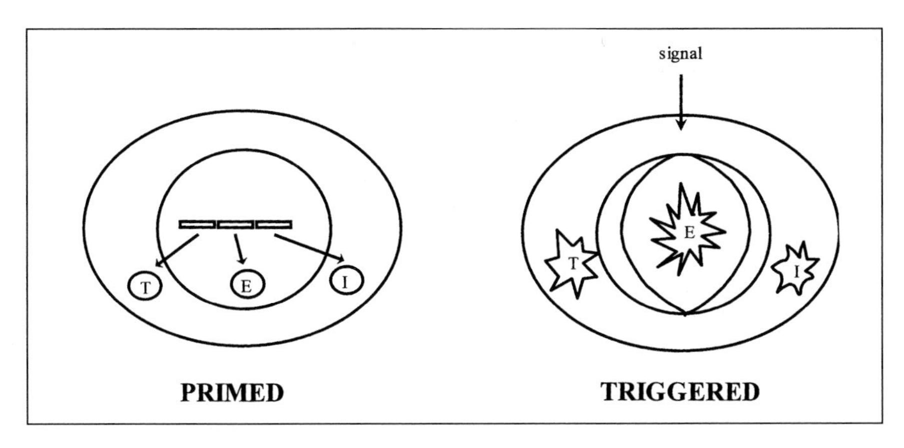

Fig. 3.3. Two stage model for priming and triggering of apoptosis. Primed cells synthesize inactive precursor effector proteins such as transglutaminases (T), endonucleases (E), ICE–like proteases (P) and other putative factors. Following specific or nonspecific signals, there is activation of the effector molecules which bring about the changes of apoptosis. ICE–like proteases may act in a cascade playing an important role in activating other precursor effectors by proteolytic cleavage at specific sites. Transglutaminases and endonucleases require calcium ions for activation and mediate crosslinking of cytoplasmic proteins and cleavage of nuclear DNA respectively.

(Fig. 3.3). Thus, whether a cell undergoes apoptosis in response to an external stimulus, such as injury due to mild or moderate ischemia/hypoxia, or treatment with cytotoxic drugs or irradiation, depends on to what extent it is primed and hence susceptible to the triggering of apoptosis.

Hypoxia can cause both apoptosis and necrosis; however, it is both the rate of development and degree of severity of hypoxia which influence the mode of death an individual cell will undergo, as well as whether that cell is primed for apoptosis. This is supported by several lines of evidence. First, in many tumors at the interface of viable tumor tissue and areas of necrosis, apoptosis has been found to be more prevalent presumably reflecting a gradient of hypoxia across this region. In one study, the extent of apoptosis at this interface has been shown to correlate broadly with other measures of susceptibility to apoptosis of the cell lines used, reflecting the level of priming for apoptosis (Fig. 3.4) [32].

Second, in the same experimental model of transformed rodent fibroblastic cell lines, those expressing mutated Ha–*ras* and grown as tumors in immune suppressed mice developed rapidly expanding tumors with high levels of mitosis, low levels of apoptosis and extensive areas of necrosis [31, 32]. Availability of endogenous endonucleases in viable cell nuclei was tested by experimental manipulation of divalent cation concentrations, known to activate this enzymic

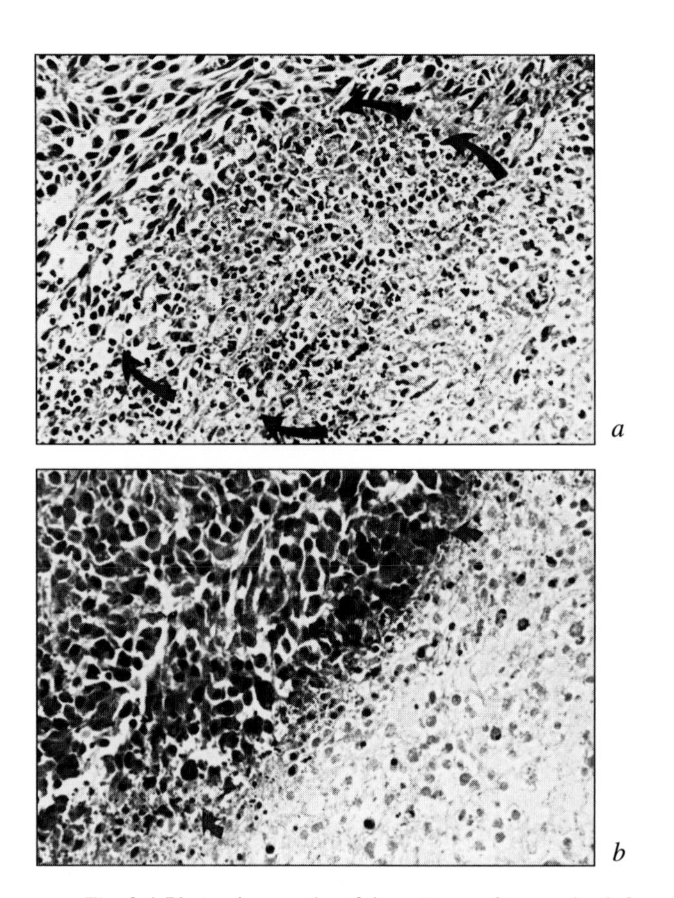

Fig. 3.4. Photomicrographs of the patterns of tumor death from (*a*) a c–myc expressing fibroblastic tumor showing a thick rim of apoptosis (between the large arrows) at the interface region with viable cells (top left corner) and a zone of necrosis (bottom right corner) in contrast to (*b*) a mutant ras expressing fibroblastic tumor displaying a large zone of confluent necrosis (bottom right corner), with only a thin rim of apoptosis (between the small arrows) at the interface with viable cells (top left corner).

activity, and no chromatin cleavage was detectable, suggesting that these cells were not primed, or had very low levels of priming for apoptosis. In contrast, similarly derived tumors from fibroblasts expressing *c–myc* showed high levels of apoptosis compared to mitosis, and grew more slowly with little necrosis; cleavage of chromatin to oligonucleosomes by endogenous endonuclease activity within viable nuclei was demonstrable, following experimental manipulation of the divalent cation concentrations, indicating that these cells were primed for apoptosis in terms of containing latent endonucleases [31, 32]. Furthermore, in this family of tumors apoptosis was found to be inversely related to necrosis

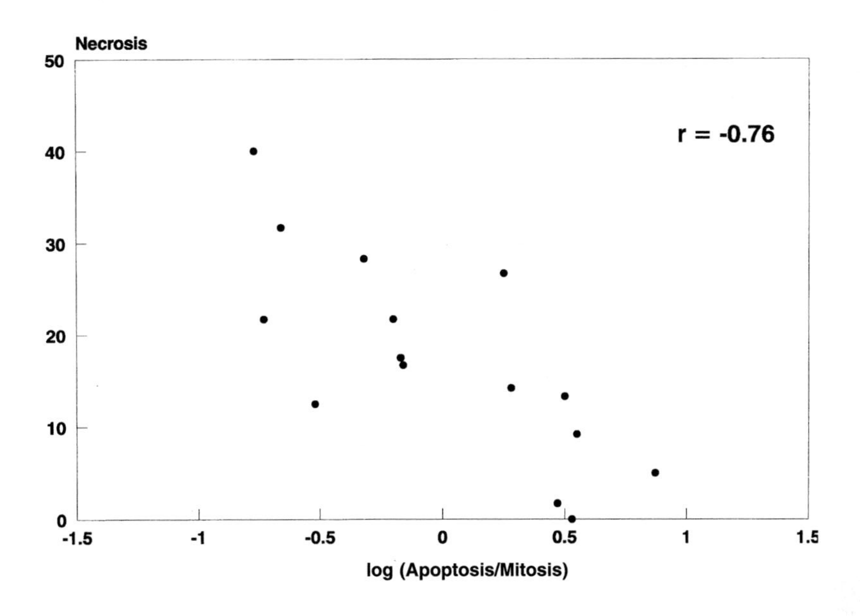

Fig. 3.5. Scatter plot of tumor necrosis scores versus log A/M ratios of fibroblastic tumors (correlation r = –0.75) (modified from Arends et al, 1994).

(Fig. 3.5) [32]. One explanation for this, that stresses the significance of priming for apoptosis, is that during growth as an expanding tumor in vivo, cells that are primed for apoptosis are liable to die by apoptosis in adverse conditions such as mild or moderate ischemia, resulting in hypoxia, reduction in availability of essential substrates and growth factors, or falling pH. In contrast, cells that are intrinsically less susceptible to apoptosis, because they have very low levels of priming for apoptosis, may be more resistant to these conditions, and so survive, at least initially, within deviant tumor microenvironments, until conditions become so adverse with severe ischemia/hypoxia, that cells die by necrosis as the only mode of death available to them.

Finally, in transplanted tumors expressing wild type p53, regions of hypoxia mapped to zones that were enriched for apoptosis, whereas in p53–deficient tumors hypoxic regions contained little apoptosis [35]. Hypoxia can induce *p53* expression apparently by a different pathway than DNA–damaging agents [36]. Hypoxia induces apoptosis in malignant cells and this appears to be largely p53–mediated, such that loss of the *p53* gene or overexpression of *bcl–2*, substantially reduces hypoxia–induced cell death [35]. Hypoxia is a stress that is commonly found in growing solid tumors (above 1–2 mm size) of diverse origin. The ability of cells to survive hypoxia therefore represents a powerful and widely

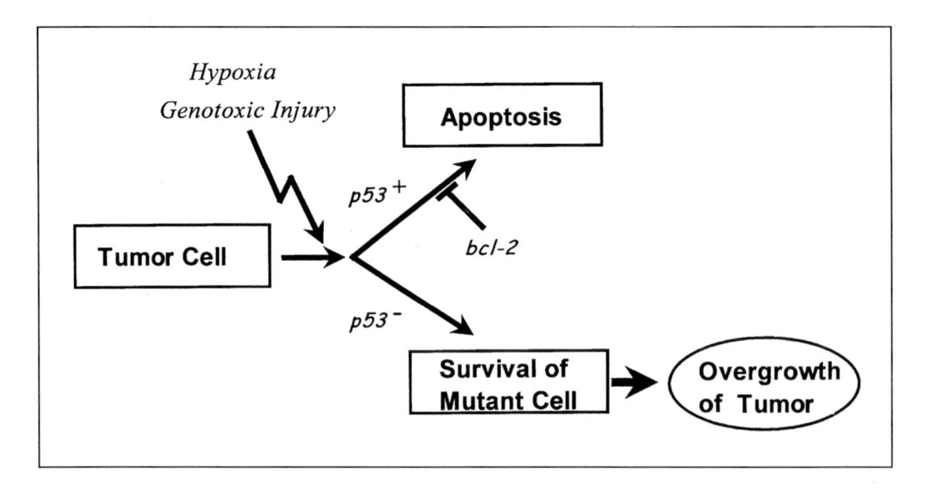

Fig. 3.6. Model of tumor cell response to hypoxia. Cellular p53 expression can be induced by hypoxia, apparently by a different pathway than by genotoxic injury, and this mediates apoptosis. However, a defective p53 pathway or overexpression of bcl–2, substantially reduces hypoxia–induced apoptosis, allowing survival of hypoxic cells that may subsequently overgrow the tumor. Such cells may also sustain DNA damage induced by a variety of agents including reactive oxygen radicals, released by adjacent necrotic cells, and thus may become mutated at key genetic loci. Hypoxia is a common stress in growing solid tumors (above 1–2 mm size) of diverse origin and the ability of cells to survive hypoxia by apoptosis–suppression mechanisms may represent an important step in neoplastic progression.

prevalent selection pressure, suggesting an explanation for why *p53* is one of the most commonly mutated genes in human cancer. Other genetic changes such as overexpression of *bcl–2* also allow survival of hypoxia–induced apoptosis and to some extent necrosis [35] and prevent activation of the ICE–like protease cascade. Furthermore, in a zone of necrosis with a surrounding rim of hypoxia within a tumor, those cells surviving the hypoxic stress at the viable cell/necrosis interface may also be exposed to reactive oxygen species, as a result of the adjacent cell death by necrosis, which may have a DNA–damaging effect on interface cells potentially resulting in mutant cells with enhanced ability to survive (Fig. 3.6).

We propose that as hypoxia develops through increasing inadequacy of blood supply to an expanding population of tumor cells, those cells that are genetically primed for it, die by apoptosis brought about by the triggering stimulus of mild to moderate ischemia. In contrast, subclones that are not primed for apoptosis have a low susceptibility to triggering of apoptosis in regions of developing hypoxia, and these cells may survive and in time overgrow the tumor. Low susceptibility to apoptosis is thus likely to associate with both rapid expansion of the tumor cell population and the development within the tumor of zones of

Table 3.1. Regulators of angiogenesis

Pro–angiogenic factors
Basic fibroblast growth factor
Interleukin–8
Platelet–derived endothelial cell growth factor
Transforming growth factor–a and b
Vascular endothelial growth factor
Anti–angiogenic factors
Angiostatin
IFN–α and β
Platelet factor 4
Protamine
Thrombospondin

hypoxic but viable tumor cells. Eventually, necrosis is the obligate mode of death for those cells not primed for apoptosis, whether provoked by the progressive imbalance between growth of the tumor and that of its supporting blood supply or by more acute vascular events (compression, thrombosis or spasm). Selection by survival of hypoxia favors apoptosis–suppression mechanisms, such as defective p53 pathway or activated bcl–2, and this may carry with it increased propensity for survival of DNA damage—"mutator phenotype"—and for induction of new vessel growth—"pro–angiogenic phenotype".

Angiogenesis in Tumors

For solid tumors to grow beyond microscopic sizes of 1–2 mm in diameter and become clinically relevant, they must induce growth of new blood vessels from the existing host vasculature and this is known as angiogenesis [37]. Before this, the size of the tumor is limited by the distance over which nutrients and oxygen can diffuse. Thus, initially tumors may remain in a dormant, balanced state with the rate of cellular proliferation being offset by an equivalent rate of cell death, mainly apoptosis [38, 39]. Following acquisition of an "angiogenic phenotype" tumors are able to grow in size, unchecked by the limits of diffusion, and embark upon the first steps of metastasis by invading into new capillaries, which have increased vascular permeability [40].

New blood vessel growth is governed by the local balance between stimulatory and inhibitory factors (Table 3.1). In non–neoplastic tissues the inhibitory influences predominate. In contrast, cancer cells that have gained the angiogenic phenotype are potently angiogenic because of a combination of increased production of inducers (which can act synergistically together) [41] and a down-regulation of inhibitors. The origin of angiogenic factors is diverse, arising from the tumors cells themselves, or from proteins mobilized from the extracellular matrix, or from cells recruited by the tumor, such as macrophages or fibroblasts, or a combination of these mechanisms [42].

Tumor cells and their supporting endothelial cells may exist in a symbiotic relationship. Tumors cells can stimulate endothelial proliferation and migration by the production of basic fibroblast growth factor (bFGF) and vascular endothelial growth factor (VEGF) and others. Endothelial cells in turn can produce tumor cell growth factors in the form of platelet–derived growth factor (PDGF), insulin–like growth factor (IGF) and granulocyte colony–stimulating factor (G–CSF).

Two of the most commonly identified angiogenic factors in tumors are bFGF and VEGF [38]. VEGF is expressed in a wide variety of solid tumors including breast, brain, kidney, ovary and colon cancer, and as well as being an endothelial mitogen, VEGF also potently increases microvessel permeability [43–45]. The expression of some angiogenic factors and inhibitors is known to be governed by certain oncogenes and tumor suppressor genes. For example, mutant K–*ras* and H–*ras* oncogenes are known to upregulate VEGF expression [46] whereas, wild type *p53* can upregulate expression of thrombospondin–1 (TSP–1), a potent inhibitor of angiogenesis, through direct binding to TSP–1 promoter sequences [47].

Hypoxia can induce cells to secrete angiogenic factors such as VEGF and bFGF [48]. Suppression of angiogenesis has been shown to maintain dormancy of micrometastases due to balanced cell proliferation and apoptosis [39]. Thus, cells surviving hypoxia due to apoptosis–suppression mechanisms, perhaps via a defective p53 pathway (Fig. 3.6), may also evolve an angiogenic phenotype, emphasizing the importance of this step in neoplastic progression.

Angiostatin, a recently discovered molecule from a variant of the murine Lewis lung carcinoma, has been shown to have dual effects on both angiogenesis and apoptosis [49]. Angiostatin, which corresponds to an internal fragment of plasminogen, is a potent inhibitor of angiogenesis that is secreted by the primary tumor and actively prevents neovascularization of metastatic tumor deposits. On removal of the primary tumor there is rapid growth of the metastatic deposits through an increase in microvessel formation and a decrease in tumor cell apoptosis (with no significant difference in cell proliferation). Whether this is a direct effect of angiostatin on apoptosis or indirect through generation of a blood supply is at present unclear.

Angiogenesis is a therapeutic target for cancer biologists and clinical trials of agents that inhibit angiogenesis at various steps are underway. These include inhibitors of angiogenic growth factors (suramin), inhibitors of endothelial cell proliferation (TNP–470) and inhibitors of matrix degradation (batimastat) [50–52]. Other compounds, including interferon alpha 2a, a potent anti–angiogenesis factor in vitro, have been successful in treating childhood hemangiomas [53]. At present the inhibition of tumor angiogenesis is an area of active ongoing research, because of the potentially favorable therapeutic window and a theoretically low incidence of drug resistance.

Prognosis

Numerous studies have tried to correlate the prevalence of apoptosis and or necrosis in tumor sections to prognosis, measured by disease–free interval or five–year survival. Tumors displaying greater cell death of either type might be expected to have a better prognosis as the cell loss factor would be higher and the tumor slower growing. However, although this simplistic view holds true for a few tumors [54, 55] in many instances the prevalence of cell death is unrelated to prognosis. With certain tumors, the reverse is true and diagnostic histopathologists are familiar with the somewhat counter–intuitive notion that some tumors with a large proportion of necrosis or apoptosis have a bad prognosis [56, 57]. This may be related to the inverse correlation between necrosis and apoptosis discussed earlier, with prominent necrosis representing a surrogate marker of low susceptibility to apoptosis and hence fast net growth and resistance to anticancer treatments. However, there is a complex interaction of multiple factors at work, in that tumors with more extensive vascular networks may also have an increased local risk of necrosis as the blood in these vessels is often slow flowing and procoagulant. Part of this phenomenon has been attributed to the production of tissue factor by tumor cells, which is a potent stimulator of coagulation [58].

An approximate measure of the rate of apoptosis in a tumor can be calculated by determining its apoptotic index. This involves counting the number of apoptotic cells or bodies seen in a given area of tumor tissue (e.g., 10 high power fields) or as a fraction of a number (e.g., 1000) of viable tumor cells. This requires reproducible identification and enumeration of apoptotic bodies, with many workers using hematoxylin, and eosin-stained thin paraffin sections and adherence to set morphological criteria, including reliance on the characteristic nuclear changes of apoptosis described earlier. Alternatively, apoptotic cells may be identified by in situ end–labeling (ISEL) of cleaved ends of DNA in apoptotic cells (or the related TUNEL method) [59]. However, these staining techniques should be used with caution as they may produce both false positive (ISEL positive viable cells with no morphological evidence of apoptosis) and false negative

(negative ISEL staining of morphologically apoptotic cells) results, and positively label nuclei containing damaged DNA following irradiation or other treatments [60]. Furthermore, apoptotic counts in tumors must only be made within regions of viable tumor cells with scattered apoptosis, as there is increased apoptosis around the edges of necrotic zones that must be avoided along with confluent areas of apoptotic cells as these may also be related to necrotic zones in adjacent tissue sections.

A number of studies have shown that the apoptotic index can be an independent marker of prognosis. For example, in neuroblastomas a high prevalence of apoptosis correlates well with low tumor grade and increased patient survival [54]. In contrast, the apoptotic index in ovarian tumors is of no predictive valve [61]. In both breast and nonsmall cell lung cancers high apoptotic indices are associated with a shorter survival [56, 62]. Further investigations are currently underway. Thus, the apoptotic index is only of prognostic value in a number of specified tumor types and is less useful in others. Furthermore, since susceptibility or resistance of tumor cells to induction of apoptosis by chemotherapeutic drugs or irradiation appears also to be genetically determined, measurements of prevalence of apoptosis in tumor sections may give an indication of the innate tumor cell susceptibility to apoptosis that may be useful in predicting response to therapy.

Accurate quantification of necrosis in tumors is technically difficult and it is often scored semiquantitatively by light microscopy into broad categories such as little or no necrosis (< 5%), moderate necrosis (< 50%) or extensive areas of necrosis (> 50%) [63]. Necrosis is often seen in rapidly growing, aneuploid tumors and its presence may denote an unfavorable outcome. However there are numerous exceptions to this rule with many aggressive tumors showing no necrosis, and in some cancers the presence of necrosis has no predictive value of outcome (e.g., nonsmall cell lung cancer) [64]. Tumor–specific scoring systems for necrosis do exist and it is a well established practice in the evaluation of prognosis of soft tissue sarcomas [57]. Quantitative analysis of new blood vessel formation in biopsy specimens can also have prognostic value in breast and ovarian cancer [65, 66]. High intratumor microvessel density is associated with a poorer prognosis in terms of survival and the occurrence of metastasis.

Cell Death and Anticancer Treatments

For years anticancer treatments have focused on targeting cell division in tumors. This emphasis has changed to tumor cell death more recently as both radiotherapy and chemotherapy are known to cause tumor regression through the induction of apoptosis. Whether a tumor cell undergoes death in response to irradiation or chemotherapy is thought to be dependent on at least three factors: the nature of the genetic controls of susceptibility to apoptosis of the cell, in terms of the changes to cancer–related genes, the dose and type of irradiation or chemotherapy and the surrounding microenvironment.

The tumor microenvironment influences the response to therapy, since hypoxia diminishes the cytotoxicity of ionizing radiation and alters the expression and efficiency of drug–metabolizing enzyme systems [67, 68]. Local permeation of systemic chemotherapeutic agents is affected by the detailed topographical relationships between tumor cells and the microcirculation. Moreover, many recent observations of cells in vitro show that cellular susceptibility to apoptosis is regulated by the expression of oncogenes and oncosuppressor genes and that this modulates their sensitivity to cytotoxic agents of various types [69, 70].

It is known that some tumors are sensitive to radiotherapy and some resistant and for a given dose of irradiation some cells in the same tumor undergo apoptosis and others do not. These phenomena may be explained by the heterogeneity of the tumor microenvironment and also of the genetic lesions in oncogenes and tumor suppressor genes seen between different subclones within a single tumor. Cells of the small and large intestine in *p53* null mice do not undergo apoptosis in response to a given dose of irradiation that in *p53* wild type animals causes abundant crypt cell apoptosis [71]. Other oncogenes can affect apoptosis with *ras* and *bcl–2* protecting against apoptosis following irradiation while *myc*–expressing cells readily enter apoptosis after the same stimulus. Thus, susceptibility to apoptosis induced by anticancer agents is modulated genetically. Interestingly, tumors showing high rates of spontaneous apoptosis prior to irradiation, suggestive of a highly primed state, are more likely to show greater regression following therapy and increased patient survival. This supports the idea that irradiation triggers apoptosis in tumor cells that are already primed for it. However, at increasing doses of irradiation all cells will eventually undergo necrosis, because of the type of cellular damage caused. It is interesting to note that the susceptibility to apoptosis can change within a tumor population following a single dose of irradiation. Following multiple, equal doses of irradiation, a subpopulation of cells that did not undergo apoptosis the first time may do so after subsequent exposures; however, this phenomenon is not well understood. The induction of apoptosis by chemotherapy closely mimics that of radiotherapy with molecular controls playing a key role, especially p53 and Bcl–2. Most common human malignancies including breast, lung and colorectal cancer, harbor mutations of *p53* and respond poorly to anticancer therapies; however, testicular cancer, Wilms tumor and acute lymphoblastic leukemia demonstrate few aberrations of *p53* and are curable even in their advanced stages [72, 73].

It is not only the tumor cells within the tumor that are affected by irradiation or chemotherapy. Endothelial cells undergo apoptosis in response to irradiation which may well lead to thrombosis with disruption of blood flow and subsequent tumor infarction. This effect can be reversed by basic fibroblast growth factor (bFGF) which is made by some tumor cells and hence accounts for the heterogeneity seen in tumor response to therapy [74]. Thus, tumor products may control the responsiveness of the surrounding endothelium to apoptosis and subsequent thrombosis. 5–fluorouracil and mitomycin C are directly cytotoxic to endothelial cells in a dose–dependent manner which is likely to augment their tumoricidal action [75]. Antihormonal chemotherapies, such as tamoxifen treatment of estrogen receptor positive breast cancers and administration of gonadotrophin releasing hormone (GnRH) analogs to patients with prostatic adenocarcinoma, have been shown to cause tumor regression by apoptosis [76]. New approaches to the treatment of cancer using gene therapy to help to bring about tumor cell apoptosis may focus on ways of priming tumor cells for apoptosis. This may not be too distant a goal with recent experiments transferring wild–type *p53* back into *p53* mutated tumor cells enhancing the sensitivity of these cells to apoptosis following irradiation or treatment with chemotherapy [77, 78].

Conclusions

1. Cell death in tumors can occur either by apoptosis or necrosis, and these can be distinguished by their characteristic morphological features. Apoptosis affects single cells surrounded by viable neighbors, whereas necrosis usually occurs in zones of contiguous cells.

2. The net growth rate of a tumor mass reflects the balance of cell gain and cell loss. The cell loss factor in many tumors is large, and an important contributor to this loss is cell death, particularly by apoptosis. There is evidence that apoptosis can act as a key regulator of net tumor growth.

3. Apoptosis and necrosis are regulated differently. Susceptibility to apoptosis is genetically determined and alterations to oncogenes and tumor suppressor genes can modulate this susceptibility, perhaps by influencing the level of priming for apoptosis. This determines the probability of triggering of apoptosis in response to various external stimuli, including hypoxia and anticancer treatments. In contrast, necrosis in untreated tumors results from marked reduction of blood supply and is not under direct genetic control. However, cells that are highly susceptible to apoptosis are more likely to die by apoptosis in a gradient of increasing hypoxia, rather than by necrosis. There appears to be an inverse relationship between susceptibility to apoptosis and necrosis in tumors. Survival of mild to moderate hypoxia may be mediated by apoptosis–suppression mechanisms and this may influence selection of certain cells, including those with higher propensities for survival of DNA damage resulting in mutation or expressing a pro–angiogenic phenotype.

4. For a tumor to grow greater than 1–2 mm in size and metastasize, it must develop its own blood supply. Angiogenesis is a tightly regulated process that can be induced by cytokines secreted by tumor cells and this influences the adequacy of blood supply to the tumor. Where there is poor development of intratumor vasculature, cell death results. Whether apoptosis or necrosis ensues depends on the genetic changes within the cell that determine the level of priming for apoptosis and the speed of onset and severity of the hypoxia.

5. In the clinical setting quantification of apoptosis and necrosis in tumor sections can have prognostic value in certain tumors. Furthermore, tumor regression following chemotherapy, radiotherapy and hormonal treatment is due to cell death, particularly apoptosis. Future research may focus on therapeutic approaches to anti–angiogenesis and new strategies to induce tumor cell apoptosis in those cells that are currently resistant to anticancer therapy. The prospect of successful gene therapy to induce tumor cell apoptosis, using *p53* for example, is an exciting possibility.

References

1 Arends MJ. How do cancer cells die? Apoptosis and its role in neoplastic progression. In: R Leake, M Gore, RH Ward, eds. The Biology of Gynaecological Cancer. RCOG Press, 1995:73–91.
2 Wyllie AH. Cell death: a new classification separating apoptosis from necrosis. In: Bowen ID, Lockshin RA, eds. Cell Death in Biology and Pathology. London: Chapman and Hall, 1981:9–34.
3 Tozer GM, Lewis S, Michalowsky A et al. The relationship between regional variations in blood flow and histology in a transplanted rat fibrosarcoma. Br J Cancer 1990; 61:250–257.
4 Thomlinson RH, Gray LH. The histological structure of some human lung cancers and the possible implications for radiotherapy. Br J Cancer 1955; 9:539–549.
5 Tannock IF. The relationship between cell proliferation and the vascular system in a transplanted mouse mammary tumor. Br J Cancer 1968; 22:258–273.
6 Franko AJ, Sutherland RM. Oxygen diffusion distance and development of necrosis in multicell spheroids. Radiation Res 1979; 79:439–453.
7 Moore JV. Cytotoxic injury to cell populations of solid tumors. In: Potten CS, Hendry JH, eds. Cytotoxic Insult to Tissue. Edinburgh: Churchill Livingstone, 1983:368–404.
8 Moore JV. Death of cells and necrosis of tumors. In: Potten CS, ed. Perspectives in Mammalian Cell Death. Oxford: Oxford University Press, 1987:295–325.
9 Kerr JFR, Wyllie AH, Currie AR. Apoptosis: a basic biological phenomenon with wide–ranging implications in tissue kinetics. Br J Cancer. 1972; 26:239–257.
10 Wyllie AH, Kerr JFR, Currie AR. Cell Death: The significance of apoptosis. Int Rev Cytol 1980; 68:251–306.
11 Wyllie AH. Apoptosis: cell death under homeostatic control. Arch Toxicol (suppl) 1987; 11:3–10.
12 Arends MJ, Wyllie AH. Apoptosis: mechanisms and roles in pathology. Int Rev Exp Pathol 1991; 32:223–254.
13 Evan GI, Wyllie AH, Gilbert CS et al. Induction of apoptosis in fibroblasts by *c–myc* protein. Cell 1992; 69:119–128.
14 Steel GG. Growth Kinetics of Tumors. 1st ed. Clarendon Press, 1977.
15 Wyllie AH. The biology of cell death in tumors. Anticancer Res 1985; 5:131–136.
16 Kerr KM, Lamb D. Actual growth rate and tumor cell proliferation in human pulmonary neoplasms. Br J Cancer 1984; 50:343–349.

17 Bowen ID, Bowen SM. Programmed Cell Death in Tumors and Tissues. London: Chapman and Hall, 1990.

18 Sarraf CE, Bowen ID. Kinetic studies on a murine sarcoma and an analysis of apoptosis. Br J Cancer 1986; 54:989–998.

19 Sarraf CE, Bowen ID. Proportions of mitotic and apoptotic cells in a range of untreated experimental tumors. Cell Tissue Kinet 1988; 21:45–49.

20 Schatten WE. An experimental study of necrosis in tumors. Cancer Res 1962; 22:286–290.

21 Folkman J. Tumor Angiogenesis. In: Mendelson J, Howley PM, Israel MA, Liotta LA, eds. The Molecular Basis of Cancer. Philadephia: W.B. Saunders, 1995:206–232.

22 Clarke AR, Purdie CA, Harrison DJ et al. Thymocyte apoptosis induced by p53—dependent and independent pathways. Nature 1993; 362:849–852.

23 Hockenbery D, Nunez G, Milliman C et al. Bcl–2 is an inner mitochondrial membrane protein that blocks programmed cell death. Nature 1990; 348:334–336.

24 Strasser A, Harris AW, Bath ML et al. Novel primitive lymphoid tumors induced in transgenic mice by cooperation between *myc* and *bcl–2*. Nature 1990; 348:331–333.

25 Strasser A, Harris AW, Cory S. *bcl–2* transgene inhibits T cell death and perturbs thymic self–censorship. Cell 1991; 67:889–899.

26 Sentman CL, Shutter JR, Hockenbery D et al. *bcl–2* inhibits multiple forms of apoptosis but not negative selection in thymocytes. Cell 1991; 67:879–888.

27 Fanidi A, Harrington EA, Evan GI. Interaction between *c–myc* and *bcl–2* proto–oncogenes: a novel paradigm for oncogene cooperation. Nature 1992; 359:554–556.

28 Bissonnette RP, Echeverri F, Mahboubi A et al. Apoptotic cell death induced by *c–myc* is inhibited by *bcl–2*. Nature. 1992; 359:552–554.

29 Clarke AR, Maandag ER, van Roon M et al. Requirement for a functional *Rb–1* gene in murine development. Nature 1992; 359:328–330.

30 Evans CA, Owen PJ, Whetton AD et al. Activation of the Abelson tyrosine kinase activity is associated with the suppression of apoptosis in hemopoietic cells. Cancer Res 1993; 53:1735–1738.

31 Arends MJ, McGregor AH, Toft NJ et al. Susceptibility to apoptosis is differentially regulated by *c–myc* and mutated Ha–*ras* oncogenes and is associated with endonuclease availability. Br J Cancer 1993; 68:1127–1133.

32 Arends MJ, McGregor AH, Wyllie AH. Apoptosis is inversely related to necrosis and determines net growth in tumors bearing constitutively expressed *myc, ras* and *HPV* oncogenes. Am J Pathol 1994; 144:1045–1057.

33 Bertrand R, Sarang M, Jenkin J et al. Differential induction of secondary DNA fragmentation by topoisomerase II inhibitors in human tumor cell lines with amplified *c–myc* expression. Cancer Res 1991; 51:6280–6285.

34 Gregory CD, Dive C, Henderson S et al. Activation of Epstein–Barr virus latent genes protects human B cells from death by apoptosis. Nature 1991; 349:612–614.

35 Graeber T, Osmanian C, Jacks T et al. Hypoxia–mediated selection of cells with diminished apoptotic potential in solid tumors. Nature 1996; 379:88–91.

36 Graeber T, Peterson JF, Tsai M et al. Hypoxia induces accumulation of p53 protein, but activation of a G_1 phase checkpoint by low oxygen conditions is independent of p53 status. Mol Cell Biol 1994; 14:6264–6277.

37 Folkman J. Tumor angiogenesis: Therapeutic implications. N Engl J Med 1971; 285:1182–1186.

38 Folkman J. Clinical applications of research on angiogenesis. N Engl J Med 1995; 333:1757–1763.

39 Holmgren L, O'Reilly MS, Folkman J. Dormancy of micrometastasis:balanced proliferation and apoptosis in the presence of angiogenesis suppression. Nat Med 1995; 1:149–153.

40 Folkman J. Angiogenesis in cancer, vascular, rheumatoid and other disease. Nat Med 1995; 1:27–31.

41 Pepper MS, Ferrara N, Orci L et al. Potent synergism between vascular endothelial growth factor and basic fibroblast growth factor in the induction of angiogenesis in vitro. Biochem Biophys Res Commun 1992; 189:824–831.

42 Folkman J. Tumor angiogenesis. In: HollandJF, Frei E III, Bast RC Jr, Kufe DW, Morton DL, Weichsbaum RR, eds. Cancer Medicine 3rd ed. Vol. 1. Philadelphia: Lea and Febiger, 1993:153–170.

43 Kolch W et al. Regulation of expression of VEGF/VPS and its receptors: role in tumor angiogenesis. Breast Cancer Res 1995; 36:139–155.
44 Dvorak HF, Brown LF, Detmar M et al. Review:vascular permeability factor/vascular endothelial growth factor, microvascular permeability, and angiogenesis. Am J Pathol 1995; 146:1029–1039.
45 Senger DR, Van De Water L, Brown LF et al. Vascular permeability factor (VPE, VEGF) in tumor biology. Cancer Metastasis Rev 1993; 12:303–324.
46 Rak J, Mitsuhashi L, Bayko J et al. Mutant *ras* oncogenes upregulate VEGF/VPF expression: implications for induction and inhibition of tumor angiogenesis. Cancer Res 1995; 55:4575–4580.
47 Dameron KM, Volpert VO, Tainsky MA et al. Control of angiogenesis in fibroblasts by p53. Regulation of thrombospondin–1. Science 1994; 265:1582–1584.
48 Shweiki D, Itin A, Soffer D et al. Vascular endothelia growth factor induced by hypoxia may mediate hypoxia–initiated angiogenesis. Nature 1992; 359:843–845.
49 O'Reilly MS, Holmgren L, Shing Y et al. Angiostatin: A novel inhibitor that mediates the suppression of metastasis by a Lewis lung carcinoma. Cell 1994; 79:315–328.
50 Myers C, Cooper M, Stein C et al. Suramin: a novel growth factor antagonist with activity in hormone–refractory metastatic prostate cancer. J Clin Oncol 1992; 10:881–889.
51 Pluda JM, Wyvill K, Figg WD et al. A phase1 study of an angiogenesis inhibitor, TNP–470(AGM–1470) administered to patients with HIV related Kaposis sarcoma. Proc Am Soc Clin Oncol 1994; 13:252.
52 Hawkins MJ. Clinical trials of antiangiogenic agents. Curr Op Oncol 1995; 7:90–93.
53 White CW, Sondheimer HM, Crouch EC et al. Therapy of pulmonary hemangiomatosis with recombinant interferon alpha–2a. N Engl J Med 1989; 320:1197–1200.
54 Gestblom C, Hoehner JC, Pahlan S. Proliferation and apoptosis in neuroblastoma: subdividing the mitosis–karyorrhexis index. European J Cancer 1995; 31A:458–463.
55 Aihara M, Scardino PT, Truong LD et al. The frequency of apoptosis correlates with the prognosis of Gleason Grade 3 adenocarcinoma of the prostate. Cancer 1995; 75:522–529.
56 Tormanen U, Eerola AK, Rainio P et al. Enhanced apoptosis predicts shortened survival in nonsmall cell lung carcinoma. Cancer Res 1995; 55:5595–5602.
57 Choong PF, Gustafson P, Willen H et al. Prognosis following locally recurrent soft–tissue sarcoma. A staging system based on primary and recurrent tumor characteristics. Int J Cancer 1995; 60:33–37.
58 Folkman J. Tumor angiogenesis and tissue factor. Nat Med 1996; 2:167–168.
59 Gavrielli Y, Sherman Y, Ben–Sasson SA. Identification of programmed cell death in situ via specific labeling of nuclear DNA fragmentation. J Cell Biol 1992; 119:493–501.
60 Coates PJ, Save V, Ansari B et al. Demonstration of DNA damage repair in individual cells using in situ end–labeling: association of p53 with sites of DNA damage. J Pathol 1995; 176:19–26.
61 Jussila T, Stenback F. Cell proliferation markers and growth factors in ovarian cancer. Ann Med 1995; 27:87–94.
62 Lipponen P, Aaltomaa S, Kosma VM et al. Apoptosis in breast cancer as related to histopathological characteristics and prognosis. Eur J Cancer 1994; 30A:2068–2073.
63 Enzinger FM, Weiss SW. General considerations. Enzinger FM, Weiss SW, eds. In: Soft Tissue Tumors. 2nd ed. St Louis: The CV Mosby Co., 1988:1–18.
64 Roberts TE, Hasleton PS, Musgrove C et al. Vascular invasion in nonsmall cell lung carcinoma. J Clin Path 1992; 45:591–593.
65 Weidner N, Semple JP, Welch WR et al.Tumor angiogenesis and metastasis–correlation in invasive breast carcinoma. N Engl J Med 1991; 324:1–8.
66 Tannock IF. Oxygen diffusion and the distribution of cellular radiosensitivity in tumors. Br J Radiol 1972; 45:515–524.
67 Hollingsworth HC, Kohn HC, Steinberg SM et al. Tumor angiogenesis in advanced stage ovarian carcinoma. Am J Pathol 1995; 147:33–40.
68 Shan X, Aw TY, Smith ER et al. Effect of chronic hypoxia on detoxication enzymes in rat liver. Biochem Pharmacol 1992; 43:2421–2426.
69 Lowe S, Ruley HE, Jacks T et al. p53–Dependent apoptosis modulates the cytotoxicity of anticancer agents. Cell 1993; 74:957–967.

70 Lotem J, Sachs L. Regulation by bcl–2, c–myc, and p53 of susceptibility to induction of apoptosis by heat shock and cancer chemotherapy compounds in differentiation–competent and –defective myeloid leukemic cells. Cell Growth Diff 1993; 4:41–47.
71 Clarke AR, Gledhill S, Hooper ML et al. p53 dependence of early apoptotic and proliferative responses within the mouse intestinal epithelium following γ–irradiation. Oncogene 1994; 9:1767–1773.
72 Hickman JA. Apoptosis induced by anticancer drugs. Cancer Metastasis Rev 1992; 11:121–140.
73 Fisher DE. Apoptosis in cancer therapy: crossing the threshold. Cell 1994; 78:539–542.
74 Fuks Z, Persaud RS, Alfieri A et al. Basic fibroblast growth factor protects endothelial cells against radiation induced programmed cell death in vitro and in vivo. Cancer Res 1994; 54:2582–2590.
75 Nuyts RM, Pels E, Greve EL. The effects of 5–fluorouracil and mitomycin C on corneal endothelium. Curr. Eye Res 1992; 11:565–570.
76 Montironi R, Magi–Galluzzi C, Muzzonigro G et al. Effects of combination endocrine treatment on normal prostate, prostatic intraepithelial neoplasia, and prostatic adenocarcinoma. J Clin Path 1994; 47:906–913.
77 Lowe SW, Bodis S, McClatchey A et al. p53 status and the efficacy of cancer therapy in vivo. Science 1994; 266:807–810.
78 Gjerset RA, Turla ST, Sobol RE et al. Use of wild type p53 to achieve complete treatment sensitization of tumor cells expressing endogenous mutant p53. Molec Carcinogenesis 1995; 14:275–285.

Chapter 4

Apoptosis and Cancer, edited by Seamus J. Martin.

Apoptosis and the Cell Cycle

Rati Fotedar,[a] Ludger Diederich[a] and Arun Fotedar[b]

[a] Institut de Biologie Structurale J.-P. Ebel, Cedex, France
[b] Sidney Kimmel Cancer Center, San Diego California, U.S.A.

Introduction

Control of cell number is determined by a delicate balance between cell death and cell proliferation. Inappropriate accumulation of cells such as those with a damaged genome, can contribute to unrestrained tumor growth, while excessive death can result in impaired development and in degenerative diseases [1]. Cell cycle progression is subject to internal controls called checkpoints that ensure the integrity of the genome through coordination of the different cell cycle events. These checkpoints for example halt the cell cycle following DNA damage or when the chromosome alignment is faulty in mitosis. Loss of cell cycle checkpoint controls can initiate the apoptotic program. Coordination between the regulatory pathways controlling the cell cycle and apoptosis is therefore important. Elucidating the mechanisms by which cell cycle control signals are coupled to apoptosis will therefore be of central importance in understanding tumor progression and designing new models of effective tumor therapy.

Regulation of Cell Cycle

Cell cycle is the process by which the cell prepares for duplication. The cell cycle is comprised of four phases: G_1, S, G_2 and M. Following division the cell cycle commences with G_1 and at a certain point in mid to late G_1 called the Restriction point, the cell becomes committed to progress through the cell cycle [2]. DNA replication occurs during S phase followed by preparation of the chromatin to enter mitosis in G_2. Finally, the chromosomes segregate in mitosis. At the end of mitosis the two daughter cells either renter the cycle or become quiescent. Cell cycle progression is controlled by a family of serine/threonine protein kinases designated cyclin dependent kinases or CDKs [3]. The CDKs require association with a distinct family of proteins, termed cyclins, for activity. In mammalian cells, the kinase activity associated with D–type cyclins (D1, D2 and D3) in association with CDK4 or CDK6 are required for G_1 progression and most likely for progression through the Restriction point [2]. Cyclin E in association with CDK2 is required for the G_1 to S phase transition [4, 5], while cyclin A in complex with CDK2 functions later, for DNA replication and progression through the S phase [6]. The CDK1–cyclin B kinase in concert with CDK2–cyclin A controls events that promote entry into mitosis [7]. CDK1 is also known as CDC2.

The timely activation of different CDK–cyclin complexes is crucial to regulation of the cell cycle. CDKs are regulated by a complex series of phosphorylation and dephosphorylation events. For example, in mammalian cells CDK2 is inactive when Thr160 is dephosphorylated or when Thr14 and Tyr15 are phosphorylated through the activation of the weel/mik1 related kinases [8]. CDK2 activation first requires association with a cyclin, and the phosphorylation of Thr160 by a CDK activating kinase, CAK. Interestingly, CAK is comprised of a complex of CDK7 with cyclin H. Final activation of CDK2 follows the dephosphorylation of Thr14 and Tyr15 probably through the action of the protein phosphatase, CDC25 [8]. Inactivation of CDKs occurs through a decrease in cyclin levels, either due to a decrease in transcription of cyclin or to its specific proteolytic degradation.

CDK–cyclin activity is modified by additional regulatory proteins termed **C**yclin Dependent **K**inase **I**nhibitors (CKIs). These proteins bind directly to either the CDK or the CDK–cyclin complex and inhibit kinase activity [9–11]. CKIs have also been described in yeast [9] and in *Xenopus* [12]. In mammalian cells two distinct families of CDK–cyclin kinase inhibitors, p21 and p16, which inhibit cell cycle progression in G_1 phase have been described. The p21 family (also known as Kip/Cip family) includes structurally related proteins, $p21^{WAF-1/Cip1}$ [13–17], $p27^{Kip1}$ [18, 19], and $p57^{Kip2}$ [20, 21], all of which inhibit a variety of CDK–cyclin kinases by binding previously activated CDK–cyclin complexes. The N–terminus of p21 binds to CDK2–cyclin complex and is sufficient to inhibit CDK–cyclin kinase activity in vitro [6, 22–27]. N–terminal residues of p21, p27 and p57 share significant homology, and the N–terminus of p27 has been shown to bind CDK2–cyclin A/E complex and to inhibit kinase activity [24]. p27, p21 and p57 differ in their C terminal regions. The C–terminal domain of p21 binds proliferating cell nuclear antigen (PCNA), a processivity factor for DNA polymerase d, and inhibits its function [6, 22, 24, 26]. Both the N–terminal domain and the C–terminal domain of p21 can function independently to block cell cycle progression and inhibit cell growth albeit via a different mechanism [6, 24]. Interestingly, p21 expression is regulated by the tumor suppressor gene *p53* upon DNA damage, although p21 can also be expressed in a *p53*–independent manner [28, 29].

The p16 (or *INK*) family is unrelated to the p21 family and its members, p14/p15^{INK4b} [30, 31], p16^{INK4a} [32], p18^{INK4c} [30], p19^{INK4d} [33, 34] preferentially inhibit CDK4–cyclin D and CDK6–cyclin D kinases. These proteins inhibit kinase activity by specifically binding to the CDK subunit and preventing the binding of D–type cyclins with CDK.

The products of tumor suppressor genes, Rb (retinoblastoma protein) [35] and p53 [36], act as negative regulators of cell growth. The *Rb* gene is mutated in several types of human cancer including retinoblastomas, some small lung carcinomas, sarcomas and bladder carcinomas [37]. *Rb* exerts its effect in G_1 phase at a defined point termed the Restriction point (R point) [35]. During early to mid G_1, Rb is in a dephosphorylated state and undergoes phosphorylation concomittant with the passage of the cell through the R point. The phosphorylation of Rb by cyclin D–CDK4, cyclin D–CDK6 and probably by cyclin E–CDK2 downregulates its growth inhibitory effects. One function of Rb is to link the cell cycle progression with the transcription machinery. Rb binds to and negatively regulates a subgroup of E2F transcription factors (E2F–1, –2, –3) whose function is to activate genes required for entry into S phase and for DNA replication [38]. Phosphorylation of Rb relieves E2F–1 from inhibition which then activates various genes associated with G_1 to S phase progression including, cyclins D [39], E [40, 41] and A [42]. Rb also plays a role in transcriptional suppression by binding to E2F that is bound to its specific DNA site [43–47]. Physiological growth inhibitory signals such as TGF–β, cyclic AMP and contact inhibition mediate arrest in G_1 phase by preventing phosphorylation of Rb. These signals activate a CKI that in turn can prevent the CDK–cyclin kinase from phosphorylating Rb [31, 48–51]. Alternatively, as in the case of TGF–β, there is dramatic reduction in the levels of CDK4 in some cell types [52].

The *p53* tumor suppressor gene is the most frequently mutated gene in human tumors [53]. Lack of functional p53 is associated with increased incidence of tumor development both in mice with homozygous deletion of *p53* gene [54, 55] and in humans carrying a heterozygous mutation in *p53* gene [56, 57]. Overexpression of *p53* in tumor cells lacking functional *p53* results in growth suppression. *p53* gene is involved in checkpoint control and its principal function appears to be in mediating a response to DNA damage by coordinating cell cycle arrest and DNA repair. p53 expression also stimulates apoptosis and this aspect of p53 function is discussed in detail later. p53 is therefore thought to function in two critical pathways: G_1/S cell cycle arrest following DNA damage (G_1 checkpoint) and apoptosis.

Transcriptional activation of target genes is an important component of p53 function. The induced genes may be involved in cell cycle arrest, cell cycle check point control or apoptosis. Following DNA damage, p53 activates transcription of p21$^{WAF-1/Cip1}$ [13], Gadd45 (a protein involved in DNA repair) [58], cyclin G (of unknown function) [59–61], bax (an inducer of apoptosis) [62] and IGF–BP3 (insulin-like growth factor binding protein 3) [63]. Other genes activated by p53 are *mdm2* (a negative regulator of p53) [64] and *Fas* receptor (triggers apoptosis upon binding Fas ligand) [65]. Gadd45 binds to PCNA [66, 67] and to p21 [68] and inhibits cell growth [69]. The interaction between Gadd45, p21 and PCNA has been proposed to play a role in DNA repair [66]. Of the known transcriptional targets, p21$^{WAF-1/Cip1}$ an inhibitor of CDK–cyclin activity [14–16], is responsible for p53–mediated cell cycle arrest [70–73]. Arrest in G_1 phase is probably mediated by binding of p21 to specific CDK–cyclin complexes and the resulting inhibition of kinase activity. Inhibition of cyclin D–CDK4/6 complexes suppresses Rb phosphorylation and prevents activation of E2F while inhibition of CDK2–cyclin E kinase decreases the protein kinase activity required for G_1 to S phase transition [74, 75].

Cell Cycle Regulators, Tumor Suppressors and Apoptosis

Cyclin Dependent Kinases and Apoptosis

Activation of cyclin dependent kinases has been observed in many forms of apoptosis. For example, apoptosis induced in a T cell hybridoma by activation with anti–CD3 or concanavalin A occurs in G_2/M phase and is accompanied by a three– to six–fold induction of CDC2–cyclin B associated kinase activity as compared to control cells [76]. Both anti–CD3 and concanavalin A stimulate T cells via surface receptors. Apoptosis induced in anti–CD3 or concanavalin A activated T cells is inhibited by antisense cyclin B oligonucleotides [76]. Apoptosis induced in target cells by fragmentin–2, a granule serine protease produced by natural killer cells, in combination with perforin, a pore–forming cytotoxic granule is also accompanied by activation of CDC2 dependent kinase activity [77]. FT–210 cells which carry a temperature sensitive CDC2 that is degraded at the restrictive temperature (39°C) were resistant to apoptosis induced by fragmentin–2 and perforin [77] at 39°C.

Apoptosis in HeLa cells blocked in S phase and treated with chemicals known to elicit premature mitosis is accompanied by elevated levels of CDK2, CDC2 and cyclin A associated kinase activity as well as elevated protein levels [78]. Expression of dominant negative mutants of CDC2 or CDK2 results in suppression of apoptosis in some cell types [79]. Apoptosis induced by overexpression of c–myc is associated with elevated levels of cyclin A mRNA [80]. Increase in cyclin A and E associated kinase activity has been observed in HIV–1 Tat-induced apoptosis in T cells [81]. Tat–induced apoptosis is inhibited by antisense oligonucleotide comprising the sequence conserved in cyclins A, B and E. Ectopic expression of *PITSLRE*, a *CDC2* related gene, induces apoptosis in Chinese hamster ovary cells [82], although *PITSLRE* has not been shown to be involved in cell cycle progression. Induction of apoptosis via the Fas pathway was shown to be associated with an increase in PITSLRE kinase activity. Fas (also known as APO–1 or CD95) is the cell surface receptor of Fas ligand and triggers apoptosis upon binding to Fas ligand [83].

Activation of cyclin dependent kinases is however not a requirement for all forms of apoptotic death. Antisense cyclin B oligonucleotides prevent apoptosis in T cells activated by anti–CD3 but have no effect on dexamethasone induced apoptosis in the same cell type [76]. CDC2–cyclin kinase activity is not involved in apoptosis of fibroblasts following serum withdrawal [84] or apoptosis of thymocytes following treatment with etoposide or dexamethasone [85]. FT–210 cells are not resistant to apoptosis induced by treatment of cells with actinomycin D (an RNA synthesis inhibitor), VP–16 (a topoisomerase II inhibitor) or ceramide (a sphingolipid implicated in several forms of apoptosis including that induced by Fas and TNF) at the restrictive temperature [86]. These results imply that CDC2 associated kinase does not regulate apoptosis by the agents listed above but leaves open the possibility that other cyclin dependent kinases may be involved.

E2F and Apoptosis

The E2F family of transcription factors plays a key role in regulation of the cell cycle. The DNA binding activity of E2F is enhanced by hetero–dimerization with members of the DP family [38]. Promoters of genes that regulate entry into S phase, such as *c–myc* [38], *cyclins D* [39], *E* [40, 41] and *A* [42] and genes involved in DNA synthesis [38, 87] contain E2F binding sites which are crucial for transcriptional activation. Activity of E2Fs as transcription factors is regulated by their interaction with Rb, p107 and p130 [35]. Rb binds to and negatively regulates E2F–1, –2, and –3. E2F–1 is the best characterized member of the E2F family. Phosphorylation of Rb in late G_1 phase results in the release of the bound E2F–1 which can then transcriptionally activate genes that enable the cell to progress into late G_1 and into S phase. Overexpression of E2F–1 is sufficient to drive the quiescent fibroblasts into S phase and induce DNA synthesis [88–91]. These studies suggest that E2F/DP transcription activity is rate limiting for entry into S phase. In addition to activating genes, E2F–1 may repress transcription when it is bound to its cognate DNA element with Rb [43–47].

Several studies indicate that in addition to making decisions leading to growth, E2F–1 may regulate apoptosis [89–93]. Overexpression of E2F–1 in serum starved or growing fibroblasts can trigger apoptosis [91]. Apoptosis under these conditions has been shown to be partially p53–dependent, that is, in the absence of a functional *p53* gene or in p53–/– fibroblasts, E2F–1 expression does not result in substantial apoptosis. Overexpression of E2F–1 may induce apoptosis by suppressing transcription of genes important for survival of the cell or by prematurely driving the cells into S phase. Coexpression of E2F–1 and p53 induces apoptosis [92]. In this case, although G_1 arrest induced by p53 is overcome by E2F–1, the conflicting signals received by the cell probably lead to the activation of the apoptotic pathway. Recently, mice lacking E2F–1 were generated [94, 95]. Loss of E2F–1 in these mice results in tumorigenesis. This is surprising given that E2F–1 has been considered primarily to be a positive regulator of cell cycle progression and overexpression of E2F–1 leads to oncogenesis [96–98]. In the context of the present discussion, it is, however, interesting to note that thymocytes from E2F–1–/– mice are less susceptible to T cell receptor–stimulated apoptosis than thymocytes from control mice [95]. It now remains to be seen whether the tumor generation in E2F–1–/– mice is the result of defects in apoptosis or in growth suppression. It is interesting to note that E2F overexpression in *Drosophila* imaginal discs leads to apoptosis and induction of the gene for reaper, a cell death inducer in *Drosophila.* [99].

p21$^{\text{WAF–1/Cip1}}$ *and Apoptosis*

p21 inhibits CDK–cyclin kinase activity and PCNA function, both of which are required for transition from G_1 to S phase. Transcriptional activity of the p21$^{WAF–1/Cip1}$ gene is under the control of p53 and this has led to the suggestion that p21 might promote p53–dependent G_1 arrest or apoptosis. p21–/– mice [70, 71] like p53–/– mice [100] exhibit defective DNA damage induced check point control. G_1 arrest following DNA damage is known to be dependent on p53 [101]. Apoptosis of thymocytes or of small intestine crypts was however unaffected by the loss of the *p21* gene suggesting that p21 is required for p53–mediated G_1 arrest but not for apoptosis [70, 71]. In contrast, *p21* transgenic mice thymocytes are hypersensitive to p53–dependent radiation induced cell death [102]. These apparently conflicting results concerning the physiological role of p21 in p53–dependent apoptosis of thymocytes may be reconciled if p21–/– mice undergo apoptosis due to defects in radiation-induced checkpoint control. A recent observation that cells of a p21–/– cell line die on exposure to DNA damage agents due to defects in checkpoint control supports this explanation [103]. The lack of checkpoint control in *p21* null mice is reflected by the induction of multiple rounds of DNA replication in the absence of mitosis. The caveat with this study is that it was performed with transformed cell lines and therefore it may not reflect what happens in vivo with untransformed cells. Future studies will clarify the physiological role of p21 in apoptosis.

Transforming Growth Factor–β1: Cell Cycle Arrest and Apoptosis

TGF–β is a multifunctional cytokine involved in controlling cell cycle progression, cell differentiation and morphogenesis [104]. In an inherited form of colon cancer known as **h**ereditary **non****p**olyposis **c**olorectal **c**ancer (HNPCC), genetic instability caused by mismatch repair has been shown to lead to mutations in TGF–β1 [105]. In certain cell types, TGF–β1 inhibits proliferation by preventing the phosphorylation of Rb by CDK–cyclin kinases leading to an arrest in G_1 phase [48]. Cell cycle arrest by TGF–β1 may be mediated through inhibition of expression of G_1 cyclins [106] or of CDK4 [52]. TGF–β1 can also increase the expression of cyclin kinase inhibitors, p14/p15^{INK4b} [31] and p21$^{WAF-1/Cip1}$ [107]. All these events result in inhibition of CDK–cyclin kinase activity. Recently, TGF–β1 has been shown to transcriptionally activate both p14/p15^{INK4b} [108] and p21 [109].

While several investigators have shown that TGF–β1 induces apoptosis, the effect of TGF–β1 on apoptosis appears to be variable [110–114]. It is not clear whether the differences observed are due to cell type or the state of transformation of the cell. We have found that TGF–β1 inhibits T cell death induced by activation via the T cell receptor [115] probably due to inhibition of cyclin associated kinase activity [76] through the induction of CDK–cyclin kinase inhibitors [107]. Recently, apoptosis mediated by anti–Fas antibody was shown to be partially inhibited in the presence of TGF–β1 [114]. A clue to why TGF–β1 has variable effects on apoptosis is suggested by the observation that TGF–β1 acts as a growth inhibitor only in the presence of Rb [116]. In the absence of Rb, TGF–β1 has growth stimulatory effects. The ability of TGF–β1 to inhibit or to induce apoptosis may therefore be determined by Rb status of the cell.

p53 Tumor Suppressor and Apoptosis

p53 induces apoptosis in skin [117, 118] and many cell types including thymocytes [119, 120] and intestinal epithelium cells [121, 122] in response to DNA damage. The intestinal cells and thymocytes in p53–/– mice are insensitive to irradiation induced death [119, 120, 122]. In contrast, glucocorticoid–mediated apoptosis of thymocytes is p53–independent in these mice [119, 120]. Apoptosis which occurs in response to withdrawal of serum factors is p53–dependent. Hematopoetic progenitors from *p53–/–* mice exhibit reduced apoptosis following deprivation of growth factors [123]. Reduction of apoptotic death upon growth factor withdrawal has been observed in leukemic cells in which p53 is inactivated [124, 125]. Similarly, p53 is involved in death due to metabolite imbalance [126] and after exposure to drugs such as PALA that do not damage DNA [127]. Apoptosis induced by E1A (viral oncoprotein) [128, 129], c–myc [130, 131], or E2F transcription factors [89, 92] also requires p53. p53–dependent apoptosis induced by expression of proteins, such as E1A, E2F and Myc, which drive cell cycle progression, suggests that a conflict of signals between growth induction and inhibition triggers apoptosis.

Several p53–inducible genes have a potential for contributing to apoptosis. p53 activates transcription of p21$^{WAF1/Cip1}$, an inhibitor of CDK–cyclin activity [13]. In this context, *p21* transgenic mice thymocytes have been found to be hypersensitive to p53–dependent radiation induced cell death [102]. p53 also transcriptionally activates *bax*, an inducer of apoptosis [132]. Irradiation of the whole mouse results in rapid expression of Bax in radiosensitive organs, followed by massive apoptosis at these sites [62]. Although Bax expression has been shown to be p53–dependent [133], p53–mediated apoptosis can occur without changes in levels of Bax [134, 135] and thymocytes from bax/– mice show a normal apoptotic response following exposure to ionizing radiation [136]. The induction of IGF–BP3 (Insulin–like growth factor binding protein 3), a protein which inhibits mitogenic signaling by Insulin–like growth factor (IGF–1), by p53 suggests a role for this protein in p53–dependent growth control [63]. Failure of mutant forms of p53 to activate IGF–BP3 transcription has been associated with impaired apoptotic function [137]. p53 activates transcription of the gene for Fas receptor thereby suggesting that p53 may play a role in Fas–Fas ligand dependent apoptosis [65]. p53 also suppresses transcription of genes from specific promoters. Of substantial interest in the context of apoptosis control, p53 inhibits the expression of Bcl–2 [138]. Bcl–2 is an inhibitor of apoptosis and has been shown to suppress p53–dependent apoptosis [139–141].

Loss of p53–mediated apoptosis has been implicated in tumor progression both in mice [142] and in humans [143]. Mice bearing an intact transgene for Simian virus 40 large T antigen develop aggressive brain tumors as a result of sequestration of both Rb and p53, while mice expressing a truncated T antigen, capable of binding Rb but not p53, develop slow growing tumors [142]. The slow growth of tumors observed in this case is p53–dependent, as the same truncated T antigen expressed in p53–/– mice elicits aggressive tumors. The p53–dependent reduction in tumor growth is accompanied by extensive apoptosis. Similarly, p53–mediated apoptosis has been shown to protect transformation of cells by E1A, another oncogenic viral product that binds Rb and inactivates its normal function [144].

The two pathways, G_1 arrest and apoptosis, represent distinct functions of p53 and the cellular response to p53 activation may depend partly on which group of genes becomes activated: the genes that lead to G_1 arrest or those that have a role in apoptosis [137, 145]. The nature of the cell's response to p53 activation also rests on the cell type in question and on environmental cues such as the presence or absence of survival factors. For example, thymocytes undergo p53–dependent apoptosis upon irradiation [120, 121]. In contrast, fibroblasts irradiated with similar doses arrest in G_1 [146]. Irradiation of murine hematopoietic cell lines in the presence of IL–3 induces G_1 arrest while in the absence of IL–3 the irradiated cells undergo apoptosis [134]. Analysis of tumor–derived *p53* point mutations (22/23 mutation in the p53 transactivation domain) that fail to activate transcription reveal that p53–mediated apoptosis requires transcription–activation function of p53 in some cell types [147, 148] but not in others [149].

While the ability of p53 to transactivate transcription has been correlated with apoptosis, p53–dependent death in the absence of transcriptional activation has also been demonstrated [131, 150]. These results have led to the proposal that p53 controls apoptosis as a result of its participation in a variety of DNA repair processes [151–154]. Alternatively, p53 may suppress transcription of survival genes.

Retinoblastoma Protein and Apoptosis

Oncogenic viral proteins, including adenovirus E1A and human papillomavirus E7, bind Rb directly and reverse its growth suppressive effects [155, 156]. The observation that E1A and E7 can induce apoptosis suggests that Rb may inhibit apoptosis [157, 158]. Overexpression of E7 in human fibroblasts leads to apoptosis upon irradiation, whereas a similar radiation exposure leads to growth arrest in the absence of E7 [158]. Further evidence that Rb may inhibit apoptosis came from the examination of lens differentiation in Rb–/– mouse embryos [159] and from inactivation of Rb function by expression of E7 in lens fiber cells in transgenic mice [160]. Rb–/– mice die in mid–gestation and there is widespread cell death in the nervous system [161–163]. In Rb–/– mouse embryos, lens fibers which normally exit the cell cycle and differentiate, continue to duplicate DNA and die by apoptosis. Apoptosis in the lens is p53–dependent since embryos that are negative for both p53 and Rb undergo significantly less death [159]. Further, embryos doubly transgenic for *E7* (which binds Rb) and *E6* (which target p53 for degradation) expression display reduced apoptosis in the lens and adult mice go on to develop lens tumors [160].

Table 4.1. Summary of the effects of cell cycle modulators on apoptosis

Gene	Effect on Cell Cycle Progression	Effect on Apoptosis	Effect on Tumor Generation
Cyclin dependent kinases	positive regulator	induces apoptosis	——[a]
Rb	negative regulator	inhibits apoptosis	tumor suppressor
p53	negative regulator	induces apoptosis	tumor suppressor
p21	negative regulator	complex effects	tumor suppressor
TGF–β1	negative regulator[b]	complex effects	variable effect
E2F–1	positive regulator	induces apoptosis	tumor suppressor
c–myc	positive regulator	induces apoptosis	oncogene

[a] Cyclins have been implicated in tumor growth [10]. Cyclin D1 has antiproliferative effects in mammary epithelium and retina [175, 176].
[b] Negative regulator under certain conditions.

Ionizing radiation induced apoptosis of SAOS–2 cells, which do not express Rb, was inhibited by the expression of Rb but not by a mutant form of Rb that fails to complex with E2F [164]. Rb exerts its growth inhibitory effect by binding to transcription factors of the E2F family that regulates progression from G_1 into S phase. Phosphorylation of Rb by CDK–cyclin kinases results in activation of E2F–1 and execution of S phase. Interestingly, the observations of Morgenbesser et al [159], and those of Pan and Griep [160] suggest that in the absence of Rb, p53 induces apoptosis. This link between Rb and p53 can be rationalized as follows. p53 induces the expression of p21, an inhibitor of cyclin dependent kinases. Possibly, in the absence of Rb, p53 attempts to arrest cell cycle progression through inhibition of CDK–cyclin kinases. In the absence of Rb the conflicting signals, stimulation of entry into S phase by E2F and inhibition of cell cycle progression by p53 may result in apoptosis [165]. The cooperation between p53 and E2F to induce apoptosis is also supported by other reports [91, 92].

c–Myc and Apoptosis

The *c–myc* proto–oncogene is implicated in control of normal cell proliferation. *c–myc* can cooperate with other oncogenes such as *ras* to induce tumors in vitro in cell lines and in vivo in transgenic mice models [166]. More recently, *c–myc* has been implicated in apoptosis. It was shown that inducible expression of *c–myc* in *rat–1* fibroblast cells lead to apoptosis [167]. Similarly, antisense

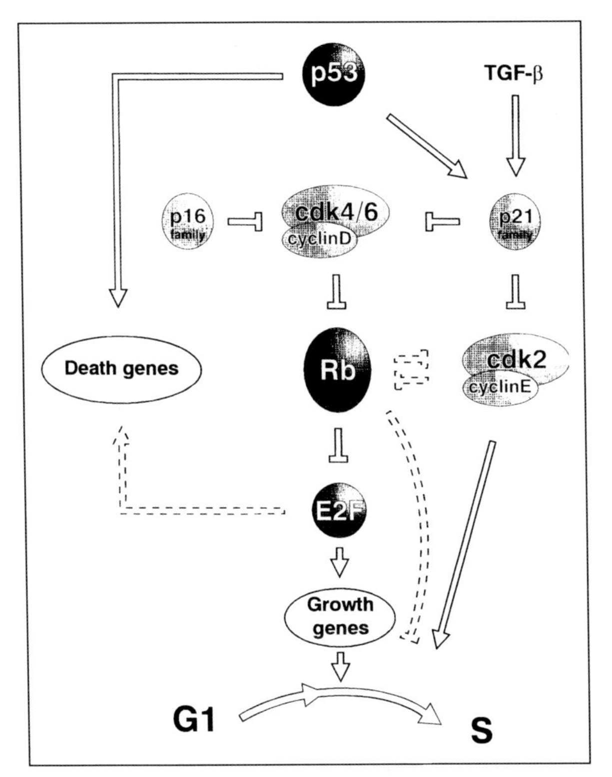

Fig. 4.1. Regulators of cell cycle and apoptosis. Mechanisms by which different cell cycle regulators may cooperate to induce apoptosis. Hypophosphorylated Rb associates with and inactivates E2F family of transcription factors which are responsible for transcription of growth related genes and entry into S phase. CDK–cyclin D and probably cyclin E phosphorylate Rb leading to its inactivation and the release of E2F. Failure to release Rb from E2F leads to a G_1 block. Presence or absence of Rb appears to be an important factor in determining whether the cell proliferates or not in response to a certain signal (such as TGFβ1) and whether the cell undergoes apoptosis or not. Overexpression of E2F induces inappropriate entry into S phase which may directly trigger apoptosis. Although E2F is a positive regulator of cell cycle, lack of the *E2F* gene leads to tumor generation and failure of some cells to undergo apoptosis. E2F may regulate expression of genes involved in apoptosis although there is no evidence at present. p53 may function by inducing p21 or by inducing the expression of genes involved in apoptosis (such as bax). p53 may also act by inhibiting the expression of survival genes or through a direct effect on DNA repair processes. A solid line represents an established correlation whereas a dotted line represents a suspected association.

oligonucleotides corresponding to *c–myc* block activation induced cell death in T cell hybridomas [168]. The ability of c–Myc to induce cell death is p53–dependent in some cell types [131, 169, 170]. c–Myc is a transcriptional factor that forms a heterodimer with another protein called Max. The complex of Myc–Max then binds to specific DNA sequences and activates transcription [171]. One target of the c–Myc transactivator is the ornithine decarboxylase enzyme, which by itself can induces apoptosis [172]. The cell cycle progression induced by c–Myc is independent of the ability of c–Myc to induce cell death [173]. Interestingly, another proto–oncogene, *CDC25*, is a transcriptional target of c–Myc [174]. CDC25 is a phosphatase which is required during the cell cycle to activate CDK1 (CDC2) kinases. Like c–Myc, the CDC25 phosphatase can also induce apoptosis in cells deprived of growth factor. Importantly, apoptosis induced by c–Myc is CDC25 dependent [174].

Conclusion

There is a consensus emerging from studies evaluating the role of cell cycle regulators on cell death and transformation (see Table 4.1 for summary). The simplistic idea that tumor suppressors induce cell death is clearly not universal. A straightforward comparison of the characteristics of such regulators reveals some correlations: a positive cell cycle regulator induces apoptosis; a negative cell cycle regulator and a tumor suppressor inhibits apoptosis; and an oncogene, a positive cell cycle regulator, induces apoptosis. The best value of such correlations, which is not necessarily a static conclusion, provides an evolving conceptual framework to evaluate this rapidly expanding literature. It is also evident that tumor progression undermines the normal cell cycle linkage to apoptosis.

It is now clear that apoptosis is regulated by inhibitors and activators of cell cycle. In this respect, it is evident that p53, Rb and E2F play a central role in regulating the cell cycle and can induce apoptosis if a checkpoint control is overridden (Fig. 4.1). It remains for future studies to provide an insight into the mechanism that links the activity of the various cell cycle components to activation of apoptosis.

Acknowledgments

The work in the authors laboratory was supported by grants from Program Région Rhone Alpes, and ARC to R.F. and by grants from American Cancer Society and NIH to A.F.

References

1 Thompson CB. Apoptosis in the pathogenesis and treatment of disease. Science 1995; 267:1456–1462.
2 Sherr CJ. G_1 phase progression: cycling on cue. Cell 1994; 79:551–555.
3 Pines J. Cyclins and cyclin–dependent kinases: take your partners. Trends Biochem Sci 1993; 18:195–197.
4 Knoblich JA, Sauer K, Jones L et al. Cyclin E controls S phase progression and its down–regulation during *Drosophila* embryogenesis is required for the arrest of cell proliferation. Cell 1994; 77:107–120.
5 Ohtsubo M, Roberts JM. Cyclin–dependent regulation of G_1 in mammalian fibroblasts. Science 1993; 259:1908–1912.
6 Fotedar R, Fotedar A. Cell cycle and control of DNA replication. In: Meijer L, Guidet S, Tung HYL, eds. Progress in Cell Cycle Research. New York: Plenum Press, 1995(1):73–89.
7 King RW, Jackson PK, Kirschner MW. Mitosis in transition. Cell 1994; 79:563–571.
8 Nigg AE. Cyclin–dependent protein kinases: key regulators of eukaryotic cell cyle. BioEssays 1995; 17:471–480.
9 Elledge SJ, Harper JW. Cdk inhibitors: on the threshold of checkpoints and development. Curr Opin Cell Biol 1994; 6:847–852.
10 Hunter T, Pines J. Cyclins and cancer. II: Cyclin D and CDK inhibitors come of age. Cell 1994; 79:573–582.
11 Sherr CJ, Roberts JM. Inhibitors of mammalian G_1 cyclin–dependent kinases. Genes Dev 1995; 9275:1149–1163.
12 Su JY, Rempel RE, Erikson E et al. Cloning and characterization of the Xenopus cyclin–dependent kinase inhibitor $p27^{XIC1}$. Proc Natl Acad Sci USA 1995; 92:10187–10191.
13 el–Deiry WS, Tokino T, Velculescu VE et al. WAF1, a potential mediator of p53 tumor suppression. Cell 1993; 75:817–825.
14 Gu Y, Turck CW, Morgan DO. Inhibition of CDK2 activity in vivo by an associated 20K regulatory subunit. Nature 1993; 366:707–710.
15 Harper JW, Adami GR, Wei N et al. The p21 Cdk–interacting protein Cip1 is a potent inhibitor of G_1 cyclin–dependent kinases. Cell 1993; 75:805–816.
16 Xiong Y, Hannon GJ, Zhang H et al. p21 is a universal inhibitor of cyclin kinases. Nature 1993; 366:701–704.
17 Noda A, Ning Y, Venable SF et al. Cloning of senescent cell–derived inhibitors of DNA synthesis using an expression screen. Exp Cell Res 1994; 211:90–98.
18 Polyak K, Lee MH, Erdjument–Bromage H et al. Cloning of $p27^{Kip1}$, a cyclin–dependent kinase inhibitor and a potential mediator of extracellular antimitogenic signals. Cell 1994; 78:59–66.
19 Toyoshima H, Hunter T. p27, a novel inhibitor of G_1 cyclin–Cdk protein kinase activity, is related to p21. Cell 1994; 78:67–74.
20 Lee MH, Reynisdottir I, Massague J. Cloning of $p57^{KIP2}$, a cyclin–dependent kinase inhibitor with unique domain structure and tissue distribution. Genes Dev 1995; 9:639–649.
21 Matsuoka S, Edwards MC, Bai C et al. $p57^{KIP2}$, a structurally distinct member of the $p21^{CIP1}$ Cdk inhibitor family, is a candidate tumor suppressor gene. Genes Dev 1995; 9:650–662.
22 Chen J, Jackson PK, Kirschner MW et al. Separate domains of p21 involved in the inhibition of Cdk kinase and PCNA. Nature 1995; 374:386–388.
23 Goubin F, Ducommun B. Identification of binding domains on the $p21^{Cip1}$ cyclin–dependent kinase inhibitor. Oncogene 1995; 10:2281–2287.
24 Luo Y, Hurwitz J, Massague J. Cell–cycle inhibition by independent CDK and PCNA binding domains in p21Cip1. Nature 1995; 375:159–161.
25 Nakanishi M, Robetorye RS, Adami GR et al. Identification of the active region of the DNA synthesis inhibitory gene $p21^{Sdi1/CIP1/WAF1}$. EMBO J 1995; 14:555–563.
26 Warbrick E, Lane DP, Glover DM et al. A small peptide inhibitor of DNA replication defines the site of interaction between the cyclin–dependent kinase inhibitor $p21^{WAF1}$ and proliferating cell nuclear antigen. Curr Biol 1995; 5:275–282.

27 Fotedar R, Fitzgerald P, Rousselle T et al. p21 contains independent binding sites for cyclin and cdk2: Both sites are required to inhibit cdk2 kinase activity. Oncogene 1996; 12:2155–2164.
28 Michieli P, Chedid M, Lin D et al. Induction of WAF1/CIP1 by a p53–independent pathway. Cancer Res 1994; 54:3391–3395.
29 Sheikh MS, Li XS, Chen JC et al. Mechanisms of regulation of *WAF1/Cip1* gene expression in human breast carcinoma: role of p53–dependent and independent signal transduction pathways. Oncogene 1994; 9:3407–3415.
30 Guan KL, Jenkins CW, Li Y et al. Growth suppression by p18, a $p16^{INK4/MTS1-}$ and $p14^{INK4B/MTS2}$– related CDK6 inhibitor, correlates with wild–type pRb function. Genes Dev 1994; 8:2939–2952.
31 Hannon GJ, Beach D. $p15^{INK4B}$ is a potential effector of TGF–beta–induced cell cycle arrest. Nature 1994; 371:257–261.
32 Serrano M, Hannon GJ, Beach D. A new regulatory motif in cell–cycle control causing specific inhibition of cyclin D/CDK4. Nature 1993; 366:704–707.
33 Hirai H, Roussel MF, Kato JY et al. Novel INK4 proteins, p19 and p18, are specific inhibitors of the cyclin D–dependent kinases CDK4 and CDK6. Mol Cell Biol 1995; 15:2672–2681.
34 Chan FK, Zhang J, Cheng L et al. Identification of human and mouse p19, a novel CDK4 and CDK6 inhibitor with homology to $p16^{ink4}$. Mol Cell Biol 1995; 15:2682–2688.
35 Weinberg RA. The retinoblastoma protein and cell cycle control. Cell 1995; 81:323–330.
36 Cox LS, Lane DP. Tumor suppressors, kinases and clamps: how p53 regulates the cell cycle in response to DNA damage. BioEssays 1995; 17:501–508.
37 Horowitz JM, Park SH, Bogenmann E et al. Frequent inactivation of the retinoblastoma anti–oncogene is restricted to a subset of human tumor cells. Proc Natl Acad Sci USA 1990; 87:2775–2779.
38 La Thangue NB. DP and E2F proteins: components of a heterodimeric transcription factor implicated in cell cycle control. Curr Opin Cell Biol 1994; 6:443–450.
39 Sala A, Nicolaides NC, Engelhard A et al. Correlation between E2F–1 requirement in the S phase and E2F–1 transactivation of cell cycle–related genes in human cells. Cancer Res 1994; 54:1402–1406.
40 Ohtani K, DeGregori J, Nevins JR. Regulation of *cyclin E* gene by transcription factor E2F1. Proc Natl Acad Sci USA 1995; 92:12146–12150.
41 Botz J, Zerfass–Thome K, Spitkovsky D et al. Cell cycle regulation of the murine *cyclin E* gene depends on an E2F binding site in the promoter. Mol Cell Biol 1996; 16:3401–3409.
42 Schulze A, Zerfass K, Spitkovsky D et al. Cell cycle regulation of the cyclin A gene promoter is mediated by a variant E2F site. Proc Natl Acad Sci USA 1995; 92:11264–11268.
43 Weintraub SJ, Prater CA, Dean DC. Retinoblastoma protein switches the E2F site from positive to negative element. Nature 1992; 358:259–261.
44 Lam EW, Watson RJ. An E2F–binding site mediates cell–cycle regulated repression of mouse *B–myb* transcription. EMBO J 1993; 12:2705–2713.
45 Qin XQ, Livingston DM, Ewen M et al. The transcription factor E2F–1 is a downstream target of RB action. Mol Cell Biol 1995; 15:742–755.
46 Weintraub SJ, Chow KN, Luo RX et al. Mechanism of active transcriptional repression by the retinoblastoma protein. Nature 1995; 375:812–815.
47 Zwicker J, Liu N, Engeland K et al. Cell cycle regulation of E2F site occupation in vivo. Science 1996; 271:1595–1597.
48 Laiho M, DeCaprio JA, Ludlow JW et al. Growth inhibition by TGF–beta linked to suppression of retinoblastoma protein phosphorylation. Cell 1990; 62:175–185.
49 Kato JY, Matsuoka M, Polyak K et al. Cyclic AMP–induced G_1 phase arrest mediated by an inhibitor ($p27^{Kip1}$) of cyclin–dependent kinase 4 activation. Cell 1994; 79:487–496.
50 Polyak K, Kato JY, Solomon MJ et al. $p27^{Kip1}$, a cyclin–Cdk inhibitor, links transforming growth factor–beta and contact inhibition to cell cycle arrest. Genes Dev 1994; 8:9–22.
51 Slingerland JM, Hengst L, Pan CH et al. A novel inhibitor of cyclin–Cdk activity detected in transforming growth factor beta–arrested epithelial cells. Mol Cell Biol 1994; 14:3683–3694.
52 Ewen ME, Sluss HK, Whitehouse LL et al. TGF beta inhibition of Cdk4 synthesis is linked to cell cycle arrest. Cell 1993; 74:1009–1020.

53 Hollstein M, Sidransky D, Vogelstein B et al. p53 mutations in human cancers. Science 1991; 253:49–53.
54 Donehower LA, Harvey M, Slagle BL et al. Mice deficient for p53 are developmentally normal but susceptible to spontaneous tumors. Nature 1992; 356:215–221.
55 Jacks T, Remington L, Williams BO et al. Tumor spectrum analysis in *p53*–mutant mice. Curr Biol 1994; 4:1–7.
56 Malkin D, Li FP, Strong LC et al. Germ line *p53* mutations in a familial syndrome of breast cancer, sarcomas, and other neoplasms. Science 1990; 250:1233–1238.
57 Srivastava S, Zou ZQ, Pirollo K et al. Germ–line transmission of a mutated *p53* gene in a cancer–prone family with Li–Fraumeni syndrome. Nature 1990; 348:747–749.
58 Zhan Q, Bae I, Kastan MB et al. The p53–dependent gamma–ray response of GADD45. Cancer Res 1994; 54:2755–2760.
59 Tamura K, Kanaoka Y, Jinno S et al. Cyclin G: a new mammalian cyclin with homology to fission yeast Cig1. Oncogene 1993; 8:2113–2118.
60 Okamoto K, Beach D. Cyclin G is a transcriptional target of the p53 tumor suppressor protein. EMBO J 1994; 13:4816–4822.
61 Zauberman A, Lupo A, Oren M. Identification of p53 target genes through immune selection of genomic DNA: the cyclin G gene contains two distinct p53 binding sites. Oncogene 1995; 10:2361–2366.
62 Kitada S, Krajewski S, Miyashita T et al. Γ–Radiation induces upregulation of Bax protein and apoptosis in radiosensitive cells in vivo. Oncogene 1996; 12:187–192.
63 Buckbinder L, Talbott R, Velasco–Miguel S et al. Induction of the growth inhibitor IGF–binding protein 3 by p53. Nature 1995; 377:646–649.
64 Barak Y, Juven T, Haffner R et al. *mdm2* expression is induced by wild type p53 activity. EMBO J 1993; 12:461–468.
65 Owen–Schaub LB, Zhang W, Cusack JC et al. Wild–type human p53 and a temperature–sensitive mutant induce Fas/APO–1 expression. Mol Cell Biol 1995; 15:3032–3040.
66 Smith ML, Chen IT, Zhan Q et al. Interaction of the p53–regulated protein Gadd45 with proliferating cell nuclear antigen. Science 1994; 266:1376–1380.
67 Hall PA, Kearsey JM, Coates PJ et al. Characterization of the interaction between PCNA and Gadd45. Oncogene 1995; 10:2427–2433.
68 Kearsey JM, Coates PJ, Prescott AR et al. Gadd45 is a nuclear cell cycle regulated protein which interacts with $p21^{Cip1}$. Oncogene 1995; 11:1675–1683.
69 Zhan Q, Carrier F, Fornace AJ, Jr. Induction of cellular p53 activity by DNA–damaging agents and growth arrest. Mol Cell Biol 1993; 13:4242–4250.
70 Brugarolas J, Chandrasekaran C, Gordon JI et al. Radiation–induced cell cycle arrest compromised by p21 deficiency. Nature 1995; 377:552–557.
71 Deng C, Zhang P, Harper JW et al. Mice lacking p21CIP1/WAF1 undergo normal development, but are defective in G_1 checkpoint control. Cell 1995; 82:675–684.
72 Waldman T, Kinzler KW, Vogelstein B. p21 is necessary for the p53–mediated G_1 arrest in human cancer cells. Cancer Res 1995; 55:5187–5190.
73 Polyak K, Waldman T, He T–C et al. Genetic determinants of p53–induced apoptosis and growth arrest. Genes Dev 1996; 10:1945–1952.
74 DeGregori J, Leone G, Ohtani K et al. E2F–1 accumulation bypasses a G_1 arrest resulting from the inhibition of G_1 cyclin–dependent kinase activity. Genes Dev 1995; 9:2873–2887.
75 Mann DJ, Jones NC. E2F–1 but not E2F–4 can overcome p16–induced G_1 cell–cycle arrest. Curr Biol 1996; 6:474–483.
76 Fotedar R, Flatt J, Gupta S et al. Activation–induced T cell death is cell cycle dependent and regulated by cyclin B. Mol Cell Biol 1995; 15:932–942.
77 Shi L, Nishioka WK, Th'ng J et al. Premature $p34^{cdc2}$ activation required for apoptosis. Science 1994; 263:1143–1145.
78 Meikrantz W, Gisselbrecht S, Tam SW et al. Activation of cyclin A–dependent protein kinases during apoptosis. Proc Natl Acad Sci USA 1994; 91:3754–3758.
79 Meikrantz W, Schlegel R. Suppression of apoptosis by dominant negative mutants of cyclin–dependent protein kinases. J Biol Chem 1996; 271:10205–10209.

80 Hoang AT, Cohen KJ, Barrett JF et al. Participation of cyclin A in Myc–induced apoptosis. Proc Natl Acad Sci USA 1994; 91:6875–6879.
81 Li CJ, Wang C, Pardee AB. Induction of apoptosis by beta–lapachone in human prostate cancer cells. Cancer Res 1995; 55:3712–3715.
82 Lahti JM, Xiang J, Heath LS et al. PITSLRE protein kinase activity is associated with apoptosis. Mol Cell Biol 1995; 15:1–11.
83 Nagata S, Golstein P. The Fas death factor. Science 1995; 267:1449–1456.
84 Oberhammer FA, Hochegger K, Froschl G et al. Chromatin condensation during apoptosis is accompanied by degradation of lamin A+B, without enhanced activation of cdc2 kinase. J Cell Biol 1994; 126:827–837.
85 Norbury C, MacFarlane M, Fearnhead H et al. Cdc2 activation is not required for thymocyte apoptosis. Biochem Biophys Res Commun 1994; 202:1400–1406.
86 Martin SJ, McGahon AJ, Nishioka WK et al. $p34^{cdc2}$ and apoptosis [letter]. Science 1995; 269:106–107.
87 DeGregori J, Kowalik T, Nevins JR. Cellular targets for activation by the E2F1 transcription factor include DNA synthesis– and G_1/S–regulatory genes. Mol Cell Biol 1995; 15:4215–4224.
88 Johnson DG, Schwarz JK, Cress WD et al. Expression of transcription factor E2F1 induces quiescent cells to enter S phase. Nature 1993; 365:349–352.
89 Qin XQ, Livingston DM, Kaelin WG, Jr. et al. Deregulated transcription factor E2F–1 expression leads to S–phase entry and p53–mediated apoptosis. Proc Natl Acad Sci USA 1994; 91:10918–10922.
90 Shan B, Lee WH. Deregulated expression of E2F–1 induces S–phase entry and leads to apoptosis. Mol Cell Biol 1994; 14:8166–8173.
91 Kowalik TF, DeGregori J, Schwarz JK et al. E2F1 overexpression in quiescent fibroblasts leads to induction of cellular DNA synthesis and apoptosis. J Virol 1995; 69:2491–2500.
92 Wu X, Levine AJ. p53 and E2F–1 cooperate to mediate apoptosis. Proc Natl Acad Sci USA 1994; 91:3602–3606.
93 Hiebert SW, Packham G, Strom DK et al. E2F–1:DP–1 induces p53 and overrides survival factors to trigger apoptosis. Mol Cell Biol 1995; 15:6864–6874.
94 Yamasaki L, Jacks T, Bronson R et al. Tumor induction and tissue atrophy in mice lacking E2F–1. Cell 1996; 85:537–548.
95 Field SJ, Tsai FY, Kuo F et al. E2F–1 functions in mice to promote apoptosis and suppress proliferation. Cell 1996; 85:549–561.
96 Johnson DG, Cress WD, Jakoi L et al. Oncogenic capacity of the E2F1 gene. Proc Natl Acad Sci USA 1994; 91:12823–12827.
97 Singh P, Wong SH, Hong W. Overexpression of E2F–1 in rat embryo fibroblasts leads to neoplastic transformation. EMBO J 1994; 13:3329–3338.
98 Xu G, Livingston DM, Krek W. Multiple members of the E2F transcription factor family are the products of oncogenes. Proc Natl Acad Sci USA 1995; 92:1357–1361.
99 Asano M, Nevins JR, Wharton RP. Ectopic E2F expression induces S phase and apoptosis in Drosophila imaginal discs. Genes Dev 1996; 10:1422–1432.
100 Kastan MB, Zhan Q, el–Deiry WS et al. A mammalian cell cycle checkpoint pathway utilizing p53 and GADD45 is defective in ataxia–telangiectasia. Cell 1992; 71:587–597.
101 Kastan MB. Signaling to p53: where does it all start? BioEssays 1996; 18:617–619.
102 Fotedar R, Brickner H, Khastilba S et al. p21 is a tumor suppressor and induces apoptosis upon DNA damage. In Preparation.
103 Waldman T, Lengauer C, Kinzler KW et al. Uncoupling of S phase and mitosis induced by anticancer agents in cells lacking p21. Nature 1996; 381:713–716.
104 Massague J, Polyak K. Mammalian antiproliferative signals and their targets. Curr Opin Genet Dev 1995; 5:91–96.
105 Markowitz S, Wang J, Myeroff L et al. Inactivation of the type II TGF–beta receptor in colon cancer cells with microsatellite instability. Science 1995; 2685:1336–1338.
106 Geng Y, Weinberg RA. Transforming growth factor beta effects on expression of G_1 cyclins and cyclin–dependent protein kinases. Proc Natl Acad Sci USA 1993; 90:10315–10319.

107 Reynisdottir I, Polyak K, Iavarone A et al. Kip/Cip and Ink4 Cdk inhibitors cooperate to induce cell cycle arrest in response to TGF–beta. Genes Dev 1995; 9:1831–1845.
108 Li JM, Nichols MA, Chandrasekharan S et al. Transforming growth factor beta activates the promoter of cyclin–dependent kinase inhibitor p15^{INK4B} through an Sp1 consensus site. J Biol Chem 1995; 270:26750–26753.
109 Datto MB, Yu Y, Wang XF. Functional analysis of the transforming growth factor beta responsive elements in the WAF1/Cip1/p21 promoter. J Biol Chem 1995; 270:28623–28628.
110 Oberhammer FA, Pavelka M, Sharma S et al. Induction of apoptosis in cultured hepatocytes and in regressing liver by transforming growth factor beta 1. Proc Natl Acad Sci USA 1992; 89:5408–5412.
111 Alam R, Forsythe P, Stafford S et al. Transforming growth factor beta abrogates the effects of hematopoietins on eosinophils and induces their apoptosis. J Exp Med. 1994; 179:1041–1045.
112 Barlat I, Henglein B, Plet A et al. TGF–beta 1 and cAMP attenuate cyclin A gene transcription via a cAMP responsive element through independent pathways. Oncogene 1995; 11:1309–1318.
113 Lomo J, Blomhoff HK, Beiske K et al. TGF–beta 1 and cyclic AMP promote apoptosis in resting human B lymphocytes. J Immunol 1995; 154:1634–1643.
114 Cerwenka A, Kovar H, Majdic O et al. Fas– and activation–induced apoptosis are reduced in human T cells preactivated in the presence of TGF–beta 1. J Immunol 1996; 156:459–464.
115 Fotedar R, Fotedar A. Unpublished observations.
116 Herrera RE, Makela TP, Weinberg RA. TGFb–induced growth inhibtion in primary fibroblasts requires the retinoblastoma protein. Mol Cell Biol 1996; 7:1335–1342.
117 Ziegler A, Jonason AS, Leffell DJ et al. Sunburn and p53 in the onset of skin cancer. Nature 1994; 372:773–776.
118 Hall PA, McKee PH, Menage HD et al. High levels of p53 protein in UV–irradiated normal human skin. Oncogene 1993; 8:203–207.
119 Clarke AR, Purdie CA, Harrison DJ et al. Thymocyte apoptosis induced by p53–dependent and independent pathways. Nature 1993; 362:849–852.
120 Lowe SW, Schmitt EM, Smith SW et al. p53 is required for radiation–induced apoptosis in mouse thymocytes. Nature 1993; 362:847–849.
121 Clarke AR, Gledhill S, Hooper ML et al. p53 dependence of early apoptotic and proliferative responses within the mouse intestinal epithelium following gamma–*irradiation*. Oncogene 1994; 9:1767–1773.
122 Merritt AJ, Potten CS, Kemp CJ et al. The role of p53 in spontaneous and radiation–induced apoptosis in the gastrointestinal tract of normal and p53–deficient mice. Cancer Res 1994; 54:614–617.
123 Lotem J, Sachs L. Hematopoietic cells from mice deficient in wild–type p53 are more resistant to induction of apoptosis by some agents. Blood 1993; 82:1092–1096.
124 Gottlieb E, Haffner R, von Ruden T et al. Down–regulation of wild–type p53 activity interferes with apoptosis of IL–3–dependent hematopoietic cells following IL–3 withdrawal. EMBO J 1994; 13:1368–1374.
125 Zhu YM, Bradbury DA, Russell NH. Wild–type p53 is required for apoptosis induced by growth factor deprivation in factor–dependent leukemic cells. Br J Cancer 1994; 69:468–472.
126 Yin Y, Tainsky MA, Bischoff FZ et al. Wild–type p53 restores cell cycle control and inhibits gene amplification in cells with mutant p53 alleles. Cell 1992; 70:937–948.
127 Almasan A, Linke SP, Paulson TG et al. Genetic instability as a consequence of inappropriate entry into and progression through S–phase. Cancer Metastasis Rev 1995; 14:59–73.
128 Debbas M, White E. Wild–type p53 mediates apoptosis by E1A, which is inhibited by E1B. Genes Dev 1993; 7:546–554.
129 Lowe SW, Ruley HE. Stabilization of the p53 tumor suppressor is induced by adenovirus 5 E1A and accompanies apoptosis. Genes Dev 1993; 7:535–545.
130 Hermeking H, Eick D. Mediation of c–Myc–induced apoptosis by p53. Science 1994; 265:2091–2093.
131 Wagner AJ, Kokontis JM, Hay N. Myc–mediated apoptosis requires wild–type p53 in a manner independent of cell cycle arrest and the ability of p53 to induce p21$^{waf1/cip1}$. Genes Dev 1994; 8:2817–2830.

132 Miyashita T, Reed JC. Tumor suppressor p53 is a direct transcriptional activator of the human bax gene. Cell 1995; 80:293–299.
133 Zhan Q, Fan S, Bae I et al. Induction of *bax* by genotoxic stress in human cells correlates with normal p53 status and apoptosis. Oncogene 1994; 9:3743–3751.
134 Canman CE, Gilmer TM, Coutts SB et al. Growth factor modulation of p53–mediated growth arrest versus apoptosis. Genes Dev 1995; 9:600–611.
135 Allday MJ, Inman GJ, Crawford DH et al. DNA damage in human B cells can induce apoptosis, proceeding from G_1/S when p53 is transactivation competent and G_2/M when it is transactivation defective. EMBO J 1995; 14:4994–5005.
136 Knudson CM, Tung KS, Tourtellotte WG et al. Bax–deficient mice with lymphoid hyperplasia and male germ cell death. Science 1995; 270:96–99.
137 Ludwig RL, Bates S, Voudsen KH. Differential activation of target cellular promoters by p53 mutants with impaired apoptotic function. Mol Cell Biol 1996; 16:4952–4960.
138 Miyashita T, Harigai M, Hanada M et al. Identification of a p53–dependent negative response element in the bcl–2 gene. Cancer Res 1994; 54:3131–3135.
139 Wang Y, Szekely L, Okan I et al. Wild type p53–triggered apoptosis is inhibited by bcl–2 in a v–myc–induced T cell lymphoma line. Oncogene 1993; 8:3427–3431.
140 Chiou SK, Rao L, White E. Bcl–2 blocks p53–dependent apoptosis. Mol Cell Biol 1994; 14:2556–2563.
141 Strasser A, Harris AW, Jacks T et al. DNA damage can induce apoptosis in proliferating lymphoid cells via p53– independent mechanisms inhibitable by Bcl–2. Cell 1994; 79:329–339.
142 Symonds H, Krall L, Remington L et al. p53–dependent apoptosis suppresses tumor growth and progression in vivo. Cell 1994; 78:703–711.
143 Bardeesy N, Beckwith JB, Pelletier J. Clonal expansion and attenuated apoptosis in Wilms' tumors are associated with p53 gene mutations. Cancer Res 1995; 55:215–219.
144 Lowe SW, Jacks T, Housman DE et al. Abrogation of oncogene–associated apoptosis allows transformation of p53–deficient cells. Proc Natl Acad Sci USA 1994; 91:2026–2030.
145 Friedlander P, Haupt Y, Prives C et al. A mutant p53 that discriminates between p53–responsive genes cannot induce apoptosis. Mol Cell Biol 1996; 16:4961–4971.
146 Di Leonardo A, Linke SP, Clarkin K et al. DNA damage triggers a prolonged p53–dependent G_1 arrest and long–term induction of Cip1 in normal human fibroblasts. Genes Dev 1994; 8:2540–2551.
147 Sabbatini P, Lin J, Levine AJ et al. Essential role for p53–mediated transcription in E1A–induced apoptosis. Genes Dev 1995; 9:2184–2192.
148 Yonish–Rouach E, Deguin V, Zaitchouk T et al. Transcriptional activation plays a role in the induction of apoptosis by transiently transfected wild–type p53. Oncogene 1996; 12:2197–2205.
149 Haupt Y, Rowan S, Shaulian E et al. Induction of apoptosis in HeLa cells by transactivation–deficient p53. Genes Dev 1995; 9:2170–2183.
150 Caelles C, Helmberg A, Karin M. p53–dependent apoptosis in the absence of transcriptional activation of p53–target genes. Nature 1994; 370:220–223.
151 Oberosler P, Hloch P, Ramsperger U et al. p53–catalyzed annealing of complementary single–stranded nucleic acids. EMBO J 1993; 12:2389–2396.
152 Bakalkin G, Yakovleva T, Selivanova G et al. p53 binds single–stranded DNA ends and catalyzes DNA renaturation and strand transfer. Proc Natl Acad Sci USA 1994; 91:413–417.
153 Brain R, Jenkins JR. Human p53 directs DNA strand reassociation and is photolabeled by 8–azido ATP. Oncogene 1994; 9:1775–1780.
154 Mummenbrauer T, Janus F, Muller B et al. p53 Protein exhibits 3'–to–5' exonuclease activity. Cell 1996; 85:1089–1099.
155 Dyson N, Howley PM, Munger K et al. The human papilloma virus–16 E7 oncoprotein is able to bind to the retinoblastoma gene product. Science 1989; 243:934–937.
156 Whyte P, Williamson NM, Harlow E. Cellular targets for transformation by the adenovirus E1A proteins. Cell 1989; 56:67–75.
157 Rao L, Debbas M, Sabbatini P et al. The adenovirus E1A proteins induce apoptosis, which is inhibited by the E1B 19–kDa and Bcl–2 proteins. Proc Natl Acad Sci USA 1992; 89:7742–7746.

158 White AE, Livanos EM, Tlsty TD. Differential disruption of genomic integrity and cell cycle regulation in normal human fibroblasts by the HPV oncoproteins. Genes Dev 1994; 8:666–677.
159 Morgenbesser SD, Williams BO, Jacks T et al. p53–dependent apoptosis produced by Rb–deficiency in the developing mouse lens. Nature 1994; 371:72–74.
160 Pan H, Griep AE. Temporally distinct patterns of p53–dependent and p53–independent apoptosis during mouse lens development. Genes Dev 1995; 9:2157–2169.
161 Lee EY, Chang CY, Hu N et al. Mice deficient for Rb are nonviable and show defects in neurogenesis and hematopoiesis. Nature 1992; 359:288–294.
162 Jacks T, Fazeli A, Schmitt EM et al. Effects of an Rb mutation in the mouse. Nature 1992; 359:295–300.
163 Clarke AR, Maandag ER, van Roon M et al. Requirement for a functional *Rb–1* gene in murine development. Nature 1992; 359:328–330.
164 Haas–Kogan DA, Kogan SC, Levi D et al. Inhibition of apoptosis by the retinoblastoma gene product. EMBO J 1995; 14:461–472.
165 Almasan A, Yin Y, Kelly RE et al. Deficiency of retinoblastoma protein leads to inappropriate S–phase entry, activation of E2F–responsive genes, and apoptosis. Proc Natl Acad Sci USA 1995; 92:5436–5440.
166 Marcu KB, Bossone SA, Patel AJ. Myc function and regulation. Annu Rev Biochem 1992; 61:809–860.
167 Evan GI, Wyllie AH, Gilbert CS et al. Induction of apoptosis in fibroblasts by c–myc protein. Cell 1992; 69:119–128.
168 Shi Y, Glynn JM, Guilbert LJ et al. Role for c–myc in activation–induced apoptotic cell death in T cell hybridomas. Science 1992; 257:212–214.
169 Wagner AJ, Small MB, Hay N. Myc–mediated apoptosis is blocked by ectopic expression of Bcl–2. Mol Cell Biol 1993; 13:2432–2440.
170 Hermeking H, Funk JO, Reichert M et al. Abrogation of p53–induced cell cycle arrest by c–Myc: evidence for an inhibitor of p21WAF1/CIP1/SDI1. Oncogene 1995; 11:1409–1415.
171 Amati B, Land H. Myc–Max–Mad: a transcription factor network controlling cell cycle progression, differentiation and death. Curr Opin Genet Dev 1994; 4:102–108.
172 Packham G, Cleveland JL. Ornithine decarboxylase is a mediator of c–Myc–induced apoptosis. Mol Cell Biol 1994; 14:5741–5747.
173 Rudolph B, Saffrich R, Zwicker J et al. Activation of cyclin–dependent kinases by Myc mediates induction of cyclin A, but not apoptosis. EMBO J 1996; 15:3065–3076.
174 Galaktionov K, Chen X, Beach D. Cdc25 cell–cycle phosphatase as a target of c–myc. Nature 1996; 382:511–517.

Chapter 5

Apoptosis and Cancer, edited by Seamus J. Martin.

Bcl–2 Family Proteins: Role in Dysregulation of Apoptosis and Chemoresistance in Cancer

John C. Reed

The Burnham Institute, Cancer Research Center, La Jolla, California, U.S.A.

Introduction

The *bcl–2* family of genes are critical regulators of the programmed cell death pathway. The expression of these genes frequently becomes altered in human cancers, thus contributing to neoplastic cell expansion by prolonging cell survival. The first identified member of this gene family, *bcl–2*, was discovered by virtue of its involvement in t(14;18) chromosomal translocations commonly found in lymphomas. Since then however, overexpression of Bcl–2 has been reported in a wide variety of cancers, including prostate, colorectal, lung, renal and other types of solid tumors and leukemias. A variety of experiments have provided conclusive evidence that elevations in Bcl–2 expression cause resistance to chemotherapeutic drugs and radiation, while decreases in Bcl–2 promote apoptotic responses to anticancer drugs and radiotherapy.

Bcl–2 family proteins share structural similarity with the pore–forming domains of certain bacterial toxins, and evidence has been obtained that some of these proteins can create channels in lipid membranes. In addition, protein–protein interactions among members of the Bcl–2 family appear to play an important role as regulators of their function. The anti–apoptotic function of the Bcl–2 protein, for example, may be dependent, at least in part, on its ability to heterodimerize with another member of this family, Bax. In contrast to Bcl–2, overexpression of the Bax protein promotes apoptosis. The relationship between Bax and apoptotic responses of cancer cells to chemotherapeutic drugs is further underscored by the findings that drug–induced damage to DNA has been shown to result in p53–dependent increases in Bax expression and that tumor cells with high levels of Bcl–2 are protected from p53–induced elevations in Bax and thus, fail to undergo apoptosis.

Recent studies of Bcl–2, Bax, and other members of the Bcl–2 protein family with regard to structure–function relationships, interactions with other nonhomologous proteins and regulation by protein phosphorylation are beginning to suggest strategies for potentially abrogating the relative resistance of many types of tumors to apoptotic stimuli such as chemotherapy and radiation.

The Bcl–2 Family of Apoptosis–Regulating Proteins

To date, 14 cellular homologs of Bcl–2 have been described, (Fig. 5.1) including the anti–apoptotic proteins Bcl–X_L, Mcl–1, A1/Bfl–1, Bcl–W, Nr–13 (avian), and Ced–9 (nematode) and the pro–apoptotic proteins Bax, Bcl–X_S, Bad, Bak, Bik and Bid [1–14]. The Bcl–X_L and Bcl–X_S protein arise through alternative mRNA splicing mechanisms from the same gene. An additional human homolog BRAG–1 and two *Xenopus* homologs have also been described whose effects on cell survival and death have not been assessed, but they are most likely suppressors of apoptosis [15,16]. In addition, four homologs of Bcl–2 have been discovered in viruses: E1b–19 kDa (adenovirus), BHRF–1 (Epstein Barr Virus), LMH–5W (African Swine Fever Virus) and ORF–16 (Herpes Saimiri Virus) [17–20].

Sequencing alignment and mutagenesis studies have identified up to four evolutionarily conserved domains within Bcl–2 family proteins which can be important for their function [21, 22]. Though various names for these domains can be found in the literature, we have recently proposed that these be termed BH1, BH2, BH3 and BH4 [21], where BH stands for Bcl–2 Homology Domain as originally suggested by Oltvai et al [2]. For historical reasons, these four domains are unfortunately not ordered sequentially along the protein from NH_2– to COOH–terminus. Most members of the Bcl–2 protein family also contain a stretch of hydrophobic amino acids near their C–termini that result in their post–translational insertion into intracellular membrane (discussed below).

Figure 5.1 depicts the structures of Bcl–2 and its cellular homologs. In the human Bcl–2 protein, the BH1, BH2, BH3 and BH4 domains reside at amino–acid positions 136–155 (BH1), 187–202 (BH2), 93–107 (BH3) and 10–30 (BH4). The transmembrane domain (TM) of Bcl–2 resides at positions 219 to 237. Of interest, the BH4 domain (also known as the A–box) is not found in most pro–apoptotic Bcl–2 family proteins, including Bax, Bak, Bik and Bad, suggesting that this domain may play a unique role in the function of the anti–apoptotic

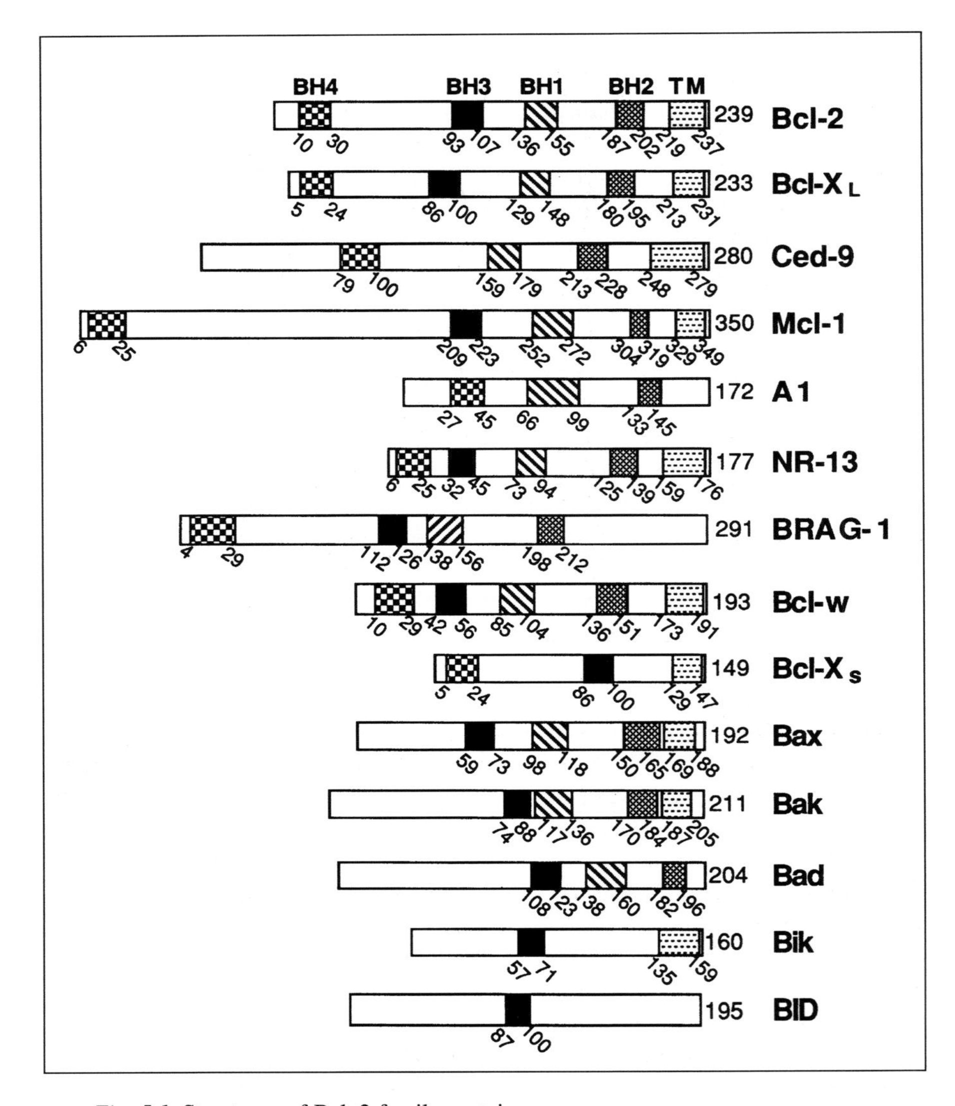

Fig. 5.1. Structures of Bcl–2 family proteins.

proteins, such as Bcl–2, Bcl–X_L, Mcl–1, A1/Bfl–1 and Ced–9. The BH4 domain however is found in the cell death promoting proteins Bcl–X_S, which suggests either that the BH4 domain is not directly involved in the anti–apoptotic function of Bcl–2 family proteins, or more likely, that the Bcl–X_S protein can function at least in part as a dominant–inhibitor of Bcl–2 and its anti–apoptotic ho-

Protein	Start																End	Apoptosis
Bax	(59)	L	S	E	C	L	K	R	I	G	D	E	L	D	S	N	(73)	+
Bak	(74)	V	G	R	Q	L	A	I	I	G	D	D	I	N	R	R	(88)	+
Bik	(57)	L	A	L	R	L	A	C	I	G	D	E	M	D	V	S	(71)	+
BID	(86)	I	A	R	H	L	A	Q	V	G	D	S	M	D	R	S	(100)	+
Hu BAD	(104)	L	W	-	A	A	Q	R	Y	G	R	E	L	-	R	R	(116)	+
Bcl-2	(93)	V	H	L	T	L	R	Q	A	G	D	D	F	S	R	R	(107)	-
Bcl-X_L	(86)	V	K	Q	A	L	R	E	A	G	D	E	F	E	L	R	(100)	-
Mcl-1	(209)	A	L	E	T	L	R	R	V	G	D	G	V	Q	R	N	(223)	-
Nr13	(32)	T	A	A	E	L	R	R	A	A	A	E	L	E	R	R	(45)	-
BRAG-1	(127)	L	G	I	D	T	Y	T	L	L	E	E	M	A	S	R	(141)	-
Bcl-W	(42)	L	H	Q	A	M	R	A	A	G	D	E	F	E	T	R	(56)	-
XR-I	(77)	L	H	S	A	M	R	A	A	G	D	E	F	E	E	R	(91)	-
XR-II	(58)	V	L	Q	A	L	L	E	A	T	E	E	F	E	L	R	(72)	-

Fig. 5.2. BH3 domains of vertebrate Bcl–2 family proteins.

mologs by competing for whatever proteins or protein domains that might normally interact with BH4. Removal of the BH4 domain from Bcl–2 has been shown to either abrogate its anti–apoptotic activity or to convert Bcl–2 from a death suppressor to a death promoter [23, 24].

Of note is the observation that two of the pro–apoptotic proteins Bik and Bid contain only the BH3 domain, implying that the BH3 domain may be uniquely important in the promotion of apoptosis. Indeed, recent deletional analysis of the Bax and Bak proteins suggest that this is the case, as will be discussed below in more detail [21, 25, 26]. Though most anti–apoptotic proteins do contain regions with homology to BH3, there exist clear sequence differences that distinguish the BH3 domains of the pro–apoptotic proteins Bax, Bak and Bik from the anti–apoptotic proteins Bcl–2, Bcl–X_L and Mcl–1 (Fig. 5.2). Moreover, it has recently been demonstrated that substituting the BH3 domain of Bax for the corresponding domain in Bcl–2 converts the Bcl–2 protein from a cell death blocker to a cell death promoter [27].

Interactions Among Bcl–2 Family Proteins

One of the most interesting aspects of Bcl–2 family proteins is their ability in many cases to interact with each other, thus constituting a network of homo– and heterodimers that regulate programmed cell death. For example, Bcl–2 can homodimerize with itself, as well as heterodimerize with Bax, Bcl–X_L, Bcl–X_S, Bad, Bik and Mcl–1. Similarly, Bcl–X_L can interact with the same proteins in addition to Bak [13, 25, 28, 29]. Functional importance for some of these homo– and heterodimerization events has been suggested by experiments where small deletions or single amino acid substitution mutants have been made that alter the ability of Bcl–2 or some of its homologs to interact with themselves or with other proteins. For example, deletion of the BH3 domain from Bax or mutations in conserved residues in this domain prevent Bax/Bax homodimerization and abrogate Bax's function as a promoter of cell death in both mammalian cells and in yeast [21, 26, 30]. Thus, for Bax to promote cell death, it appears that homodimerization is required. However, complicating interpretation of the results, at least in mammalian cells, is the observation that these same mutations that abolish Bax/Bax homodimerization also prevent Bax/Bcl–2 heterodimerization. Consequently, it could be that for Bax to kill, it must be capable of binding to and neutralizing Bcl–2. Given the observations that Bax also promotes cell death in yeast and that yeast seem to lack Bcl–2 homologs, it appears likely that Bax/Bax homodimerization rather than Bax/Bcl–2 heterodimerization is the critical event. Of course, it is entirely possible that Bax promotes cell death through more than one mechanism, i.e., both directly as a Bax/Bax homodimer and indirectly by binding to and preventing Bcl–2 from fulfilling its mission as a death blocker. Indeed, experiments using truncation mutants of Bax that represent little more than the BH3 domain but which retain cytotoxic function in mammalian cells argue that at least one mechanism by which Bax functions is by neutralizing Bcl–2 via its BH3 domain [21, 26, 30]. Similar results have been obtained for Bak, which seems to prefer Bcl–X_L as its heterodimerization partner rather than Bcl–2 [25], though less is known about the effects of BH3 domain mutations on Bak/Bak homodimerization. Nevertheless, these and other studies have implicated the BH3 domains of Bax and Bak as functionally important structures that mediate homo– and heterodimerization events. One caveat with the data suggesting an important functional role for the BH3 domain of Bax is a report that Bax (DBH3) mutants can still retain their ability to antagonize Bcl–X_L–mediated protection from chemotherapy–induced apoptosis under some circumstances [31] in contrast to most other studies where deletion of the BH3 domain abrogated the pro–apoptotic function of Bax [21, 25, 26]. Moreover, the observation that some tumors develop mutations in the BH3 domain proves further argument that this domain is of importance for Bax function [32].

Single amino–acid substitution mutants within the BH1 and BH2 domains of Bcl–2 can prevent binding to Bax and abrogate Bcl–2's protective action against apoptosis but do not impair homodimerization with endogenous wild–type Bcl–2 [33, 34]. The simplest interpretation of these results is that for Bcl–2 to function as a suppressor of cell death, it must be able to form heterodimers with Bax. However, it cannot be entirely excluded that mutants of Bcl–2 that bind to endogenous wild–type Bcl–2 are functioning as dominant inhibitors of Bcl–2, forming dysfunctional Bcl–2/Bcl–2 heterodimers (mutant X wild–type) that cannot prevent cell death. Studies where Bcl–2 and Bax have been expressed in yeast argue in favor of heterodimerization of Bcl–2 with Bax, rather than homodimerization of Bcl–2, as the more significant interaction. For example, when expressed in yeast, Bax induces cell death which can be suppressed by Bcl–2 but not by mutants of Bcl–2 that fail to bind to Bax, whereas expression of Bcl–2 mutants in yeast by themselves have no apparent effect on cell viability [34–36]. Similarly, mutations in the BH1 and BH2 domains of Bcl–X_L have been shown to abrogate the ability of Bcl–X_L to heterodimerize with Bax and to suppress apoptosis [29]. However, some specific mutations in these domains also apparently diminish interactions with Bax as well as with Bak, but do not impair the ability of Bcl–X_L to suppress cell death [31, 37]. Though the effects of this latter type of BH1 and BH2 domain mutation on Bcl–X_L homodimerization or interactions with other Bcl–2 family proteins were not tested, the simplest interpretation of these results is that Bcl–X_L does not necessarily need to dimerize with Bax or Bak to suppress apoptosis. It should be noted, however, that it is currently unclear whether these BH1 and BH2 domain mutants of Bcl–X_L lack all Bax– and Bak–binding activity, versus retaining some ability to bind these pro–apoptotic proteins, albeit with lower affinity.

Other interactions among Bcl–2 family proteins appear to be more straight forward than discerning the role of Bcl–2/Bax and Bcl–X_L/Bak heterodimers. For instance, the pro–apoptotic proteins Bcl–X_S and BAD cannot homodimerize with themselves and do not bind to Bax, Bak or other pro–apoptotic proteins, but can bind to the anti–apoptotic protein Bcl–2 and Bcl–X_L [8, 29, 38]. These proteins appear to function as dominant–inhibitors of Bcl–2 and Bcl–X_L, preventing these anti–apoptotic proteins from either homodimerizing with themselves or from heterodimerizing with Bax, Bak or other Bcl–2 family proteins [8]. One report involving studies of Bcl–X_S, however, has suggested an alternative mechanism, which implies that Bcl–X_S may compete for binding to other types of proteins with which Bcl–2 and Bcl–X_L needs to interact rather than by binding to and preventing Bcl–2 or Bcl–X_L from binding to Bax [39].

Recently the three–dimensional structure of the Bcl–XL protein has been determined by both x–ray crystallographic and NMR–solution methods [40]. The structure of Bcl–X_L provides insights into the probable mechanisms of homo–

and heterodimerization among Bcl–2 family proteins. Specifically, a hydrophobic pocket was found on the surface of the Bcl–X_L protein which is lined by residues that are known from mutagenesis studies to be important for dimerization. Moreover, the BH3 domain was shown to correspond to the second amphiphatic α–helix in the protein. Since the BH3 domains of Bax and Bak are known to be necessary and sufficient for interacting with anti–apoptotic Bcl–2 family proteins, it is speculated that the hydrophobic pocket on the surface of Bcl–X_L binds the hydrophobic side of the amphipathic helix that constitutes the BH3 domain. Interestingly, the BH1, BH2 and BH3 domains of Bcl–X_L make significant contributions to the borders of the pocket that presumably binds to the BH3 domain of Bax or Bak. Thus, an asymmetry exists to the dimerization, with one partner providing the pocket ("receptor") and the other the BH3 domain ("ligand") needed for interaction.

The implication therefore is that for one of the two partners to play the receptor role, it must undergo a conformational change that allows the second α–helix (BH3 domain) to rotate away from the rest of the protein and present its hydrophobic surface (ligand) to the pocket of the corresponding partner protein. Though highly speculative at this time, the propensity to adopt either a "receptor" or "ligand" conformation may dictate whether a member of the Bcl–2 family functions as an anti–apoptotic versus pro–apoptotic protein. If true, then changes in protein conformation may switch the phenotype of a Bcl–2 family protein, from death blocker to death inducer or vice versa, thus potentially providing an explanation for reports of paradoxical behavior among Bcl–2 family proteins such as the blocking of cell death by Bax and Bak seen in some circumstances [10, 41]. Using conformation to get two functions from the same protein could also potentially explain why the nematode *C. elegans* is evidently able to cope with only one homolog of Bcl–2 (ced–9) and yet achieve physiologically appropriate regulation of programmed cell death.

Bcl–2 Family Proteins as Determinants of Chemoresponses and Chemoresistance

A variety of investigations have shown that essentially all currently available anticancer drugs induce tumor cell death by triggering apoptosis (reviewed in 42–46). Thus, while specific chemotherapeutic drugs may have unique primary mechanisms by which they damage tumor cells, presumably they all share the ability to activate cell suicide pathways which culminate in apoptotic cell death.

Elements of the cell death pathway upon which anticancer drugs depend for killing tumor cells is conserved throughout evolution. Components of this pathway have been identified, for example, in the free–living nematode, *C. elegans* [47]. In *C. Elegans*, all 131 cell deaths that normally occur during development of this organism can be blocked by gain of function mutations in the gene *ced–9* (48). The *ced–9* gene is a homolog of Bcl–2 [7]. The finding of homologs of Bcl–2 in viruses also suggests a distant evolutionarily conserved origin for this family of proteins.

Because Bcl–2 is a blocker of programmed cell death [49–52], overexpression of Bcl–2 or its related anti–apoptotic homologs prevents or markedly delays the normal turnover of cells in vivo, thus contributing to neoplastic cell growth by prolonging cell survival rather than by accelerating cell division. The relevance of dysregulated cell death to human cancer can be readily appreciated by consideration of the follicular non–Hodgkin's B cell lymphomas, in which the *bcl–2* gene was first discovered because of its involvement in t(14;18) chromosomal translocations that fuse the *bcl–2* gene from chromosome 18 with the immunoglobulin gene heavy chain locus on chromosome 14 resulting in overexpression of *bcl–2* [53, 54]. The low–grade follicular lymphomas represent the most common type of non–Hodgkin's lymphomas, with ~20,000 new cases reported annually in the United States alone. Cell cycle analysis of these tumors has shown that the malignant cells are almost entirely G_0/G_1–phase resting B cells, which accumulate in patients not because they are dividing more rapidly than normal but because they are failing to turnover at appropriate rates by programmed cell death mechanisms (reviewed in ref. 55). This ability of overexpression of Bcl–2 to produce abnormal expansions of resting B cells by prolonging cell survival has been confirmed in transgenic mouse experiments, where Bcl–2 expression was driven under the influence of B cell–specific immunoglobulin heavy–chain enhancer elements [56–58].

Overexpression of Bcl–2 contributes not only to the origins of cancer but also to difficulties in treating it. This is because Bcl–2 can block or markedly impair the induction of apoptosis by essentially every currently known anticancer drug. In experiments where gene transfer methods have been used to produce elevations in Bcl–2 protein, it has been shown that Bcl–2 renders various types of tumor and leukemia cell lines resistant to induction of apoptosis by a wide variety of anticancer drugs including cyclophosphamide, cisplatin, etoposide (VP16), mitoxantrone, adriamycin, cytosine arabinoside (Ara–C), methotrexate, dexamethasone, nitrogen mustards (BCNU), vincristine, taxol, fludarabine, 2–chlorodeoxyadenosine, 5–fluorouracil and radiation [59–64]. Similar data have recently been obtained for Bcl–X_L, a close cousin of Bcl–2 that also suppresses apoptosis [65, 66]. These findings further strengthen arguments that currently existing drugs have activation of apoptosis as their final common mechanism of cytotoxicity and are consistent with other data which have suggested that Bcl–2 and its homologs regulate a distal step in the evolutionarily conserved cell death pathway [43].

Several groups have performed experiments using antisense technology, demonstrating that downregulation of Bcl–2 protein levels can reverse chemoresistance, rendering malignant cells more sensitive to the cytotoxic effects of conventional anticancer drugs. For example, antisense oligonucleotides and expression plasmids have been shown to markedly increase the sensitivity of human B cell lymphoma cell lines to drugs such as dexamethasone, Ara–C and methotrexate [67]. Similarly, *bcl–2* antisense oligonucleotides can increase the sensitivity of freshly isolated AML blasts to Ara–C [68]. In addition, *bcl–2* antisense oligomers can improve cytotoxic responses of human prostate carcinoma cell lines to VP–16 [69]. Also, *bcl–2* antisense expression plasmids markedly improved the cytotoxic responses of human breast cancer cell lines to adriamycin [70]. In addition to experiments in which Bcl–2 protein levels were downregulated using antisense approaches, chemosensitization has also been achieved by overexpression of pro–apoptotic Bcl–2 family proteins such as Bcl–X_S and Bax [71, 72]. Taken together, these findings strongly suggest that pharmacological agents that impair Bcl–2 function or that enhance the actions of Bcl–2 antagonists such as Bax can potentially have a major impact on the treatment of cancer. By removing Bcl–2 as a barrier to apoptosis, theoretically, it should be possible to allow the apoptotic signals generated by currently available anticancer agents to much more effectively drive tumor cells into and through the cell suicide process.

In all cases examined to date, overexpression of Bcl–2 has had no effect on the entry of drugs into tumor cells. Drugs are also still able to interact with their primary molecular targets in Bcl–2–overexpressing cells, inducing damage to DNA or other macromolecules, and causing cell cycle arrest. Bcl–2 therefore defines a novel type of drug–resistance mechanism, one which is distinctly different from other classical mechanisms of chemoresistance involving problems with drug accumulation in tumor cells (mdr–1), amelioration of drug–induced damage (DNA repair enzymes) or reduced amounts of drug–mediated injury (glutathione overproduction). Rather than blocking the primary injury mediated by anticancer drugs, what Bcl–2 does is to prevent drug–induced damage from being effectively translated into signals for cell death (reviewed in ref. 73). For this reason, Bcl–2 essentially converts anticancer drugs from cytotoxic to cytostatic. Since tumor cells that express Bcl–2 are not readily killed and therefore remain viable, the idea is that these cells may have opportunities to attempt repair of drug–induced damage after drugs are withdrawn or to develop additional secondary genetic changes that result in acquired, classical drug resistance through some of the mechanisms mentioned above.

Dysregulation of Bcl–2 Protein Production in Human Malignancies

Though the *bcl–2* gene was first discovered because of its involvement in the t(14;18) chromosomal translocations commonly found in B cell lymphomas, high levels or aberrant patterns (i.e., differentiation–stage inappropriate expression) of Bcl–2 protein production have been reported in a wide variety of human cancers, including a high percentage of hormone–independent adenocarcinomas of the prostate, as well as significant proportions of colorectal adenocarcinomas, small cell and nonsmall cell lung cancers, and many other types of solid tumors and hematogenous malignancies [74–89]. In some cases, these changes in *bcl–2* gene expression appear to occur as early events in the progression of tumors, such as in colorectal adenocarcinomas where changes in the normal patterns of Bcl–2 protein production along the crypt–villus axis are demonstrable in early stage adenomatous polyps or alternatively, as late events such as in prostate cancers where the onset of *bcl–2* expression has been associated with progression to androgen–independent, metastatic disease. Based on currently available results, it can be estimated that high levels of Bcl–2 protein production occur in about one–half of all human cancers, suggesting that dysregulation of programmed cell death as a result of changes in the expression of Bcl–2 represents a fundamental step in human carcinogenesis.

Unlike the chromosomal translocations seen in B cell lymphomas, the *bcl–2* gene is not grossly altered in its structure in solid tumors and most types of leukemia, and the mechanisms responsible for high levels of *bcl–2* gene expression remain largely unknown. One potential explanation, however, is the loss of p53 function which is estimated to occur in about half of human cancers. In this regard, the tumor suppressor p53 has been shown to function as a repressor of *bcl–2* gene expression in some types of tumor cell lines in vitro and some tissues in vivo [90, 91]. Loss of p53 therefore may represent one mechanism that contributes to *bcl–2* gene deregulation in cancer, by relieving *bcl–2* from the transcriptional repression of the wild–type p53 protein. The connection between p53 and regulation of *bcl–2* further emphasizes the important role that *bcl–2* plays as a determinant of chemoresponse, given that p53 has been shown to have major influences on chemo– and radioresponses of tumors and is of prognostic significance for patients with several types of cancer.

Recently, homozygous inactivating mutations have been reported in the *bax* genes of some common types of tumors, suggesting that *bax* is a bone fide tumor suppressor gene [92]. The second exon of the *bax* gene contains a stretch of eight guanosines, creating a homopolymeric track in the coding region of the gene that appears to be subject to DNA polymerase slippage during DNA replication. DNA mismatch repair enzymes normally are responsible for repairing such errors in the DNA but, in tumors with the microsatellite instability phenotype, these mismatch repair mechanisms are faulty and the result is increased frequency of mutations in such repeated stretches of DNA. The resulting +1 or –1 nucleotide insertions/deletions result in frame–shift mutations that abolish Bax protein production. Interestingly, while *bax* gene mutation in the G_8 track do commonly occur in tumors with microsatellite instability, such as the colon cancers associated with Lynch syndrome, they are also found in variety of tumor cell lines that do not have generalized microsatellite instability [92] (and our unpublished data).

Abnormally high levels of Bcl–XL protein have been reported in several types of cancer, including poorly differentiated colon adenocarcinomas and gastric cancers, as well as advanced prostate adenocarcinomas [79, 93, 94]. Conversely, marked reductions in the levels of the pro–apoptotic protein Bak have been observed in adenocarcinomas of the colon and stomach [93]. The molecular explanation for these changes in the expression of these Bcl–2 family proteins has not yet been determined.

Prognostic Significance of Bcl–2 Family Proteins in Cancer

Bcl–2 has been shown to be of prognostic significance in at least some types of cancer patients, including subgroups of patients with lymphomas, leukemias and prostate cancer in which high levels of *bcl–2* expression have been associated with poor responses to chemotherapy, shorter disease–free survival, faster times to relapse, shorter overall survival or other end–points that generally are associated with poor clinical outcome [74, 95–98].

In some types of cancer, however, *bcl–2* expression has not correlated with poor clinical outcome and has even been paradoxically associated with favorable outcome for patients [83, 99–104]. Recent studies of other members of the Bcl–2 protein family, particularly Bax, however, suggest a potential explanation for these seemingly paradoxical observations.

Though most studies to date have focused exclusively on Bcl–2, in fact, it is the ratio of anti–apoptotic proteins, such as Bcl–2, to pro–apoptotic proteins, such as Bax, that defines the relative sensitivity or resistance of tumor cells to apoptotic stimuli, such as chemotherapeutic drugs and radiation. Consequently, tumor cells can also become resistant to therapy by reducing their expression of

Bax, as opposed to increasing their levels of Bcl–2. This is precisely what has been recently observed for metastatic breast cancer [105] and progressive CLLs [106]. Reductions in the levels of Bax proteins, for example, have been reported to occur in about one–third of breast cancers and have been associated with poor responses to therapy, faster time to tumor progression and shorter overall survival in women with metastatic disease who were treated with combination chemotherapy [105]. Interestingly, these same tumors with reduction in Bax also had reduced Bcl–2 protein levels, potentially explaining why reduced Bcl–2 has been previously associated with unfavorable outcome for some subgroups of women with breast cancer [105, 107]. Thus, the reduced levels of Bcl–2 were offset by reductions in Bax, presumably resulting on balance in a net survival advantage for these chemoresistant cancers (reviewed in ref. 108). In this regard, reduced levels of Bax protein in human breast cancer lines have also been associated with increased resistance to a variety of apoptotic stimuli [109]. Moreover, gene transfer–mediated increases in Bax in human breast cancer cell lines have been reported to restore sensitivity to apoptosis and to impair tumor formation in SCID mice [71, 110].

Factoring in the Bcl–2/Bax ratio may also explain some other clinical observations that have previously been difficult to reconcile with Bcl–2's documented function as a suppressor of apoptosis induced by chemotherapeutic drugs and radiation. For example, despite the high levels of Bcl–2 protein found in $> 85\%$ of follicular B cell lymphomas as the result of t(14;18) chromosomal translocations, most patients with this disease respond well, at least initially, to therapy and can be induced into a partial or complete remission. Recently, however, it has been shown that genotoxic stress stimulates marked increases in Bax protein production in lymphoid cells in vivo in normal and in *bcl–2* transgenic mice, presumably because of the ability of p53 to directly transactivate the *bax* gene promoter [111, 112]. Thus, by inducing increases in Bax protein levels, genotoxic stress produced by DNA–damaging drugs or radiation may partially overcome the high levels of Bcl–2 caused by t(14;18) translocations in follicular lymphomas.

Still another possible explanation for paradoxical associations of Bcl–2 with better clinical outcome in some types of cancer may be attributable to compensatory increases in Bcl–X_L. For example, in colorectal cancers, Bcl–2 tends to be present at high levels in early–stage well–differentiated tumors but often declines during progression to more aggressive undifferentiated tumors. In contrast to the well–differentiated tumors, these undifferentiated colorectal tumors tend to express Bcl–X_L at high levels rather than Bcl–2 [79]. Similar observations have made for gastric carcinomas [93]. Moreover, it has been reported that Bcl–X_L levels become elevated during progression of prostate cancers to high–grade primary (Gleason stage 8–10) and metastatic disease [94]. Thus, Bcl–X_L may substitute for Bcl–2 in some types of advanced cancers. 

Taken together, these observations illustrate the complexity of attempting to predict clinical outcome based on measurements of Bcl–2 alone and emphasize the importance of systematically screening tumors for alterations in the expression of other members of the Bcl–2 family. Each particular type of cancer may have a certain member of the Bcl–2 family that predominates in terms of its prognostic power. The challenge then is to determine which members of the Bcl–2 family are the most relevant to predicting responses to chemotherapy or overall survival in specific subgroups of patients. It is also important to bear in mind, that essentially all efforts to utilize Bcl–2 family proteins as prognostic indicators have relied upon sampling the tumors at one particular time prior to institution of therapy. However, the regulation of the levels of Bcl–2 and its homologs is likely to be a dynamic process in many tumors, as illustrated by the effects that p53 can have on expression of Bax and Bcl–2. Thus, there exist inherent limitations in the approaches that are typically taken for assessing the value of Bcl–2 family proteins as prognostic indicators.

Potential Functions of Bcl–2 Family Proteins

The biochemical mechanism by which Bcl–2 and its relatives regulate cell death remains unknown to date. The predicted amino–acid sequences of Bcl–2 and its homologs, as deduced from cDNA cloning, share no significant homology with other proteins that have a defined biochemical or enzymatic function. Thus, it has been a challenge to understand in biochemical terms how Bcl–2 family proteins modulate cell life and death.

In the simplest sense, Bcl–2 family proteins can be thought of as regulators of a distal step in the cell death pathway. The idea is that multiple stimuli that initiate the cell death process funnel their signals somehow through a final common pathway or pathways that are regulable, at least in part, by Bcl–2 and its homologs (Fig. 5.3). The available information suggests that Bcl–2 family proteins somehow either amplify or squelch these poorly defined signaling events that initiate apoptosis, with the relative ratios of the pro–apoptotic and anti–apoptotic members of the family dictating whether cells remain viable or trigger the apoptotic cell death machinery when confronted with a given cell death signal. The primary effectors of apoptosis appear to be a family of cysteine proteases that cleave their target proteins after Asparatic acid residues ("caspases" for cysteine **asp**artic acid prote**ases**) [113]. The caspases are produced as inactive zymogens (proproteins) in cells and become activated by proteolytic cleavage (reviewed in ref. 114). Overexpression of Bcl–2 or Bcl–X_L, for example, prevents the cleavage and activation of various caspases under circumstances where these Bcl–2 family proteins prevent apoptosis [115–120]. Conversely,

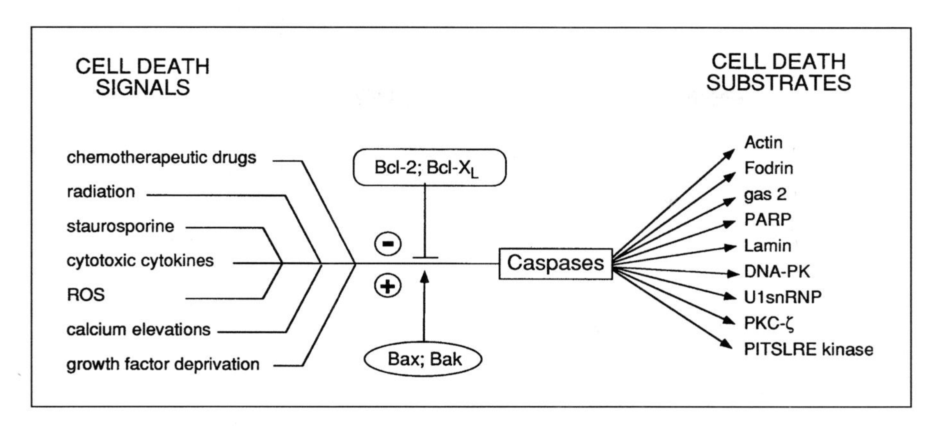

Fig. 5.3. Model for the apoptotic cell death pathway.

overexpression of Bax induces processing and activation of certain members of the caspase family [121]. This relation of Bcl–2 to the proteases has it roots in the genetic analysis of cell death in the nematode, *C. elegans*, where it was first established that the worm homolog of Bcl–2 (Ced–9) suppresses the function of the worm Caspase (Ced3) [122].

Though the model presented in Figure 5.3 provides a conceptual framework for thinking about Bcl–2, what have we learned after 10 years of effort about the biochemical mechanism by which Bcl–2 prevents and Bax enhances protease activation? First, any discussion of the functions of Bcl–2 family proteins must take into consideration the locations of these proteins in cells. As mentioned above, the human Bcl–2 protein contains a stretch of hydrophobic amino–acids at its C–terminus that allows it to post–translationally insert into intracellular membranes, primarily the outer mitochondrial membrane, nuclear envelope and endoplasmic reticulum (ER) [123, 124]. Most other members of the Bcl–2 family also contain a transmembrane domain near the C–termini and, where examined to date, appear to reside within approximately the same intracellular membrane compartments but may favor one over the others. For example, Mcl–1 appears to reside more in the ER/nuclear envelope compartment than the mitochondrial, whereas Bax appears to be located primarily in association with mitochondria [30, 125]. The adenovirus E1b–19 kDa protein resides primarily in association with the nuclear envelope and has been reported to bind to the nuclear lamins that provide structural support for the nuclear envelope [126].

Several theories have been advanced as to how Bcl–2 and its homologs control cell life and death. Certainly, Bcl–2 family proteins could potentially regulate the caspases directly, though to date there is no support for this idea. In contrast, data have been presented which argue both in favor and against the possibility of an effect of Bcl–2 on an anti–oxidant pathway in cells [127–130]. It has also been suggested that Bcl–2 may regulate the homeostasis of Ca^{2+} in cells, based on experiments which have shown an ability of Bcl–2 overexpression to (a) influence the sequestration of Ca^{2+} within the ER, preventing its release into the cytosol during apoptosis and delaying release induced by thapsigargin, a specific inhibitor of the ER's Ca^{2+}–ATPase; (b) prevent entry of Ca^{2+} into the nucleus after treatment of cells with thapsigargin or chemotherapeutic drugs; (c) inhibit the release of Ca^{2+} from mitochondria after exposure of cells to uncouplers of oxidative phosphorylation and (d) potentiate maximal Ca^{2+} uptake capacity of mitochondria [131–135]. Evidence has also been presented suggesting that Bcl–2 can control the transport of proteins across biological membranes, particularly the nuclear envelope [135–137]. In this regard, electron microscopic studies have demonstrated the presence of Bcl–2 protein in association with what appear to be nuclear pore complexes [123]. Finally, the association of Bcl–2 with the kinase Raf–1 and possibly with the GTPase R–Ras has raised the possibility that Bcl–2 may control a signal transduction pathway that is focused on the intracellular membrane compartments where Bcl–2 resides rather than the plasma membrane, where such enzymes are associated with growth factor receptor [22, 138–140]. In no case, however, has a direct cause–and–effect relationship been demonstrated between Bcl–2 and these processes, and at this point we do not know whether the effects of Bcl–2 on redox state, Ca^{2+} compartmentalization, protein transport and protease activation represent direct effects of Bcl–2 versus downstream events that are hundreds of steps removed from Bcl–2.

Bcl–2 Family Proteins as Regulators of Mitochondrial Permeability Transition

Recently, evidence has been accumulating that Bcl–2 family proteins can regulate the phenomenon of permeability transition (PT) in mitochondria. PT results from the opening of a large megachannel located at the contact sites in mitochondria where the inner and outer membranes abut and where various transport processes involving ions and proteins occur (reviewed in ref. 141). It can be measured in isolated mitochondria based on an increase in their light–absorbance caused by mitochondria swelling in isotonic medium or by the failure of such mitochondria to take up cationic fluorescent dyes which rely upon an intact electrochemical gradient for entry into mitochondria. Interestingly, electron microscopic studies suggest that Bcl–2 is concentrated at these contact sites in the outer membrane of mitochondria [142].

The biochemical composition of the megachannel is poorly–defined at present, but among its components appears to be the voltage–dependent anion channel (which is also a peripheral benzodiazepine receptor) located in the outer membrane, porin in the outer membrane and the adenine nucleotide translocator located in the inner membrane [142]. A mitochondria–specific cyclophilin located in either the intermembrane space or the matrix also interacts with the megachannel and cyclosporin analogs that inhibit its peptidylproyl cis–trans isomerase activity can prevent the induction of PT under some circumstances [143].

The induction of PT results in several potentially lethal events in cells, including (a) dissipation of the electrochemical gradient and a subsequent shut–down in oxidative phosphorylation, resulting secondarily in the generation of reactive oxygen species (ROS) because of interrupted flow in the electron transport chain at the level of cytochrome c; (b) dumping of Ca^{2+} from mitochondria into the cytosol; and (c) release of mitochondrial proteins into the cytosol [144, 145]. Among the proteins released are cytochrome c and an unidentified ~50 kDa apoptogenic protein termed AIF for Apoptosis Inducing Factor [146, 147]. Both cytochrome c and AIF reside normally in the intermembrane space of mitochondria. When added to cytosolic extracts derived from healthy nonapoptotic cells, both cytochrome c and AIF can induce rapid activation of ICE–family proteases (caspases) and trigger apoptosis–like destruction of native nuclei added to these extracts [146, 147]. Thus, with effects on ROS, Ca^{2+} and caspase activation, if Bcl–2 does regulate mitochondrial PT, then this could unify several of the phenomena described for Bcl–2 into a single mechanism.

The evidence that Bcl–2 can regulate PT comes from experiments using both intact cells and isolated mitochondria. In intact cells, a wide variety of apoptotic insults can induce mitochondrial PT, as defined by reduced uptake of cationic fluorescent dyes into mitochondria and subsequent generation of reactive oxygen species [40, 148–150]. Overexpression of Bcl–2 prevents the loss of the electrochemical gradient across mitochondria and suppresses the subsequent production of ROS and release of apoptogenic proteins under conditions where Bcl–2 also prevents apoptosis, but not when the apoptotic stimulus is such that Bcl–2 fails to prevent cell death [145, 147]. In experiments employing isolated mitochondria where the mitochondria are derived from Bcl–2 overexpressing cells or their control transfected counterparts that have low levels of Bcl–2, it has been shown that Bcl–2 can prevent the induction of PT by oxidants, Ca^{2+} and atractyloside (an inhibitor of the adenine nucleotide translocator). Though examined only recently within the context of apoptotic cell death, mitochondrial PT has been studied for decades with regards to mechanisms of necrotic cell death particularly during ischemia and reperfusion injury [141]. It is tempting therefore to speculate that the ability of Bcl–2 to prevent PT induction by a wide variety of insults, including elevated cytosolic Ca^{2+} and oxidative injury, may provide an explanation for reports that Bcl–2 can also prevent necrotic cell death under some circumstances [129, 151–153].

Major questions at this point for the field of apoptosis research are whether all the effects of Bcl–2 can be explained by regulation of mitochondrial PT and whether all pathways to apoptosis go through a mitochondria–dependent step. The finding that Bcl–2 protects against cell death even in cells that lack mitochondrial DNA (rho–zero cells) and that are therefore incapable of oxidative phosphorylation at first glance appears to argue against the PT hypothesis [154], but even cells that cannot execute oxidative phosphorylation maintain an electrochemical gradient across their mitochondria due to reverse function of the ADP/ATP antiporter, which under times of anaerobic metabolism transports ATP made in the cytosol from glycolysis into the mitochondria for sustenance of various mitochondrial functions that are essential for cell viability. Moreover, even rho–zero cells can be induced to undergo apoptosis in association with triggering of mitochondrial permeability transition [155].

However, it has also been shown that Bcl–2 can suppress apoptosis induced by some types of stimuli in certain types of cells when targeted to the endoplasmic reticulum by replacement of the usual C–terminal membrane anchoring domain of Bcl–2 with an ER targeting transmembrane domain from cytochrome b5 [150]. Though it is possible that a small portion of this chimeric Bcl–2/cyto–b5 protein manages to find its way onto the mitochondria surface, this result suggests that Bcl–2 may have other functions besides inhibiting mitochondrial PT. Moreover, the aforementioned experiments demonstrating effects of Bcl–2 overexpression on Ca^{2+} and protein transport across the nuclear envelope also speak to potential nonmitochondrial functions for Bcl–2.

Bcl–2 Family Proteins as Channel Formers

A milestone in our understanding of Bcl–2 family protein function has come from the three–dimensional structure of the Bcl–X_L protein, which has revealed striking structural similarity with the pore–forming domains of the bacterial toxins, diphtheria toxin (DT) and the colicins [40]. The structures of Bcl–X_L and the pore–forming domains DT and colicins consist entirely of α–helices connected by variable length loops. Each structure contains a pair of core hydrophobic helices that are long enough to penetrate the lipid bilayer and which are shielded from the aqueous environment by the other five to seven amphipathic helices that orient their hydrophobic surfaces towards the central core helices and their hydrophilic surfaces outward. Studies of the bacterial toxins suggest that under conditions of low pH, acidic lipid membranes and creation of a voltage gradient, the central hydrophobic helices efficiently insert through the lipid bilayer as one step in the process of pore formation [156–160]. True to its structural similarity to these bacterial toxins, recent data indicate that recombinant Bcl–X_L protein, as well as Bcl–2 and Bax, can form pores in liposomes in a pH– and acidic lipid membrane–dependent fashion [161, 162] (S. Korsmeyer, personal communications). Moreover, single channel recordings in planar bilayers which provide an exquisitely sensitive method for monitoring pore formation indicate that even at neutral pH, Bcl–2, Bcl–X_L and Bax can form discrete ion–conducting channels.

Since both anti–apoptotic (Bcl–2; Bcl–X_L) and pro–apoptotic (Bax) proteins are capable of forming channels in membranes, it remains unclear at present how this pore–forming activity relates to the bioactivities of these proteins. In addition, many other issues remain unresolved, such as the diameter of the channels and how many Bcl–2, Bcl–X_L or Bax proteins it takes to create an aqueous channel in membranes. In the case of DT, the channels are evidently large enough to transport a protein, since the primary function of DT is thought to be transport of the ADP–ribosylation factor subunit of the toxin from lysosomes and endosomes into the cytosol [163]. In contrast, the bacterial colicins transport ions [159, 164]. With respect to some of the cellular phenomena that Bcl–2 family proteins have been reported to control such as mitochondrial PT, transporting either ions (such as Ca^{2+}) or proteins (such as cytochrome C and AIF) could fit nicely with the structural and electrophysiological evidence of pore formation. Indeed, it could even be argued that Bcl–2 family proteins create the mysterious megachannel that causes mitochondrial PT. Arguing against this possibility however are data indicating that even yeast mitochondria exhibit megachannel behavior, and yet no homologs of Bcl–2 evidently exist in yeast.

Bcl–2 Family Proteins as Adaptor Proteins

Though Bcl–2 family proteins are best known for their ability to interact with each other, Bcl–2 has also been reported to bind several other nonhomologous proteins, including the kinase Raf–1, the GTPases R–Ras and H–Ras, the p53–binding protein p53–BP2, the prion protein Pr–1 and several novel proteins including BAG–1, Nip–1, Nip–2 and Nip–3 [138, 140, 165–169]. In most cases, neither the functional significance of these interactions with Bcl–2 nor the regions on the Bcl–2 protein required for binding to these proteins has been explored.

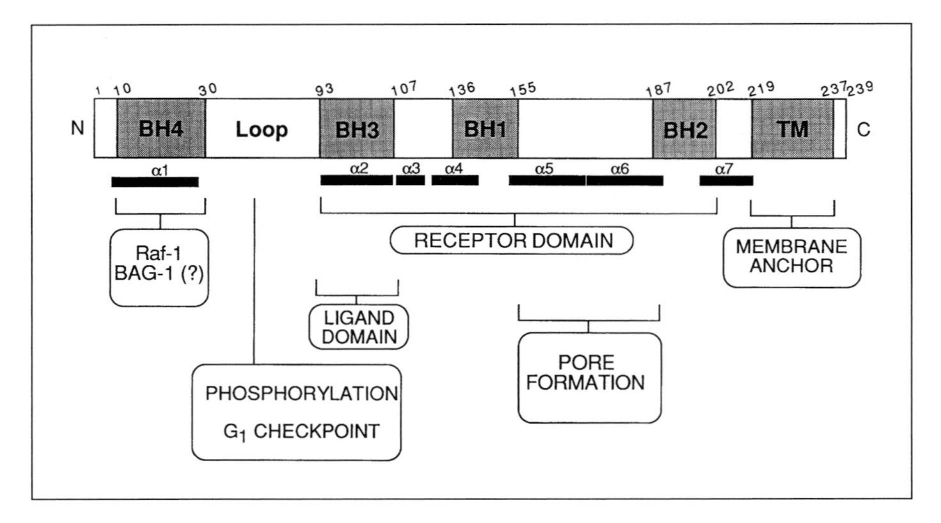

Fig. 5.4. Human Bcl–2 protein.

Recently, however, it has been reported that Raf–1 and possibly BAG–1 bind to Bcl–2 in a BH4–dependent manner. These proteins can also cooperate with Bcl–2 in the suppression of cell death [22, 138, 168, 170]. The BH4 domain of Bcl–2 corresponds to the first α–helix in the structure of Bcl–X_L. This amphipathic helix is 16 residues in length and is separated from the rest of the protein by a flexible ~50 amino acid long loop that is rich in proline residues, which presumably interfere with its formation of higher–order secondary structures (Fig. 5.4). That the BH4 domain is tethered to the rest of the protein by a long flexible loop implies that it may be able to swing away from the body of the protein for the purpose of interacting with other proteins such as BAG–1 and Raf–1. In the crystal structure of Bcl–X_L, the BH4 domain is shown inserted on the back–side of the protein relative to the hydrophobic pocket, with its hydrophobic face interacting with the core hydrophobic helices that are presumably responsible for penetration of the lipid bilayer. Upon insertion into the membrane or adopting an open conformation that resembles the membrane integrated state, therefore, the BH4 domain would be anticipated to swing free—thus having the opportunity to bind to other proteins. In addition to interacting with BAG–1 and Raf–1, the liberated BH4 domain also appears to be able to interact with other Bcl–2, Bcl–X_L and Mcl–1 molecules based on yeast two–hybrid and in vitro GST–fusion protein binding experiments which have shown that the region of Bcl–2 where the BH4 domain resides can bind to the distal portions of the Bcl–2 downstream of the proline rich loop where the other six helices reside.

Thus, the BH4 domain could potentially mediate a different type of dimerization event that involves the first helix of Bcl–2 binding to the opposite side of a partner protein relative to the location of the hydrophobic pocket thought to be involved in BH3 domain binding. Interestingly, protein–protein interaction studies indicate that this type of BH4–mediated dimerization occurs primarily among anti–apoptotic members of the family, including Bcl–2, Bcl–X_L and Mcl–1, but is not seen for the pro–apoptotic protein Bax. In this regard, inspection of the predicted first helix of Bax reveals that it both lacks significant primary amino-acids sequence similarity to the BH4 domains of the anti–apoptotic members of the Bcl–2 family and is nonamphipathic. Moreover, the first predicted "helix" of Bax contains three glycine residues that might be expected to produce kinks in helices. Thus, it seems likely that the structure of the first helix in the pro–apoptotic protein Bax is quite different from the BH4 domains of Bcl–2, Bcl–X_L, Mcl–1 and other anti–apoptotic Bcl–2 family proteins.

How does the BH4–dependent interaction of Bcl–2 with Raf–1 and BAG–1 enhance protection from apoptosis? Recently it has been shown that BAG–1 not only binds to Bcl–2 but also can specifically interact with Raf–1. Moreover, BAG–1 can activate this kinase, presumably through a protein–protein interaction that involves binding of BAG–1 to the catalytic domain of Raf–1 [170]. This activation of Raf–1 occurs in a Ras–independent fashion. Thus, the concept emerges of Bcl–2 as a docking or adaptor protein around which "signal–transduction" events occur. In the case of Raf–1 and BAG–1, dimers of Bcl–2 could be viewed as a docking site that allows Raf–1 and BAG–1 to meet each other in the cells, interact and result in transient reversible activation of the kinase Raf–1 locally in the vicinity of Bcl–2 on the surface of mitochondria, ER or nuclear envelope membranes. A further implication is that a similar mechanism may be employed to bring a serine/threonine phosphatase into the vicinity of Bcl–2, since a mechanism for reversing the effects of Raf–1 would also be needed. Once activated locally in the vicinity of Bcl–2, the kinase Raf–1 appears to mediate directly or indirectly phosphorylation of the pro–apoptotic protein BAD [22]. The unphosphorylated BAD protein can bind to anti–apoptotic proteins, such as Bcl–2 and Bcl–X_L, preventing them from dimerizing with Bax and abrogating their function as blockers of cell death [8]. Studies in which phosphorylation of BAD was induced by stimulating a hemopoeitic cell line with the growth factor interleukin–3 (IL–3) suggest that once phosphorylated, BAD no longer binds to Bcl–2 and Bcl–X_L [171]. Moreover, the two serines that become phosphorylated in response to IL–3 stimulation reside within consensus binding sites for 14–3–3, an abundant cytosolic protein that sequesters the phosphorylated BAD protein in the cytosol where it cannot interact with Bcl–2 or Bcl–X_L. In this regard, it is important to note that BAD represents one of the few members of the Bcl–2 family that does not possess a C–terminal membrane anchoring

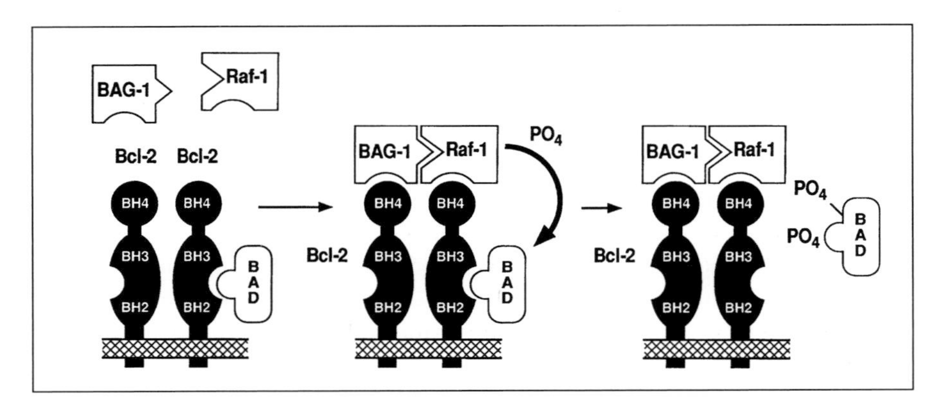

Fig. 5.5. Model for BCL–2/RAF–1 mediated phosphorylation of BAD.

domain (see Fig. 5.1). Though it remains to be determined whether the same two serines in BAD become phosphorylated within the context of Raf–1 targeting to Bcl–2 at the surface of mitochondria or other organelles as opposed to Raf–1 activation at the plasma membrane via Ras–dependent mechanisms after growth factor stimulation, the implication is that the interaction of Bcl–2 with Raf–1 provides a mechanism for eradicating the suppression of BAD (Fig. 5.5). Thus, while perhaps not central to the purported pore forming functions of Bcl–2 family proteins, this role of Bcl–2 as adaptor protein may nevertheless provide a means of modulating the activity of Bcl–2 family proteins to either maintain cell survival or promote cell death.

Additional examples of this adaptor protein function may explain the observation that Bcl–2 overexpression can prevent p53 translocation from cytosol to nucleus in some types of cells but not others. In this regard, the p53–BP2 interactions with Bcl–2 could potentially provide a mechanism for tethering Bcl–2 to p53. Though highly speculative, since Bcl–2 appears to be concentrated at nuclear pore complexes [123], p53–BP2–mediated bridging of Bcl–2 to p53 could hypothetically provide a convenient means for blocking the entry of p53 into the nucleus which would be cell–type dependent in the sense that only in those cells which express p53–BP2 would this phenomenon of Bcl–2–dependent blockage of p53 translocation be exhibited. Though the domains within Bcl–2 needed for binding to p53–BP2 have yet to be explored, a reasonable expectation is that the proline rich loop in Bcl–2 will be involved, given that p53–BP2 contains a SH3 domain. Moreover, inasmuch as Bax and the other pro–apoptotic members of the Bcl–2 family lack this long proline–rich loop, one could imagine specificity for interactions with p53–BP2 among the anti–apoptotic members of the Bcl–2 family proteins such as Bcl–2, Bcl–X_L and Mcl–1.

Post–translational Modifications of Bcl–2 Family Proteins

The loop region of Bcl–2 appears to contain a phosphorylation site(s) or is necessary for binding of a kinase that induces phosphorylation of Bcl–2 [172]. Phosphorylation can evidently be a mechanism for inactivating the Bcl–2 protein, at least in part by impairing its ability to heterodimerize with Bax [173], though a direct cause–and–effect relationship between specific sites of phosphorylation and altered behavior of Bcl–2 has yet to be demonstrated. The kinases and phosphatases that regulate phosphorylation of Bcl–2 are currently unknown. However, it has been shown that antimicrotubule drugs, such as taxol, vincristine and nocodazole, can induce phosphorylation of Bcl–2, suggesting the involvement of either a microtubule associated kinase (MAP kinase) or a cyclin–dependent kinase that becomes active during mitosis [173, 174]. Though it was suggested that Raf–1 may be responsible based on use of a pharmacological inhibitor that results in Raf–1 protein degradation through disrupting its interaction with Hsp90 [175], recent data using activated or dominant–negative versions of Raf–1 suggest otherwise [176].

Interestingly, mutations have been identified in the loop region of Bcl–2 in lymphomas containing a t(14;18) chromosomal translocation involving the *bcl–2* gene, sometimes occurring in association with transformation from low–grade follicular lymphoma to aggressive diffuse large–cell non–Hodgkin's lymphoma [177, 178]. Though little has been done so far to explore the function of Bcl–2 family proteins that contain loop mutations, alterations at one common mutation site (proline 59) have been associated with reduced Bcl–2 phosphorylation in response to taxol [176], and enhanced cell proliferation [179] (and our unpublished data). In this regard, evidence that Bcl–2 can promote a G_1–S phase cell cycle block has been obtained for several but not all types of cells [180–183]. Thus, the enhanced proliferation seen in cells transfected with a Bcl–2 (pro 59) mutant relative to the wild–type Bcl–2 protein potentially may reflect a loss of interactions of Bcl–2 with other proteins responsible for this cell cycle slowing phenomenon seen in some types of cells that have been engineered to overexpress Bcl–2. A failure of such mutants to undergo phosphorylation theoretically could also be involved and might provide a teleological explanation for why Bcl–2 becomes phosphorylated after stimulation of some types of hematopoietic cells by IL–3 [184].

Candidate proteins that can potentially speed cell cycle rates and whose function may be inhibited by overexpression of Bcl–2 include the transcription factors NF–kB and NF–AT. Inhibition of entry into the nucleus and/or successful transactivation of target reporter genes by NF–kB and NF–AT has been reported in Bcl–2 overexpressing cells [183, 185]. It should be noted, however, that the wild–type Bcl–2 protein does not cause a G_1–S delay in all types of cells, and indeed, has been shown to enhance rather than inhibit cell proliferation in some cellular contexts [186, 187]. Nevertheless, the potential for Bcl–2 to modulate the cell cycle, causing a slowing at G_1–S, may represent a protective function that allows cells more time, for example, to repair damaged DNA prior to S–phase. It may also underlie the observation that Bcl–2 overexpression in some types of solid tumors, such as adenocarcinomas of the breast and colon, tends to be found in the earlier stage, well–differentiated malignancies which have lower mitotic rates as opposed to the aggressive, highly proliferative tumors.

Little is known about whether other members of the Bcl–2 family of proteins become phosphorylated and what the functional ramifications of such phosphorylation events are if they occur. As mentioned above, some growth factors can induce phosphorylation of BAD on two serines, thus preventing BAD from dimerizing with Bcl–X_L. It has been suggested that growth factor stimulation of cell survival may be attributable to induction of phosphorylation of the pro–apoptotic BAD protein. However, thus far, this has only been documented for one type of growth factor (interleukin–3) and in one cell line (a murine pro–B cell, FL5.12) [171]. Moreover, while IL–3 does induce phosphorylation of exogenous BAD protein when expressed in FL5.12 cells [171], these cells do not naturally contain BAD protein and thus IL–3–induced phosphorylation of the BAD protein cannot account for how this lymphokine provides survival signals under normal circumstances in these cells.

Strategies for Inhibiting Bcl–2 Function in Cancers

To a large extent, the failure to find cures for cancer lies with the lack of any new drugs that attack the problem in truly novel ways. Essentially all anticancer drugs currently available to clinical oncologists target some aspect of the cell division machinery in an effort to arrest the growth of dividing cells. These agents, for example, may target enzymes required for nucleotide precursor synthesis, induce damage directly to DNA in an effort to slow DNA–replication and lethally damage cells, or interfere with the function of microtubules thereby arresting cell division during mitosis. While the mechanisms by which most currently available anticancer drugs arrest cell proliferation are well understood, it has only been in recent years that the molecular details of how anticancer drugs actually kill tumor cells have begun to be delineated and the central role that apoptosis plays in this process appreciated.

The central role that Bcl–2 family proteins play as regulators of a distal step in this pathway positions them as ideal targets for impacting cancer therapy in truly significant ways. Several strategies can be envisioned for abrogating the effects of Bcl–2 or other anti–apoptotic members of the Bcl–2 family, such as

Bcl–X_L in human cancers. These include small molecule blockers of Bcl–2/Bax and Bcl–X_L/Bak dimerization, as well as blockers of the interactions of Bcl–2 with accessory proteins, such as BAG–1 and Raf–1, that promote cell survival in collaboration with Bcl–2. In addition, once the pore–like functions of Bcl–2 family proteins are better elucidated, and in particular their biological relevance becomes better understood, this may also create opportunities for screening for small molecules that alter the channel activities of these proteins by analogy to current cardiovascular drugs. Another approach focuses on the idea that phosphorylation of Bcl–2 could be a mechanism for inactivating the protein. Clearly, therefore, if more were known about the kinases and phosphatases that control phosphorylation of Bcl–2, then this could suggest additional strategies for counteracting Bcl–2 in cancer cells. In this regard, while taxol and other antimicrotubule drugs can induce phosphorylation of Bcl–2, typically less than half of the Bcl–2 protein molecules become phosphorylated and, moreover, the phosphorylation is dependent on cells cycling through M–phase—a potential limitation given the generally low proliferative fractions of most solid tumors. Antisense oligonucleotides targeted against Bcl–2 have also been employed both in vitro and in animal models to inducing apoptosis of malignant cells or sensitizing them to conventional chemotherapeutic drugs. Thus, if inherent problems in the uptake and compartmentalization of these rather large highly charged compounds can be overcome, this approach could also prove viable. Finally, the recognition that expression of Bcl–2 family genes can be regulated by various biological response modifiers such as retinoids, estrogens, cytokines and growth factors, suggests opportunities for modulating the ratios of anti– and pro–apoptotic members of the Bcl–2 protein family in ways that might sensitize cancer cells to conventional cytotoxic drugs. Though the mechanisms of action of Bcl–2 and its homologs remain unclear, improved knowledge about the pathways that control the expression of these genes and the biochemical mechanisms of action of their encoded proteins offers the promise of improved strategies for the treatment of human cancers and other diseases that arise in whole or in part because of dysregulation of the physiological cell death pathway.

An important question when considering the idea of targeting Bcl–2 for cancer therapy is that of toxic side–effects. Though Bcl–2 and its homologs are expressed in a wide variety of normal tissues, tumor cells may be more dependent on the survival–promoting functions of members of this family of apoptosis–regulating proteins. First, tumor cells typically contain alterations in the expression of *c–myc* and other cell–cycle genes that promote apoptosis. This deregulated proto–oncogene expression not only creates an incessant drive for cell proliferation but also sets up a clash in signals that would normally induce apoptosis, were it not for the actions of anti–apoptotic genes such as *bcl–2* [188–190]. Given that normal cells do not have these contradictory signals caused by oncogene activation, drugs that target Bcl–2 or its homologs should be relatively specific for cancer cells, since cancer cells will presumably be more dependent on proteins such as Bcl–2 for their survival. Second, normal epithelial cells are highly dependent on cues from their surrounding environment for their survival. At least in part, these cues come from cell contact with the extracellular matrix via integrins, cell–cell interactions and local production of growth factors [191, 192]. Tumor cells are able to invade locally and metastasize to distant sites partly because they are less dependent on these environmental cues, probably as the result of deregulated expression of genes such as *bcl–2*. Hence, because of their localization to foreign environments, the survival of metastatic cells potentially may be severely comprised by drugs that disrupt the function of Bcl–2 or Bcl–X_L, whereas the viability of normal cell will not be. Third, the genetic instability that occurs in many solid tumors creates signals for apoptosis that would normally kill cells by triggering the endogenous program for cell death, were it not for the actions of anti–apoptotic proteins such as Bcl–2. Indeed, disturbances in the balance of pro– and anti–apoptotic protein that result in an anti–apoptotic state may be a prerequisite for the accumulation of genetic defects [92]. Finally, the reason why existing chemotherapeutic drugs display some selectivity for tumor cells over normal cells can be attributed at least in part to the loss of cell cycle check point controls in cancer cells [193]. These checkpoint controls normally cause cells with damaged DNA, for example, to arrest until repair mechanisms can correct drug–induced injury before resuming attempts at DNA–replication. The lack of cell cycle checkpoint control mechanisms in tumor cells drives these malignant cells to cycle despite the presence of drug–induced damage, thus creating signals that induce apoptosis. The problem with all too many tumors, however, is that these apoptotic signals are blocked from fulfilling their intended mission, presumably because of the actions of Bcl–2, Bcl–X_L or other anti–apoptotic proteins. Agents or strategies that counteract the effects of Bcl–2 and Bcl–X_L, therefore, should release these barriers to apoptosis in drug–damaged tumor cells and promote synergistic tumor responses to conventional chemotherapeutic drugs. Unfortunately, in the absence of effective compounds that interfere with the functions of Bcl–2, Bcl–X_L or other anti–apoptotic members of the Bcl–2 protein family, one can only speculate about whether a workable therapeutic index is possible.

Summary

After 10 years of effort, we are beginning to understand the biochemical functions of Bcl–2 family proteins and to develop strategies for attempting to manipulate them for the treatment of cancer. Bcl–2 and several of its homologs seem likely to be multifunctional proteins, that not only form channels in membranes but that also have important interactions with other types of proteins in their capacity of adapter or docking proteins. Further knowledge about the structural and functional details of these two facets of Bcl–2 family proteins will contribute additional insights that can be applied in efforts to promote apoptosis selectively in tumors.

Acknowledgments

I thank Heidi Gallant for manuscript preparation, and Hongbin Zha and Muizz Hasham for some artwork.

References

1 Boise LH, Gonzalez–Garcia M, Postema CE et al. *bcl–x*, a *bcl–2*–related gene that functions as a dominant regulator of apoptotic cell death. Cell 1993; 74:597–608.
2 Oltvai Z, Milliman C, Korsmeyer SJ. Bcl–2 heterodimerizes in vivo with a conserved homolog, Bax, that accelerates programmed cell death. Cell 1993; 74:609–619.
3 Kozopas KM, Yang T, Buchan HL, Zhou P, Craig R. *mcl–1*, a gene expressed in programmed myeloid cell differentiation, has sequence similarity to *bcl–2*. Proc Natl Acad Sci USA 1993; 90:3516–3520.
4 Lin EY, Orlofsky A, Berger MS, Prystowsky MB. Characterization of A1, a novel hemopoietic–specific early–response gene with sequence similarity to bcl–2. J Immunol 1993; 151:1979–1988.
5 Gibson L, Holmgreen SP, Huang DCS et al. Bcl–w, a novel member of the bcl–2 family, promotes cell survival. Oncogene 1996; 13:665–675.
6 Gillet G, Guerin M, Trembleau A, Brun G. A Bcl–2–related gene is activated in avian cells transformed by the Rous sarcoma virus. EMBO J 1995; 14:1372–1381.
7 Hengartner MO, Horvitz HR. *C. elegans* cell survival gene *ced–9* encodes a functional homolog of the mammlian proto–oncogene bcl–2. Cell 1994; 76:665–676.
8 Yang E, Zha J, Jockel J et al. Bad: a heterodimeric partner for Bcl–XL and Bcl–2, displaces bax and promotes cell death. Cell 1995; 80:285–291.
9 Chittenden T, Harrington EA, O'Connor R, Flemington C, Lutz RJ, Evan GI, Guild BC. Induction of apoptosis by the Bcl–2 homolog Bak. Nature 1995; 374:733–736.
10 Kiefer MC, Brauer MJ, Powers VC et al. Modulation of apoptosis by the widely distributed Bcl–2 homolog Bak. Nature 1995; 374:736–739.
11 Choi SS, Park I–C, Yun J W et al. A novel *Bcl–2* related gene, *Bfl–1*, is overexpressed in stomach cancer and preferentially expressed in bone marrow. Oncogene 1995; 11:1693–1698.
12 Farrow SN, White JHM, Martinou I et al. Cloning of a bcl–2 homolog by interaction with adenovirus E1B 19K. Nature 1995; 374:731–733.
13 Boyd JM, Gallo GJ, Elangovan B et al. Bik, a novel death–inducing protein shares a distinct sequence motif with Bcl–2 family proteins and interacts with viral and cellular survival–promoting proteins. Oncogene 1995; 11:1921–1928.
14 Wang K, Yin W–M, Chao DT et al. BID: a novel BH3 domain–only death agonist. Genes & Development 1996; 10:2859–2869.
15 Das R, Reddy EP, Chatterjee D, Adrews DW. Identification of a novel *Bcl–2* related gene, *BRAG–1*, in human glioma. Oncogene 1996; 12:947–951.

16 Cruz–Reyes J, Tata JR. Cloning, characterization and expression of two Xenopus *bcl–2*–like cell–survival genes. Gene 1995; 158:171–179.
17 Rao L, Debbas M, Sabbatini P, Hockenbery D, Korsmeyer S, White E. The adenovirus E1A proteins induce apoptosis, which is inhibited by the E1B 19–kDa and bcl–2 proteins. Proc Natl Acad Sci USA 1992; 89:7742–7746.
18 Henderson S, Huen D, Rowe M et al. Epstein–Barr virus–coded BHRF1 protein, a viral homolog of Bcl–2, protects human B cells from programmed cell death. Proc Natl Acad Sci USA 1993; 90:8479–8483.
19 Neilan JG, Lu Z, Afonso CL et al. An African swine fever virus gene with similarity to the proto–oncogene *bcl–2* and the Epstein–Barr virus gene BHRF1. J Virol 1993; 67:4391–4394.
20 Smith CA. A novel viral homolgue of Bcl–2 and Ced–9. Trends Cell Biol 1995; 5:344–345.
21 Zha H, Aime–Sempe C, Sato T, Reed JC. Pro–apoptotic protein Bax heterodimerizes with Bcl–2 and homodimerizes with Bax via a novel domain (BH3) distinct from BH1 and BH2. J Biol Chem 1996; 271:7440–7444.
22 Wang H–G, Rapp UR, Reed JC. Bcl–2 targets the protein kinase Raf–1 to mitochondria. Cell 1996; 87:629–638.
23 Hunter JJ, Bond BL, Parslow TG. Functional dissection of the human Bcl–2 protein: sequence requirements for inhibition of apoptosis. Mol Cell Biol 1996; 16:877–883.
24 Borner C, Martinou I, Mattmann C et al. The protein bcl–2α does not require membrane attachment, but two conserved domains to suppress apoptosis. J Cell Biol 1994; 126:1059–1068.
25 Chittenden T, Flemington C, Houghton AB et al. A conserved domain in Bak, distinct from BH1 and BH2, mediates cell death and protein binding functions. EMBO J 1995; 14:5589–5596.
26 Han J, Sabbatini P, Perez D et al. The E1B 19K protein blocks apoptosis by interacting with and inhibiting the p53–inducible and death–promoting Bax protein. Genes & Dev 1996; 10:461–477.
27 Hunter JJ, Parslow TG. A peptide sequence from Bax that converts Bcl–2 into an activator of apoptosis. J Biol Chem 1996; 271:8521–8524.
28 Sato T, Irie S, Krajewski S, Reed JC. Cloning and sequencing of cDNA encoding rat bcl–2 protein. Gene 1994; 140:291–292.
29 Sedlak TW, Oltvai ZN, Yang E et al. Multiple Bcl–2 family members demonstrate selective dimerizations with Bax. Proc Natl Acad Sci USA 1995; 92:7834–7838.
30 Zha H, Fisk HA, Yaffe MP et al. Structure–Function comparisons of the pro–apoptotic protein bax in yeast and mammalian cells. Moll Cell Biol 1996; 16(11):6494–6508.
31 Simonian PL, Grillot DAM, Merino R, Nunez G. Bax can antagonize Bcl–XL during etoposide and cisplatin–induced cell death independently of its heterodimerization with Bcl–XL. J Biol Chem 1996; 271(37):22764–72.
32 Meiijerink JPP, Smetsers TFCM, Slöetjes A et al. Bax mutations in cell lines derived from hematological malignancies. Leukemia 1995; 9:1828–1832.
33 Yin XM, Oltvai ZN, Korsmeyer SJ. BH1 and BH2 domains of bcl–2 are required for inhibition of apoptosis and heterodimerization with bax. Nature 1994; 369:321–333.
34 Hanada M, Aimé–Sempé C, Sato T, Reed J C. Structure–function analysis of bcl–2 protein: identification of conserved domains important for homodimerization with bcl–2 and heterodimerization with bax. J Biol Chem 1995; 270:11962–11968.
35 Badley AD, McElhinny JA, Leibson P et al. Upregulation of Fas ligand expression by Human Immunodeficiency Virus in human macrohpages mediates apoptosis of uninfected T lymphocytes. J Virol 1996; 70:199–206.
36 Bodrug SE, Aimé–Sempé C, Sato T et al. Biochemical and functional comparisons of Mcl–1 and Bcl–2 proteins: evidence for a novel mechanism of regulating Bcl–2 family protein function. Cell Death Differ 1995; 2:173–182.
37 Cheng EH–Y, Levine B, Boise LH et al. Bax–independent inhibition of apoptosis by Bcl–XL. Nature 1996; 379:554–556.
38 Sato T, Hanada M, Bodrug S et al. Interactions among members of the bcl–2 protein family analyzed with a yeast two–hybrid system. Proc Natl Acad Sci USA 1994; 91:9238–9242.
39 Minn AJ, Boise LH, Thompson CB. Bcl–XS antagonizes the protective effects of Bcl–XL. J Biol Chem 1996; 271:6306–6312.

40 Muchmore SW, Sattler M, Liang H et al. X–ray and NMR structure of human Bcl–XL, an inhibitor of programmed cell death. Nature 1996; 381:335–341.
41 Middleton G, Nunez G, Davies AM. Bax promotes neuronal survival and antagonizes the survival effects of neurotrophic factors. Development 1996; 122:695–701.
42 Eastman A. Apoptosis: A product of programmed and unprogrammed cell death. Toxicol Appl Pharmacol 1993; 121:160–164.
43 Reed JC. Bcl–2 and the regulation of programmed cell death. J Cell Biol 1994; 124:1–6.
44 Reed JC. Bcl–2 family proteins: regulators of chemoresistance in cancer. Toxicol Lett 1995; 82/83:155–158.
45 Reed JC. Regulation of apoptosis by Bcl–2 family proteins and its role in cancer and chemoresistance. Curr Opin Oncol 1995; 7:541–546.
46 Patel T, Gores GJ, Kaufmann SH. The role of proteases during apoptosis. FASEB J 1996; 10:587–597.
47 Horvitz HR, Shaham S, Hengartner MO. The genetics of programmed cell death in the nematode *Caenorhabditis elegans*. Cold Spring Harb Symp Quant Biol 1994; 59:377–385.
48 Hengartner MO, Ellis RE, Horvitz HR. *Caenorhabditis elegans* gene ced–9 protects cells from programmed cell death. Nature 1992; 356:494–499.
49 Vaux DL, Cory S, Adams JM. *Bcl–2* gene promotes hemopoietic cell survival and cooperates with *c–myc* to immortalize pre–B cells. Nature 1988; 335:440–442.
50 Hockenbery DM, Nunez G, Milliman C, Schreiber RD, Korsmeyer SJ. Bcl–2 is an inner mitochondrial membrane protein that blocks programmed cell death. Nature 1990; 348:334–336.
51 Reed J, Stein C, Haldar S et al. Antisense–mediated inhibition of *bcl–2* proto–oncogene expression and leukemic cell growth: Comparisons of phosphodiester and phosphorothioate oligodeoxynucleotides. Cancer Res 1990; 50:6565–6570.
52 Reed J, Haldar S, Cuddy M et al. Bcl–2–mediated tumorigenicity in a T–lymphoid cell line: Synergy with C–MYC and inhibition by Bcl–2 antisense. Proc Natl Acad Sci USA 1990; 87:3660–3664.
53 Tsujimoto Y, Cossman J, Jaffe E, Croce C. Involvement of the *bcl–2* gene in human follicular lymphoma. Science 1985; 228:1440–1443.
54 Tsujimoto Y, Croce CM. Analysis of the structure, transcripts, and protein products of *bcl–2*, the gene involved in human follicular lymphoma. Proc Natl Acad Sci USA 1986; 83:5214–5218.
55 Reed JC. In: Bertino JR, ed. Encyclopedia of Cancer. Vol 1, 3 vols. San Diego: Academic Press, 1997:125–143.
56 McDonnell TJ, Deane N, Platt F et al. Bcl–2–immunoglobulin transgenic mice demonstrate extended B cell survival and follicular lymphoproliferation. Cell 1989; 57:79–88.
57 Katsumata M, Siegel R, Louie D et al. Differential effects of bcl–2 on B– and T– lymphocytes in transgenic mice. Proc Natl Acad Sci USA 1992; 89:11376–11380.
58 McDonnell TJ, Nunez G, Platt F et al. Deregulated *bcl–2*–immunoglobulin transgene expands a resting but responsive immunoglobulin M and D–expressing B cell population. Mol Cell Biol 1990; 10:1901–1907.
59 Miyashita T, Reed JC. *Bcl–2* gene transfer increases relative resistance of S491 and WEHI72 lymphoid cells to cell death and DNA fragmentation induced by glucocorticoids and multiple chemotherapeutic drugs. Cancer Res 1992; 52:5407–5411.
60 Miyashita T, Reed JC. Bcl–2 oncoprotein blocks chemotherapy–induced apoptosis in a human leukemia cell line Blood 1993; 81:151–157.
61 Walton WI, Whysong D, O'Connor PM et al. Constitutive expression of human bcl–2 modulates mitrogen mustard and camptothecin induced apoptosis. Cancer Res 1993; 53:1853–1861.
62 Kamesaki S, Kamesaki H, Jorgensen T et al. Bcl–2 protein inhibits etoposide–induced apoptosis through its effects on events subsequent to topoisomerase II–induced DNA strand breaks and their repair. Cancer Res 1993; 53:4251–4256.
63 Fisher TC, Milner AE, Gregory CD et al. Bcl–2 modulation of apoptosis induced by anticancer drugs: resistance to thymidylate stress is independent of classical resistance pathways. Cancer Res 1993; 53:3321–3326.
64 Tang C, Willingham MC, Reed JC et al. High levels of p26BCL–2 oncoprotein retard taxol–induced apoptosis in human pre–B leukemia cells. Leukemia 1994; 8:1960–1969.

65 Minn AJ, Rudin CM, Boise LH, Thompson CB. Expression of Bcl–XL can confer a multidrug resistance phenotype. Blood 1995; 86:1903–1910.
66 Ibrado AM, Huang Y, Fang G, Bhalla K. Bcl–x_L overexpression inhibits taxol–induced yama protease activity and apoptosis. Cell Growth & Diff 1996; 7:1087–1094.
67 Kitada S, Takayama S, DeRiel K, Tanaka S, Reed JC. Reversal of chemoresistance of lymphoma cells by antisense–mediated reduction of *bcl–2* gene expression. Antisense Res Dev 1994; 4:71–79.
68 Campos L, Sabido O, Rouault J–P, Guyotat D. Effects of Bcl–2 antisense oligodeoxynucleotides on in vitro proliferation and survival of normal marrow progenitors and leukemic cells. Blood 1994; 84:595–600.
69 Berchem GJ, Bosseler M, Sugars LY et al. Androgens induce resistance to bcl–2–mediated apoptosis in LNCaP prostate cancer cells. Cancer Res 1995; 55:735–738.
70 Teixeira C, Reed JC, Pratt MAC. Estrogen promotes chemotherapeutic drug resistance by a mechanism involving *Bcl–2* protooncogene expression in human breast cancer cells. Cancer Res 1995; 55:3902–3907.
71 Bargou, RC, Wagener C, Bommert K et al. Overexpression of the death–promoting gene *bax–α* which is downregulated in breast cancer restores sensitivity to different apoptotic stimuli and reduces tumor growth in SCID mice. J Clin Invest 1996; 97:2651–2659.
72 Sumantran VN, Ealovega MW, Nuñez G et al. Overexpression of Bcl–XS sentizes MCF–7 cells to chemotherapy–induced apoptosis. Cancer Res 1995; 55:2507–2510.
73 Reed JC. Bcl–2: Prevention of apoptosis as a mechanism of drug resistance. Hematology/Oncology Clinics of North America 1995; 9:451–474.
74 McDonnell T, Troncoso T, Brisbary P et al. Expression of the protooncogene *BCL–2* in the prostate and its association with emergence of androgen–independent prostate cancer. Cancer Res 1992; 52:6940–6944.
75 Furuya Y, Krajewski S, Epstein JI, Reed JC, Isaacs JT. Enhanced expression of Bcl–2 and the progression of human and rodent prostatic cancers. Clin Cancer Res 1996; 2:389–398.
76 Colombel M, Symmans F, Gil S et al. Detection of the apoptosis–suppressing oncoprotein bcl–2 in hormone–refractory human prostate cancers. Am J Pathol 1993; 143:390–400.
77 Hague A, Moorghen M, Hicks D, Chapman M, Paraskeva C. Bcl–2 expression in human colorectal adenomas and carcinomas. Oncogene 1994; 9:3367–3370.
78 Bedi A, Pasricha PJ, Akhtar AJ et al. Inhibition of apoptosis during development of colorectal cancer. Cancer Res 1995; 55:1811–1816.
79 Krajewska M, Moss S, Krajewski S et al. Elevated expression of Bcl–X and reduced Bak in primary colorectal adenocarcinomas. Cancer Res 1996; 56:2422–2427.
80 Flohil CC, Janssen PA, Bosman FT. Expression of bcl–2 protein in hyperplastic polyps, adenomas, and carcinomas of the colon. J Pathol 1996; 178:393–397.
81 Bronner M, Culin C, Reed JC, Furth EE. *Bcl–2* proto–oncogene and the gastrointestinal epithelial tumor progression model. Am J Pathol 1995; 146:20–26.
82 Ben–Ezra JM, Kornstein MJ, Grimes MM, Krystal G. Small cell carcinomas of the lung express the bcl–2 protein. Am J Pathol 1994; 145:1036–1040.
83 Fontanini G, Vignati S, Bigini D et al. Bcl–2 protein: a prognostic factor inversely correlated to p53 in nonsmall–cell lung cancer. Br J Cancer 1995; 71:1003–1007.
84 Jiang S–X, Sato Y, Kuwao S, Kameya T. Expression of *bcl–2* oncogene protein is prevalent in small cell lung carcinomas. J Pathol 1995; 177:135–138.
85 Pezzella F, Turley H, Kuzu I et al. Bcl–2 protein in nonsmall–cell lung carcinoma. N Engl J Med 1993; 329:690–694.
86 Chandler D, El–Naggar AK, Brisbay S, Redline RW, McDonnell TJ. Apoptosis and expression of the *bcl–2* proto–oncogene in the fetal and adult human kidney: evidence for the contribution of bcl–2 expression to renal carcinogenesis. Hum Pathol 1994; 25:789–796.
87 Krajewski S, Chatten J, Hanada M, Womer R, Reed JC. Immunohistochemical analysis of the Bcl–2 oncoprotein in human neuroblastomas. Lab Invest 1995; 71:42–54.
88 Castle V, Heidelberger KP, Bromberg J et al. Expression of the apoptosis–suppressing protein bcl–2, in neuroblastoma is associated with unfavorable histology and N–myc amplification. Am J Pathol 1993; 143:1543–1550.

89 Tron VA, Krajewski S, Klein–Parker H et al. Immunohistochemical analysis of Bcl–2 protein regulation in cutaneous melanoma. Am J Pathol 1995; 146(3):643–650.
90 Miyashita T, Krajewski S, Krajewska M et al. Tumor suppressor p53 is a regulator of *bcl–2* and *bax* in gene expression in vitro and in vivo. Oncogene 1994; 9:1799–1805.
91 Miyashita T, Harigai M, Hanada M, Reed JC. Identification of a p53–dependent negative response element in the *bcl–2* gene. Cancer Res 1994; 54:3131–3135.
92 Rampino N, Yamamoto H, Ionov Y, Li Y, Sawai H, Reed JC, Perucho M. Frequent framshift somatic mutations in the pro–apoptotic gene *bax* in colon cancer of the microsatellite mutator phenotype. Science (in press).
93 Krajewska M, Fenoglio–Preiser CM, Krajewski S et al. Immunohistochemical analysis of Bcl–2 family proteins in adenocarcinomas of the stomach. Amer J Pathol 1996; 149(5):1449–1457.
94 Krajewski, M, Krajewski S, Epstein, JI, Shavaik A, Sauvageot, J, Song K, Kitada S, Reed JC Immunohistochemical analysis of bcl–2, bax, bcl–X, and mcl–1 expression in prostate cancers. Am J Pathol 1996; 148,1567–1576.
95 Hermine O, Haioun C, Lepage E et al. Prognostic significance of Bcl–2 protein expression in aggressive non–Hodgkin's Lymphoma. Blood 1996; 87:265–272.
96 Hill ME, MacLennan KA, Cunningham DC et al. Prognostic significance of BCL–2 expression and bcl–2 major breakpoint region rearrangement in diffuse large cell non–Hodgkin's lymphoma: a british national lymphoma investigation study. Blood 1996; 88:1046–1051.
97 Campos L, Roualult J–P, Sabido O et al. High expression of bcl–2 protein in acute myeloid leukemia cells is associated with poor response to chemotherapy. Blood 1993; 81:3091–3096.
98 Maung ZT, MacLean FR, Reid M et al. The relationship between bcl–2 expression and response to chemotherapy in acute leukemia. Br J Hematol 1994; 88:105–109.
99 Gasparini G, Barbareschi M, Doglioni C et al. Expression of Bcl–2 protein predicts efficacy of adjuvant treatments in operable node–positive breast cancer. Clin Cancer Res 1995; 1:189–198.
100 Hellemans P, van Dam PA, Weyler J et al. Prognostic value of bcl–2 expression in invasive breast cancer. Br J Cancer 1995; 72:354–360.
101 Joensuu H, Pylkkänen L, Toikkanen S. Bcl–2 protein expression and long–term survival in breast cancer. Am J Pathol 1994; 145:1191–1198.
102 Silvestrini R, Veneroni S, Daidone MG et al. The bcl–2 protein: a prognostic indicator strongly related to p53 protein in lymph node–negative breast cancer patients. J Natl Cancer Inst 1994; 86:499–504.
103 Silvestrini R, Benini E, Veneroni S et al. p53 and bcl–2 expression correlates with clinical outcome in a series of node–positive breast cancer patients. J Clin Oncol 1996; 14:1604–1610.
104 Ohsaki Y, Toyoshima E, Fujiuichi S. Bcl–2 and p53 protein expression in nonsmall cell lung cancers: correlation with survival time. Clin Cancer Res 1996; 2:915–920.
105 Krajewski S, Blomvqvist C, Franssila K et al. Reduced expression of pro–apoptotic gene *Bax* is associated with poor response rates to combination chemotherapy and shorter survival in women with metastatic breast adenocarcinoma. Cancer Res 1995; 55:4471–4478.
106 Aguilar–Santelises M, Rottenberg ME, Lewin N et al. Bcl–2, Bax, and p53 expression in B cell in relation to in vitro survival and clinical progression. Int J Cancer 1996; 69:114–119.
107 Krajewski S, Thor A, Edgerton S et al. Immunohistochemical analysis of Bax and Bcl–2 in p53–immunopositive breast cancers. Lab Invest (in press).
108 Reed JC. Balancing cell life and death: Bax, apoptosis, and breast cancer. J Clin Invest 1996; 97:2403–2404.
109 Bargou RC, Daniel PT, Mapara MY et al. Expression of the *bcl–2* gene family in normal and malignant breast tissue: low bax–a expression in tumor cells correlates with resistance towards apoptosis. Int J Cancer 1995; 60:854–859.
110 Wagener C, Bargou RC, Daniel et al. Induction of the death–promoting gene *bax–alpha* sensitizes cultured breast–cancer cells to drug–induced apoptosis. Int J Cancer 1996; 67(1):138–141.
111 Kitada S, Krajewski S, Miyashita T et al. Gamma–radiation induces upregulation of Bax protein and apoptosis in radiosensitive cells in vivo. Oncogene 1996; 12:187–192.
112 Miyashita T, Reed JC. Tumor suppressor p53 is a direct transcriptional activator of human *bax* gene. Cell 1995; 80:293–299.

113 Alnemri ES, Livingston DJ, Nicholson DW, Salvesen G, Thornberry NA, Wong WW, Yuan J. Human ICE/CED–3 protease nomenclature. Cell 1996; 87:171.
114 Martin SJ, Green DR. Protease activation during apopotosis: death by a thousand cuts? Cell 1995; 82:349–352.
115 Boulakia CA, Chen G, Ng FWH et al. Bcl–2 and adenovirus E1B 19 kDA protein prevent E1A–induced processing of CPP32 and cleavage of poly(ADP–ribose) polymerase. Oncogene 1996; 12:29–36.
116 Monney B, Otter I, Olivier R et al. Bcl–2 overexpression blocks activation of the death protease CPP32/Yama/Apopain. Biochem Biophys Res Commun 1996; 221:340–345.
117 Srinivasan A, Foster LM, Testa M–P et al. Bcl–2 expression in neural cells blocks activation of ice/ced–3 family proteases during apoptosis. J Neurosci 1996; 16:5654–5660.
118 Ibrado AM, Huang Y, Fang G et al. Overexpression of Bcl–2 or Bcl–xl inhibits Ara–C–induced Cpp32/Yama protease activity and apoptosis of human acute myelogenous leukemia HL–60 cells. Cancer Research 1996; 56:4743–4748.
119 Smyth MJ, Perry DK, Zhang J et al. prICE: a downstream target for ceramide–induced apoptosis and for the inhibitory action of Bcl–2. Biochem J 1996; 316:25–28.
120 Mebmer UK, Reimer DM, Reed JC, Brune B. Nitric oxide induced poly(ADP–ribose) polymerase cleavage in RAW 2647 macrophage apoptosis is blocked by Bcl–2. FEBS Lett 1996; 384:162–166.
121 Jürgensmeier JM, Krajewski S, Armstrong R, Wilson GM, Oltersdorf T, Fritz LC, Reed JC, Ottilie S. Bax– and Bak–induced cell death in the fission yeast Schizosaccharomyces pombe. Mol Biol Cell 1997; 8:(in press).
122 Hengartner MO, Horvitz HR. Programmed cell death in *Caenorhabditis elegans*. Curr Opin Genet Dev 1994; 4:581–586.
123 Krajewski S, Tanaka S, Takayama S, Schibler MJ, Fenton W, Reed JC. Investigations of the subcellular distribution of the bcl–2 oncoprotein: residence in the nuclear envelope, endoplasmic reticulum, and outer mitochondrial membranes. Cancer Res 1993; 53:4701–4714.
124 Lithgow T, van Driel R, Bertram JF, Strasser A. The protein product of the oncogene *bcl–2* is a component of the nuclear envelope, the endoplasmic reticulum, and the other mitochondrial membrane. Cell Growth Differ 1994; 3:411–417.
125 Yang T, Kozopas KM, Craig RW. The intracellular distribution and pattern of expression of Mcl–1 overlap with, but are not identical to, those of Bcl–2. J Cell Biol 1995; 128:1173–1184.
126 White E, Cipriani R. Specific disruption of intermediate filaments and the nuclear lamina by the 19–kDa product of the adenovirus E1B oncogene. Proc Natl Acad Sci USA 1989; 86:9886–9890.
127 Kane DJ, Sarafin TA, Auton S et al. Bcl–2 inhibition of neural cell death: decreased generation of reactive oxygen species. Science 1993; 262:1274–1276.
128 Hockenbery D, Oltvai Z, Yin X–M et al. Bcl–2 functions in an antioxidant pathway to prevent apoptosis. Cell 1993; 75:241–251.
129 Shimizu S, Eguchi Y, Kosaka H et al. Prevention of hypoxia–induced cell death by Bcl–2 and BclxL. Nature 1995; 374:811–813.
130 Jacobson MD, Raff MC. Programmed cell death and Bcl–2 protection in very low oxygen. Nature 1995; 374:814–816.
131 Distelhorst CW, Lam M, McCormick TS. Bcl–2 inhibits hydrogen peroxide–induced ER Ca^{2+} pool depletion Oncogene 1996; 12:2051–2055.
132 Baffy G, Miyashita T, Williamson JR, Reed JC. Apoptosis induced by withdrawal of Interleukin–3 [IL–3] from an IL–3–dependent hematopoietic cell line associated with repartitioning of intracellular calcium and is blocked by enforced Bcl–2 oncoprotein production. J Biol Chem 1993; 268:6511–6519.
133 Lam M, Dubyak G, Chen L et al. Evidence that Bcl–2 represses apoptosis by regulating endoplasmic reticulum–associated Ca2+ fluxes. Proc Natl Acad Sci USA 1994; 91:6569–6573.
134 Murphy AN, Bredesen DE, Cotopassi G et al. Bcl–2 potentiates the maximal calcium uptake capacity of neural cell mitochondria. Neurobiology 1996; 93:9893–9898.
135 Marin CM, Fernandez A, Bick RJ et al. Apoptosis suppression by bcl–2 is correlated with the regulation of nuclear and cytosolic Ca^{2+}. Oncogene 1996; 12:2259–2266.

136 Ryan JJ, Prochownik E, Gottlieb CA et al. c–myc and bcl–2 modulates p53 function by altering p53 subcellular trafficking during the cell cycle. Proc Natl Acad Sci USA 1994; 91:5878–5882.

137 Meikrantz W, Gisselbrecht S, Tam SW, Schlegel R. Activation of cyclin A–dependent protein kinases during apoptosis. Proc Natl Acad Sci USA 1994; 91:3754–3758.

138 Wang H–G, Miyashita T, Takayama S et al. Apoptosis regulation by interaction of bcl–2 protein and Raf–1 kinase. Oncogene 1994; 9:2751–2756.

139 Wang H–G, Millan JA, Cox AD et al. R–ras promotes apoptosis caused by growth factor deprivation via a Bcl–2 suppressible mechanism. J Cell Biol 1995; 129:1103–1114.

140 Fernandez–Sarbia MJ, Bischoff JR. Bcl–2 associates with the ras–related protein R–ras p23. Nature 1993; 366:274–275.

141 Bernardi P, Broekemeier KM, Pfeiffer DR. Recent progress on regulation of the mitochondrial permeability transition pore; a cyclosporin–sensitive pore in the inner mitochondrial membrane. Journal of Bioenergetics and Biomembranes 1994; 26:509–517.

142 De Jong D, Prins FA, Mason DY et al. Subcellular localization of the bcl–2 protein in malignant and normal lymphoid cells. Cancer Res 1994; 54:256–260.

143 Nicolli A, Basso E, Petronilli V et al. Interactions of cyclophilin with the mitochondrial inner membrane and regulation of the permeability transition pore, a clyclosporin A–sensitive channel. J Biol Chem 1996; 271:2185–2192.

144 Igbavboa U, Zwinzinski CW, Pfeiffer DR. Release of mitochondrial matrix proteins through a Ca^{2+}–requiring, cyclosporin–sensitive pathway. Biochem Biophys Res Commun 1989; 161, 619–625.

145 Kroemer G, Zamzami N, Susin SA. Mitochondrial control of apoptosis. Immunol Today 1996; (in press).

146 Liu X, Kim CN, Yang J et al. Induction of apoptotic program in cell–free extracts: requirement for dATP and Cytochrome C. Cell 1996; 86:147–157.

147 Susin SA, Zamzami N, Castedo M et al. Bcl–2 inhibits the mitochondrial release of an apoptogenic protease. J Exp Med 1996; 184:1331–1342.

148 Zamzami N, Marchetti P, Castedo M et al. Inhibitors of permeability transition interfere with the disruption of the mitochondrial transmembrane potential during apoptosis. FEBS Lett 1996; 384:53–57.

149 Marchetti P, Hirsch T, Zamzami N et al. Mitochondrial permeability transition triggers lymphocyte apoptosis. J Immunol (in press).

150 Zhu W, Cowie A, Wasfy L, Leber B, Andrews D. Bcl–2 mutants with restricted subcellular location reveal spatially distinct pathways for apoptosis in different cell types. EMBO J 1996; 15:4130–4141.

151 Shimizu S, Eguchi Y, Kamiike W et al. Retardation of chemical hypoxia–induced necrotic cell death by Bcl–2 and ICE inhibitors: possible involvement of common mediators in apoptotic and necrotic signal transductions. Oncogene 1996; 12:2045–2050.

152 Shimizu S, Eguchi Y, Kamiike W et al. Induction of apoptosis as well as necrosis by hypoxia and predominant prevention of apoptosis by bcl–2 and bcl–XL. Cancer Res 1996; 56:2161–2166.

153 Kane DJ, Örd T, Anton R, Bredesen DE. Expression of Bcl–2 inhibits necrotic neural cell death. J Neurosci Res 1995; 40:269–275.

154 Jacobson MD, Burne JF, King MP et al. Apoptosis and bcl–2 protein in cells without mitochondrial DNA. Nature 1993; 361:365–368.

155 Marchetti P, Susin SA, Decaudin D et al. Apoptosis–associated derangement of mitochondrial function in cells lacking mitochondrial DNA. Cancer Res 1996; 56:2033–2038.

156 Parker MW, Postma JPM, Pattus F, Tucker AD, Tsernoglou D. Structure of the pore–forming domain of colicin A at 24 Å resolution. J Mol Biol 1992; 224:639–657.

157 Duche D, Izard J, Gonzales–Manas JM et al. Membrane topology of the colicin a pore–forming domain analyzed by disulfide bond engineering. J Biol Chem 1996; 271:15401–15406.

158 Merrill AR, Cohen FS, Cramer WA. On the nature of the structural change of the colicin E1 channel peptide necessary for its translocation–competent state. Biochemistry 1990; 29:5829–5836.

159 Cramer WA, Zhang Y–L, Schendel S et al. Dynamic properties of the colicin E1 ion channel. FEMS Microbiology Immunol 1992; 105:71–82.

160 Shin Y–K, Levinthal C, Levinthal F, Hubbell WL. Colicin E1 binding to membranes: time–resolved studies of spin–labeled mutants. Science 1993; 259:960–963.

161 Minn AJ, Valez PR, Schendel SL et al. BclX_L forms an ion channel in synthetic lipid membranes. Nature 1996; (in press).
162 Schendel SL, Xie Z, Montal MO, Matsuyama S, Montal M, Reed JC. Channel formation by anti–apoptotic protein, Bcl–2. (submitted).
163 Donovan JJ, Simon MI, Draper RK, Montal M. Diphtheria toxin forms transmembrane channels in planar lipid bilayers. Proc Natl Acad Sci USA 1981; 78:172–176.
164 Konisky J. Colicins and other bacteriocins with established modes of action. Annu Rev Microbiol 1982; 36:125–144.
165 Chen C–Y, Faller DV. Phosphorylation of Bcl–2 protein and association with p21Ras in Ras–induced apoptosis. J Biol Chem 1996; 271:2376–2379.
166 Naumovski L, Cleary ML. The p53–binding protein 53BP2 also interacts with Bcl–2 and impedes cell cycle progression at G_2/M. Mol Cell Biol 1996; 16:3884–3892.
167 Kurschner C, Morgan JI. The cellular prion protein (PrP) selectively binds to Bcl–2 in the yeast two–hybrid system. Mol Brain Res 1995; 30:165–168.
168 Takayama S, Sato T, Krajewski S et al. Cloning and functional analysis of BAG–1: a novel Bcl–2 binding protein with anti–cell death activity. Cell 1995; 80:279–284.
169 Boyd JM, Malstrom S, Subramanian T et al. Adenovirus E1B 19 kDa and bcl–2 proteins interact with a common set of cellular proteins. Cell 1993; 79:341–351.
170 Wang H–G, Takayama S, Rapp UR, Reed JC. Bcl–2 interacting protein, BAG–1, binds to and activates the kinase Raf–1. Proc Natl Acad Sci USA 1996; 93:7063–7068.
171 Zha J, Harada H, Yang E et al. Serine phosphorylation of death agonist BAD in response to survival factor results in binding to 14–3–3 not BCL–X_L. Cell 1996; 87:619–628.
172 Chang BS, Minn AJ, Muchmore SW et al. Identification of a novel regularoty domain in Bcl–X(L) and Bcl–2. EMBO J. 1997;16:969–977.
173 Haldar S, Chintapallli J, Croce, CM. Taxol induces bcl–2 phosphorylation and death of prostate cancer cells. Cancer Res 1996; 56:1253–1255.
174 Blagosklonny M, Schulte T, Nguyen P, Trepel J, Neckers L. Taxol–induced apoptosis and phosphorylation of Bcl–2 protein involves c–Raf–1 and represents a novel c–Raf–1 transduction pathway. Cancer Res 1996; 56:1851–1854.
175 Blagosklonny MV, Schulte TW, Nguyen P et al. Taxol induction of p21WAF1 and p53 requires c–raf–1. Cancer Res 1995; 55:4623–4626.
176 Aimé–Sempé C, Kitada S, Reed JC. Investigations of Taxol–mediated phosphorylationn of bcl–2. Blood 1996; 88(10 Supp 1):106.
177 Tanaka S, Louie DC, Kant JA, Reed JC. Frequent somatic mutations in translocated *BCL2* genes of non–Hodgkin's lymphomas patients. Blood 1992; 79:229.
178 Matolcsy A, Casali P, Warnke RA, Knowles DM. Morphologic transformation of follicular lymphoma is associated with somatic mutation of the translocated *Bcl–2* gene. Blood 1996; 88(10):3937–3944.
179 Reed JC, Tanaka ST. Somatic Point mutations in translocated *bcl–2* alleles of non–Hodgkin's lymphomas and lymphocytic leukemias: Implications for mechanisms of tumor progression. Leuk Lymphoma 1993; 10:157–163.
180 Pietenpol JA, Papadopoulos N, Markowitz S et al. Paradoxical inhibition of solid tumor cell growth by bcl–2. Cancer Res 1994; 54:3714–3717.
181 Borner C. Diminished cell proliferation associated with the death–protective activity of Bcl–2. J Biol Chem 1996;271:12695–12698.
182 Mazel S, Burtrum D, Petrie HT. Regulation of cell division cycle progression by bcl–2 expression: a potential mechanism for inhibition of programmed cell death. J Exp Med 1996; 183:2219–2226.
183 Linette GP, Li Y, Roth K, Korsmeyer SJ. Cross talk between cell death and cell cycle progression: BCL–2 regulates NFAT–medicated activation. Proc Natl Acad Sci USA 1996; 93:9545–9552.
184 May WS, Tyler PG, Ito T et al. Interleukin–3 and bryostatin–1 mediate hyperphosphorylation of Bcl–2 alpha in association with suppression of apoptosis. J Biol Chem 1994; 269:26865–26870.
185 Grimm S, Bauer MKA, Baeuerle P, Schulze–Osthoff K. Bcl–2 down–regulates the activity of transcription factor NF–kB induced apoptosis. J Cell Biol 1996; 134:13–23.

186 Reed JC, Talwar HS, Rapp UR et al. Mitochondrial protein p26 bcl–2 reduces growth factor requirements of NIH3T3 fibroblasts. Exp Cell Res 1991; 195:277–283.
187 Miyazaki T, Liu Z–J, Kawahara A et al. Three distinct IL–2 signaling pathways mediated by bcl–2, c–myc, and lck cooperate in hematopoietic cell proliferation. Cell 1995; 81:223–231.
188 Bissonnette RP, Echeverri F, Mahboubi A, Green DR. Apoptotic cell death induced by c–myc is inhibited by bcl–2. Nature 1992; 359:552–554.
189 Fanidi A, Harrington EA, Evan GI. Cooperative interaction between c–*myc* and *bcl–2* proto–oncogenes. Nature 1992; 359:554–556.
190 Kranenburg O, van der Eb AJ, Zantema A. Cyclin D1 is an essential mediator of apoptotic neuronal cell death. EMBO J 1996; 15:46–54.
191 Raff MC, Barres BA, Burne JF, Coles HS, Ishizaki Y, Jacobson MD. Programmed cell death and the control of cell survival: lessons from the nervous system. Science 1993; 262:695–700.
192 Frisch SM, Francis H. Disruption of epithelial cell–matrix interactions induces apoptosis. J Cell Biol 1994; 124:619–626.
193 Elledge SJ. Cell cycle checkpoints: preventing an identity crisis. Science 1996; 274:1664–1667.

Chapter 6

Apoptosis and Cancer, edited by Seamus J. Martin.

Abl Tyrosine Kinase and the Control of Apoptosis

Gustavo P. Amarante–Mendes and Douglas R. Green

Division of Cellular Immunology, La Jolla Institute for Allergy and Immunology, San Diego, California, U.S.A.

Introduction

The process of oncogenesis has been considered for many years as a deregulation of cellular proliferation leading to the expansion of a particular cell population. This view is supported by the observations that many oncogene products are homologous either to growth factors, growth factor receptors or molecules that participate in growth factor signal transduction. However, to be more precise, the expansion of a tumor (or any tissue) is directly related to the difference of the rates of proliferation and cell death [1]. This idea suggests that defects in the apoptosis machinery may contribute to the malignant phenotype found in some forms of cancer. Indeed, it was recently demonstrated that tumors with a higher incidence of apoptosis have a better prognosis than those with lower rates of apoptotic cell death and/or high rates of necrotic death [2]. Thus, two possible strategies can be used by oncogenes: cellular transformation, which we will consider only in terms of deregulated proliferative events, including autocrine growth, anchorage independence and increased mitotic index, and anti–apoptotic signaling. Here we will review data demonstrating that the oncogenic forms of Abl trigger both events and will discuss the possibility that different domains of the molecule are responsible for each of these events. Also, we will suggest that oncogenic forms of Abl are among the most potent mammalian repressors of apoptosis.

The Abl Family of Tyrosine Kinases

The Abl family of tyrosine kinases consists of five members, v–Abl, c–Abl, Bcr–Abl, TEL–Abl and Arg (Fig. 6.1). The first described member of this family was v–Abl. It was identified as the viral oncogene encoded by the Abelson murine leukemia virus (A–MuLV) [3], which was later found to be generated by a recombination between the Moloney murine leukemia virus (Mo–MuLV) and the proto–oncogene *c–abl* [4–6]. v–Abl occurs mainly in three isoforms, p160, p120 and p90, and is composed of the *gag* coding sequence of Mo–MuLV linked to a deleted form of the mouse c–Abl that lacks the region extending from the amino terminus through the SH3 domain. A different form of v–Abl was described in which the SH3 domain is preserved but part of the Abl carboxy–terminal sequence is deleted and the truncated gene is then fused between *gag* and *pol* in the Hardy–Zuckerman–2 feline sarcoma virus (HZ2–FSV) [7, 8].

Cellular homologs of v–Abl were found in human, mouse, cat, fruit fly and in the nematode *C. elegans* [4, 6, 9–13]. c–Abl is a nonreceptor tyrosine kinase localized preferentially to the nucleus [14], but is also found in the cytoplasm where it is associated with the actin filaments [15]. It is composed of multiple domains, some of them shared with proteins involved in signal transduction. Two isoforms of c–Abl, type I and IV (or 1a and 1b in humans), are produced by alternative splicing of the first two exons [13, 16] and are ubiquitously expressed in mammalian cells [17]. In contrast to c–Abl 1a, c–Abl 1b contains a myristoylation site at its amino–terminus. In addition, Abl–family kinases contain **S**rc–**h**omology 3 (SH3), SH2, SH1 (kinase domain), DNA binding (DB) and actin binding (AB) domains, as well as a nuclear translocation signal (NTS), two sites for protein kinase C (PKC) phosphorylation and a proline–rich sequence. Among these domains the SH1 is the most conserved throughout evolution and assumed to be essential for coupling the pathways that lead to cellular proliferation and resistance to apoptosis.

The fourth member of the Abl–family is Bcr–Abl, the outcome of a translocation between chromosomes 9 and 22 found in Philadelphia (Ph^1) chromosome–positive chronic myelogenous leukemia (CML) and acute lymphocytic leukemia (ALL) [18–21]. This genetic event replaces the amino–terminal end of c–Abl by sequences encoded by the *bcr* gene. In CML, *bcr–abl* produces a chimeric tyrosine kinase of 210 kDa whereas in ALL Bcr–Abl is expressed as a 185 kDa protein. This difference reflects the position of the breakpoint in the *bcr* gene. The 185 kDa form contains *bcr* sequences coding for a coiled–coil domain, a putative serine/threonine kinase, a Grb2 binding site and a SH2 binding domain [22]. The 210 kDa protein shares all of these features and has an additional domain homologous to DBL, VAV, ECT2, CDC24 and CDC25 [22].

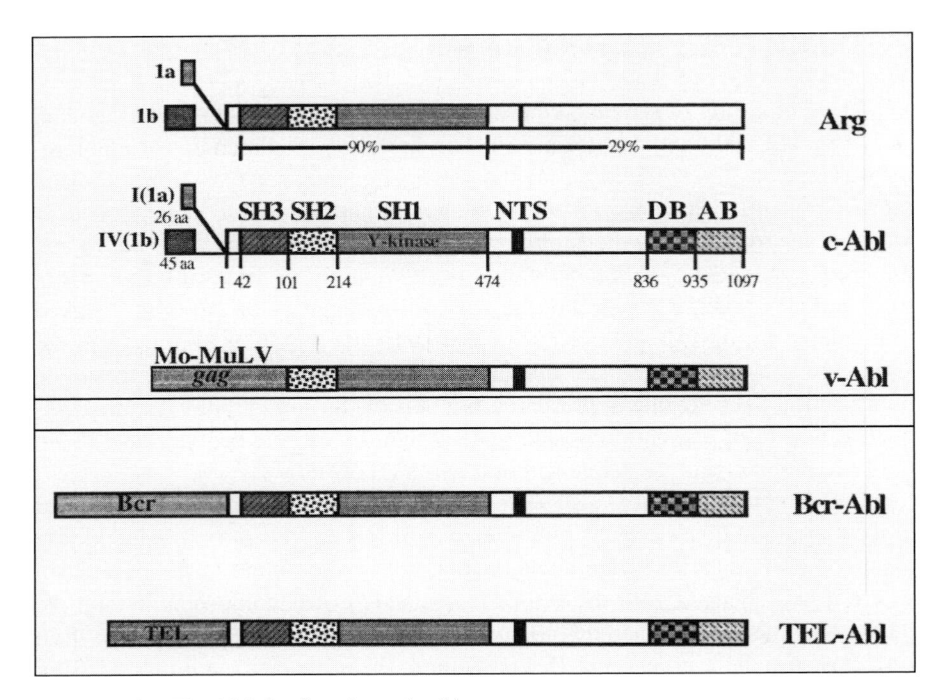

Fig. 6.1. The Ab1 family of tyrosine kinases.

Another member of the Abl–family of tyrosine kinases was identified by hybridization of genomic clones containing sequences related to *c–abl* [23] and, therefore, designated *arg* (for **A**belson–**r**elated **g**ene). The characterization of this gene revealed a high level of identity (around 90%) between Arg and c–Abl SH3/SH2/SH1 domains but weak similarity (29%) between the carboxy–terminal domains [24]. Nevertheless, several important features of the carboxy–terminal domains are conserved in these two proteins. These include the proline–rich sequence, part of the NTS, and at least one of the PKC phosphorylation sites [24]. In addition, like *c–abl, arg* produces two alternative spliced mRNAs. Although Arg and c–Abl amino–terminal sequences have diverged, Arg type 1b shares the myristoylated site found in c–Abl type 1b.

The last member of this family was recently cloned and characterized as a chromosomal translocation associated with acute myeloid leukemia, which generates the TEL–Abl fusion protein [25, 26]. As in the case of *bcr–abl*, a truncated form of *TEL* (nt 1–1033) is fused in–frame to exon 2 of *c–abl*. TEL–Abl retains the helix–loop–helix domain but not the DNA–binding domain present in the wild–type TEL. The biological activity of TEL–Abl was shown to be similar to that of Bcr–Abl, including constitutive activation of the Abl kinase and tyrosine phosphorylation of the molecule, cytoskeleton localization and dependence on oligomerization for transformation [26]. Moreover, proteins known to be substrates for Bcr–Abl were also found to be tyrosine phosphorylated in response to TEL–Abl [27].

Abl and Cellular Transformation

The oncogenic features of activated Abl proteins have been extensively described in the last 10–20 years and are not the focus of this chapter. However, we will summarize some of the most important findings that may have implications for the anti–apoptotic effect of these tyrosine kinases.

The Kinase Domain

The first suggestion that the tyrosine kinase activity of Abl is important for the transformed phenotype was provided by the observation that both oncogenic forms, v–Abl and Bcr–Abl, are constitutively activated. Furthermore, the 185 kDa isoform of Bcr–Abl has a higher tyrosine kinase activity when compared to the 210 kDa Bcr–Abl and much greater transforming activity in both hematopoietic cells and fibroblasts [28]. Temperature–sensitive mutants of these molecules were capable of transforming cells at the permissive temperature but not at the nonpermissive temperature, at which the molecules display marked reduction in their tyrosine kinase activity [29–33]. Finally, kinase deficient mutants (e.g., K671R) have no tumorigenic activity and are unable to confer growth factor independence [34].

The SH3 Domain

Deletion of the amino–terminal sequences that include the SH3 domain was shown to be sufficient to activate the transforming potential of c–Abl [35, 36]. In addition, a point mutation in this domain which abolishes the binding to proline–rich ligands also activates its transforming ability [37]. These observations suggest that c–Abl is tightly regulated by the interaction with a proline–rich cellular inhibitor (see also ref. 38). One such protein could be 3BP–1, which was shown to have a high specificity for the Abl SH3 sequences [39]. Recently, two novel and highly related proteins were identified, Abi–1 and Abi–2 (for **Abl**–**i**nteractor), that bind to both Abl SH3 and Abl–carboxy–terminal domains [40,

41] (see Fig. 6.2). Most importantly, a truncated form of Abi–2 that lacks the Abl SH3–binding site but still interacts with Abl–carboxy–terminal sequence activates c–Abl tyrosine kinase and its transforming potential. This suggests that the full–length Abi–2 is a repressor of c–Abl. In addition, Abi–1 was found to suppress v–Abl transforming activity. Taken together, these results point to a new family of inhibitory proteins involved in the control of cellular proliferation—and maybe cell death—mediated by c–Abl. These molecules are similar to proteins involved in Src–mediated signaling, such as HS1, LckBP1 and cortactin. Interestingly, HS1–knock–out mice have a defect in the lymphocyte compartment in respect to antigen–receptor induced proliferative and apoptotic responses [42]. It is important to note that the SH3 domain is preserved in the Bcr–Abl molecule, and thus the Bcr sequences are presumably able to overcome the inhibitory effect of Abl SH3–binding proteins or hinder their association. Phosphoserine/phosphothreonine residues present on the amino–terminus of Bcr can associate with the Abl SH2 domain in the Bcr–Abl molecule and this interaction is necessary for transformation, raising the possibility that this interaction shelters the Abl SH3 domain [43] (see Fig. 6.2).

The SH2 Domain

In contrast to the SH3 domain, Abl SH2 domain has a positive effect in transformation. Deletion of this domain or mutation in the residue responsible for binding phosphotyrosine impairs the transforming potential of v–Abl [44]. The same mutation in the p185 isoform of Bcr–Abl was shown to impair its ability to transform fibroblasts [45]. However, the SH2 domain is not required for transformation of hematopoietic cells and induction of growth factor independence by Bcr–Abl [15, 46].

Subcellular Localization

A very significant characteristic that is shared by every oncogenic form of Abl is the cytoplasmic localization of these proteins. This finding is intriguing, given the fact that c–Abl is predominantly a nuclear protein, and it has at least two important implications. First, Abl substrates that participate in normal cellular regulation may be different from the ones involved in transformation and protection from apoptosis. Second, it reinforces the idea that the central mechanism of apoptosis is localized primarily in the cytoplasm. The mechanism responsible for changing the subcellular localization of Abl is not yet completely defined. Amino–terminal myristoylation, an event known to target proteins to the

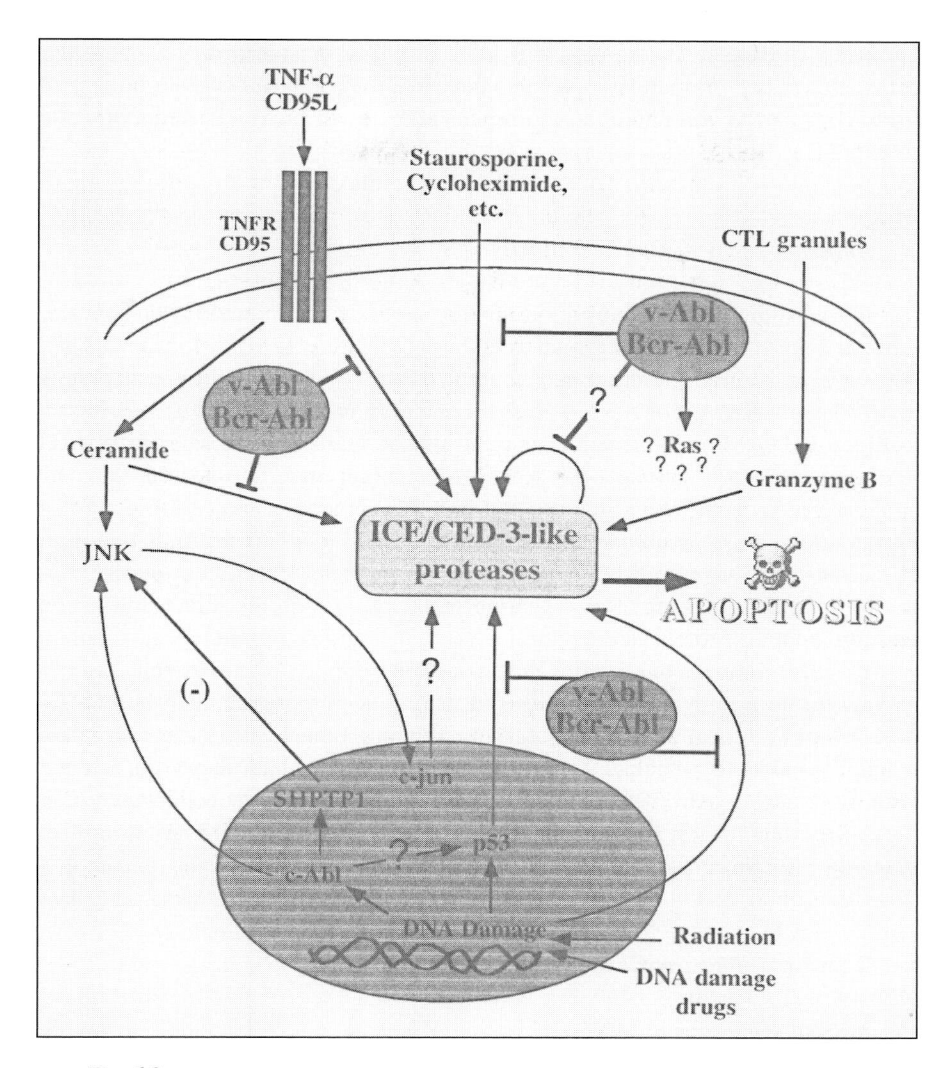

Fig. 6.2.

cellular membrane, may participate in some instances but is not a requirement since Bcr–Abl does not have such a translational modification. Another possible mechanism involves the carboxy–terminal end of the Abl molecule, which contains a domain that specifically interacts with actin filaments [15, 47]. The binding of Abl to actin is strongly increased by the first 63 amino acids of Bcr and correlates with the degree of transformation by Abl–family tyrosine kinases.

Deletion of the actin–binding domain impairs the association with the cytoskeleton and dramatically reduces the ability of Bcr–Abl both to transform fibroblasts and to abrogate the requirement for growth factors in hematopoietic cells [15].

Influence of the Cellular Context

Perhaps one of the most important aspects to be considered in the study of Abl–mediated cellular transformation is the fact that this tyrosine kinase can initiate different signaling pathways depending on the cellular context. One example involves the activation of Ras. Expression of a dominant–negative *ras* was shown to block Bcr–Abl transformation of mouse bone marrow cells whereas normal *ras* accelerated this activity [48]. A Bcr–Abl mutant unable to couple the Grb2 adapter protein to the Ras signaling pathway failed to transform fibroblasts [49]. Remarkably, the same mutant caused growth factor independence in hematopoietic cells [50]. This effect was found to be achieved in this context by the connection of Bcr–Abl to the Ras signaling pathway through the direct interaction with the adapter protein SHC.

Another example of the relevance of cellular context is provided by the observation that nonmyristoylated Abl transforms hematopoietic cells but not fibroblasts [51]. Given that N–terminal myristoylation provides a means for intracellular proteins to associate with the plasma membrane and, therefore, interact with molecules localized in this subcellular compartment, it is conceivable that the Abl partners implicated in the first steps of cellular transformation are restricted to the plasma membrane in fibroblasts.

The Anti–Apoptotic Effect of Abl

Abl in Development

Some of the most compelling evidence for a role of the Abl–family kinases in the control of development—perhaps regulating the life span of lymphocytes—is provided by targeted disruption of *c–abl* in mouse embryonic stem cells [52, 53]. Mice homozygous for the *c–abl* null mutation are runted and display increased perinatal mortality, along with abnormal head, eye, spleen and thymus development and an associated lymphopenia. Although c–Abl is expressed ubiquitously, only the lymphocyte compartment is severely affected in these mice, and thus it is conceivable that a redundant mechanism, perhaps represented by the *arg* gene product, acts in other cell types.

A second set of observations supporting a role for Abl in development come from work in *Drosophila* [10]. A series of transgenic flies were generated in which the *D–abl* gene was mutated or part of it replaced by the corresponding sequence in human, mouse or virus *abl*. In this model, disruption of *D–abl* was associated with pupal lethality and defective eye development. Most interestingly, this aberrant phenotype could be rescued by a kinase–inactive form of Abl and required an intact carboxy–terminal domain which was responsible for a proper subcellular localization. It is worth noting that the kinase activity of D–Abl became necessary in a more severe genetic background, in which the flies carried only one copy of the *disabled* (*dab*) gene. These observations indicate that a second gene product may complement the kinase activity of D–Abl. We wonder, at this point, whether such a complementary mechanism exists in the mammalian Abl network.

The Response to DNA Damage

Recently, important observations have connected c–Abl to the jun kinase/ stress activated protein kinase (JNK/SAPK) pathway [54, 55]. Increased tyrosine kinase activity of c–Abl was observed after treatment with different DNA–damaging agents including ionizing radiation, mitomycin C, *cis*–platinum and 1–β–D–arabinofuranosylcytosine (ara–C), all known to trigger the JNK/SAPK pathway. Interestingly, these same stimuli failed to activate JNK/SAPK in c–Abl–deficient cells. In comparison, TNF–α was not able to activate the Abl tyrosine kinase but did engage the JNK/SAPK cascade in Abl–negative cells. These data showed for the first time that c–Abl is activated in response to a extracellular stimulus and is required for linking DNA damage events with activation of stress responses. Moreover, the same group showed recently that c–Abl is constitutively associated with SHPTP1 and mediate tyrosine phosphorylation of this protein tyrosine phosphatase after ionizing radiation [56]. In cotransfection experiments, a wild–type SHPTP1 blocked, whereas a dominant negative SHPTP1 stimulated, irradiation–induced activation of JNK/SAPK, suggesting a regulatory circuit in which c–Abl activates both JNK/SAPK and SHPTP1, the latter acting as a negative regulator of the former.

It is important to mention that the JNK/SAPK pathway has been shown to be activated by the same mechanisms that culminate in apoptosis. Whether this is just a bystander phenomenon or the JNK/SAPK pathway is indeed involved in apoptosis is not clear. Recent observations suggested that different stress–inducing agents, such as x–rays, heat, H_2O_2, TNF–α and γ–irradiation, use ceramide to initiate the apoptosis cascade through a JNK/SAPK pathway [57]. This view was recently challenged by experiments showing that JNK/SAPK activation after ligation of the TNF type 1 receptor (TNFR1) is mediated through a signaling pathway distinct from the one leading to apoptosis [58]. Apoptosis in this case is most likely to be initiated via direct activation of the caspases [59], cysteine proteases belonging to the ICE/CED–3 family known to be part of the central machinery of apoptosis. Interestingly enough, as mentioned before, TNF–α was shown to trigger the JNK/SAPK pathway through a c–Abl–independent pathway [60]. The mechanism through which DNA–damage agents trigger the caspase cascade is unknown. Whether this mechanism involves the activation of c–Abl and subsequent activation of the JNK/SAPK pathway remains elusive.

Abl as a Regulator of Cell Growth

Observations that c–Abl is differentially phosphorylated by cdc–2 or cdc–2–like kinases at several serine/threonine residues during the cell cycle [61] and that cdc–2–mediated phosphorylation controls the binding of c–Abl to DNA [62] suggest a role for c–Abl in cell cycle regulation. In addition, it was shown that the tumor suppressor retinoblastoma (Rb) protein, which is implicated in the control of the G_1–S transition, interacts with c–Abl and inhibits its tyrosine kinase activity [63]. Interestingly, overexpression of c–Abl in fibroblasts inhibits growth due to cell cycle arrest in the G_1 phase by a mechanism dependent on the tumor suppresser protein p53 [64]. This growth inhibitory characteristic of c–Abl is shared with v–Abl, which also causes G_1 arrest in a majority of NIH3T3 cells following transfection [65]. It is important to note that v–Abl does induce cell growth in some NIH3T3 cells, probably due to a permissive cell context [65]. In comparison, a kinase–defective mutant of c–Abl acts as a dominant negative molecule and has a stimulatory effect on cell growth [64]. Whether the anti–apoptosis signaling triggered by v–Abl and Bcr–Abl is influenced by the cell cycle regulatory function of c–Abl is yet to be determined.

The fact that growth suppressor and oncogenic behaviors may reside in the same molecule makes Abl of considerable interest. More intriguingly, recent observations suggest that different domains of Abl may be responsible for the transformation and anti–apoptotic phenotype observed in cells expressing oncogenic forms of this tyrosine kinase [66]. These domains would initiate different signaling pathways which may cooperate with each other or be responsible for only the transformation or the anti–apoptotic phenotype. In this regard, oncogenic forms of Abl were associated with increased phosphatidylinositol 3–kinase (PI3K) activity [67], which was shown to be required for the growth of Bcr–Abl–expressing cells [68]. Using HL–60 or Ba/F3 cells expressing $p185^{bcr-abl}$ or the CML line K562, we observed that PI3K activity is not involved in the anti–apoptotic effect of Bcr–Abl [69]. Treatment of these cells with wortmannin, a PI3K specific inhibitor, blocked the activity of this enzyme without affecting their resistance to VP–16– or CHX–induced apoptosis. Thus, PI3K activity seems to be implicated in transformation but not in the anti–apoptotic phenotype of Bcr–Abl–positive cells.

CML, A Myeloaccumulative Disorder

Chronic myelogenous leukemia is characterized by a triphasic course. The disease evolves from an indolent benign or chronic phase to a more aggressive accelerated phase and ends in blast crisis [70, 71]. Most studies have found that the relative rate of proliferation is not increased in CML [72–74]. This is in accordance with the fact that Bcr–Abl was shown to be unable to activate the mitogenesis in fibroblasts but rendered these cells anchorage independent [75]. Given that Bcr–Abl has a very strong anti–apoptotic effect, the above mentioned observations suggest that the myeloid expansion in CML is due to a prolongation of cell survival rather than an excessive proliferation, providing support for strategies that activate the central apoptotic machinery of CML cells during the chronic phase of the disease. A very efficient approach would be the activation of a specific cytotoxic cell response. The mechanism behind this response will be discussed in more detail in a following section.

Abl–Mediated Repression of Apoptosis

The first indications that the oncogenic forms of Abl tyrosine kinase act as repressors of cell death came from work with growth factor dependent cell lines. In this regard, both v–Abl and Bcr–Abl were shown to induce growth factor independence in a variety of cell populations otherwise fated to die after being deprived of such factors [30, 76–81]. It is important to note, however, that it is still not clear whether stimulation of growth factor receptors protects cells from a constitutive apoptotic signal or simply provides a complementary mechanism for cell cycle progression resulting in the completion of cellular proliferation instead of mitotic catastrophe. Interestingly, despite the fact that IL–3 and Bcr–Abl trigger common signaling events, the same level of anti–apoptotic phenotype observed in Bcr–Abl positive cells cannot be provided by IL–3 [82].

The K562 line, which is derived from a CML patient and expresses $p210^{bcr-abl}$, was shown to be resistant to apoptosis when compared to other leukemia cell lines [83]. Later, Bcr–Abl was proved to be the molecule responsible for this effect since antisense oligonucleotides (AS) directed to this gene reduced Bcr–Abl levels and reverted the resistant phenotype of K562 cells, without having any direct effect on the cell viability [84]. Similar results using different AS were obtained in two IL–3–dependent cell lines induced to constitutively express p210*bcr–abl* as well as cells from CML patients [85]. Care should be taken when interpreting these AS experiments since it was suggested by Vaerman et al that the antiproliferative effect of Bcr–Abl AS may not be attributed to an antisense mechanism [86]. Also, Smetsers et al (87) detected cell death after AS treatment of BV173 and LAMA–84 cells without concomitant reduction of Bcr–Abl protein levels.

Further proof that the oncogenic forms of Abl tyrosine kinase confer resistance to apoptosis came from studies using ectopic expression of either Bcr–Abl or v–Abl and treatment of the cells with chemotherapeutic drugs. Thus, it was shown that the mouse mast cell line IC.DP becomes resistant to melphalan or hydroxyurea upon expression of v–Abl [82, 88]. Using the HL–60 cell line, we observed that both Bcr–Abl and v–Abl induce resistance to a wide variety of apoptotic stimuli, including anti–CD95 antibodies, ceramide, U.V. irradiation, staurosporine, VP–16, VM–25, actinomycin D and cycloheximide [89–91].

How the oncogenic forms of Abl confer such a wide resistance to apoptosis remains largely obscure. Bcr–Abl and v–Abl are among the most potent anti–apoptotic molecules and nevertheless share no structural similarity with any of the known death repressors, such as Bcl–2, Bcl–x_L, CrmA and p35. There is no clear evidence for an obligate interaction between anti–apoptotic Abl and Bcl–2 family members although some studies have been suggestive. In this regard, transfection of Bcr–Abl–positive BaF3 cells with a plasmid containing mouse *bcl–2* in antisense orientation was shown to decrease *bcl–2* mRNA down to undetectable levels and revert the IL–3 independent phenotype of these cells, suggesting that Bcl–2 may participate in the anti–apoptotic and/or transformation effect of Bcr–Abl [92]. In contrast, both wild–type and the apoptosis–resistant, v–Abl–positive IC.DP cells were shown to exhibit similar levels and subcellular localization of the Bcl–2 protein, suggesting that v–Abl acts through a Bcl–2–independent mechanism [88]. Moreover, Bcl–2 protein and mRNA levels were surprisingly downregulated by the function of a temperature–sensitive mutant of v–Abl (v–Abl^{ts}) in HL–60 cells [91]. To further support a Bcl–2–independent mechanism of the anti–apoptotic effect of v–Abl, a Bcl–2–negative HL–60.v–Abl^{ts} clone was found to be resistant to apoptosis at the permissive temperature [91]. In addition, both wild–type and v–Abl^{ts}–positive HL–60 cells expressed the same amount of Bcl–x. Taken together, these results suggests that Bcr–Abl and v–Abl can act through distinct mechanisms, probably not dependent upon Bcl–2.

Interestingly though, Bax, another member of the Bcl–2 family, was also found to be downregulated in HL–60.v–Abl^{ts} cells in a temperature–dependent manner. This effect was necessary for v–Abl to protect HL–60 cells since ectopic overexpression of Bax in HL–60.v–Abl^{ts} cells restored the susceptibility to apoptosis found in the wild–type HL–60 counterparts [91]. Nevertheless, downregulation of Bax is not sufficient to block apoptosis in these cells and overexpression of Bax by itself does not induce apoptosis in HL–60 cells.

The possibility that oncogenic forms of Abl act through Bcl–2 family members raises two important questions. First, is the interaction with Bcl–2 members the only mechanism involved in Abl–mediated resistance to apoptosis? Second, is this a direct interaction or does it require intermediate steps in order to couple to the Bcl–2–family pathway? In the case of the downregulation of Bax by v–Abl, we observed a negative effect of Abl on the Bax promoter, as assessed by a transient transfection assay using a Bax promoter reporter construct [91]. Indirect or alternative pathways may involve Ras, as suggested by experiments in which a dominant negative v–H–Ras were transfected into K562 cells resulting in cell death [93]. Another possibility involves the translocation of protein kinase C (PKC) to the nucleus as suggested by experiments in IC.DP cells [94]. In the latter case, these authors observed that treatment of IC.DP.v–Abl cells with calphostin C, a PKC inhibitor, blocked translocation of PKCβII to the nucleus and restored apoptosis in these cells, which suggests that a nuclear event is required in v–Abl–mediated resistance to apoptosis.

This prompted us to investigate whether v–Abl and Bcr–Abl could induce their anti–apoptotic effect in enucleated cells. Cytoplasts obtained from HL–60 cells expressing either v–Abl or Bcr–Abl are much more resistant to the "spontaneous" apoptosis induced during the process of nuclear extrusion. Apoptosis in HL–60 cytoplasts was accelerated by treatment with staurosporine or anti–CD95. However, this was not observed in HL–60 cytoplasts expressing either of the Abl oncogenes (unpublished observations). Thus, our observations indicate that the anti–apoptotic effect of both v–Abl and Bcr–Abl occurs in the absence of nucleus.

Cytotoxic Lymphocytes Overcome Bcr–Abl–Mediated Protection

To date, the most efficient mechanism for inducing apoptosis in Bcr–Abl–positive cells is the attack of cytotoxic cells. These cells utilize two possible mechanisms to induce apoptosis in their targets: the first involves expression of CD95 ligand which then acts on CD95–positive targets and the second mechanism involves a discharge of cytotoxic granules containing perforin, a pore–forming protein, and a cocktail of proteases. Among these enzymes, granzyme B deserves special attention since it has been shown to be able to initiate apoptosis in a cell–free system by itself, probably acting as a proteolytic processor and activator of the caspases [95, 96]. While Bcr–Abl has been shown to block anti–CD95–induced apoptosis [89], it may not interfere with this second mechanism.

One of the most traditional targets for cytotoxic cell killing is the K562 cells previously mentioned to be largely resistant to apoptosis–inducing stimuli. Further, bone marrow transplantation is a very efficient treatment for CML patients, possibly by inducing a T cell dependent graft–versus–leukemia type of reaction. Thus, it is likely that cytotoxic cells are able to overcome Bcr–Abl–mediated resistance to apoptosis. Indeed, Fuchs et al [97] showed that BaF3 and FDC–P1 cells expressing p210$^{bcr-abl}$ are as susceptible to cytotoxic T cell killing as their Bcr–Abl–negative counterparts. In comparison, these cells showed a marked resistance to VP–16 and radiation–induced apoptosis. In addition, Roger et al [98] demonstrated that NK and LAK cells also induce apoptosis in Bcr–Abl–positive targets and that inhibition of granule exocytosis interferes with this process. Taken together, these observations suggest that cytotoxic cells overcome Bcr–Abl protection from apoptosis through a cytotoxic granule–mediated mechanism.

Where Does Abl Block the Apoptotic Process?

Given the fact that the oncogenic forms of Abl confer resistance to a wide range of apoptosis–inducing stimuli, it is likely that these molecules operate close to the central apoptotic machinery. This machinery, referred to as the apoptotic "executioner" [99], is composed of members of the caspases. They share the conserved QACXG active site and have the distinguishing property of cleaving at aspartic acid (Asp) residues [100, 101]. One or multiple caspases account for the specific cleavage of certain proteins, such as poly(ADP–ribose) polymerase (PARP) [102, 103], the 70 kDa subunit of U1snRNP [104], fodrin [105], lamins [106–108], topoisomerase I [107], DNA–dependent protein kinase [109] and others, during apoptosis. The consequences of the cleavage of each specific substrate is not yet defined but it is conceivable that the proteolysis of some of them contribute to the morphological manifestations of the apoptotic process. Investigation on the relationship between oncogenic forms of Abl tyrosine kinase and the caspases is, in our opinion, one of the most important issues to be pursued in the future. Whether Bcr–Abl or v–Abl can directly associate with and/or phosphorylate the caspases and, as a consequence, influence their catalytic property is unknown at this time. We envision two models to account for the effects of oncogenic Abl in the apoptotic process. In the first one, Bcr–Abl and v–Abl would act downstream of the "executioner," by inactivating the caspases. Alternatively, Abl could act upstream of the "executioner" by preventing the activation of the caspases. We would like to support the second model based on the observations mentioned earlier in this chapter that maybe the only insult capable of overcoming the anti–apoptotic effect of Abl is cytotoxic cell killing. At this point, this seems to be mediated only by the discharge of granzyme–containing granules. Given the fact that granzyme B directly activates caspase 3 (CPP32/Yama/Apopain) [95, 96], it is likely that once these proteases are activated Bcr–Abl can no longer protect cells from apoptosis. Studies to test this idea are underway.

The realization that oncogenic forms of Abl protect cells from apoptosis has given us new insights into how they promote some forms of cancer, and suggest new strategies for therapy. Thus, downregulation of Abl expression and/or function must be coupled with apoptosis–inducing agents for maximum effect. Continued investigations into the mechanism by which oncogenic Abl blocks cell death is also likely to give us new approaches to understanding the central mechanism of apoptosis as well.

Acknowledgments

The authors thank Jean Wang, Owen Witte, Daniel Afar, Anne McGahon, Deborah Finucane, Tom Cotter and Caroline Dive for their ongoing discussions and collaborations on the function of Abl in regulating apoptosis. Our research in this area has been supported by the American Cancer Society (CB–82). G.P.A.–M. is a Brazilian Research Council (CNPq) Fellow.

References

1 Bresciani F, Paoluzi R, Benassi M et al. Cell kinetics and growth of squamous cell carcinomas in man. Cancer–Res 1974; 34:2405–15.
2 Arends MJ, McGregor AH, Wyllie AH. Apoptosis is inversely related to necrosis and determines net growth in tumors bearing constitutively expressed *myc*, *ras*, and HPV oncogenes. J Pathol 1994; 144:1045–1057.
3 Abelson HT, Rabstein LS. Lymphosarcoma: Virus–induced thymic–independent disease in mice. Cancer Res. 1970; 30:2213–2222.
4 Goff S, Gilboa E, Witte ON et al. Structure of the Abelson murine leukemia virus genome and the homologous cellular gene: Studies with cloned viral DNA. Cell 1980; 22:777–785.
5 Dale B, Ozanne B. Characterization of mouse cellular DNA homologous to Abelson murine leukemia virus–specific sequences. Mol Cell Biol 1981; 1:731–742.
6 Wang JYJ, Ledley F, Goff S et al. The mouse *c–abl* locus: Molecular cloning and characterization. Cell 1984; 36:349–356.
7 Besmer P, Hardy WD Jr, Zuckerman EE et al. The Hardy–Zuckerman 2–FeSV, a new feline retrovirus with oncogene homology to Abelson–MuLV. Nature 1983; 303:825–828.
8 Bergold PJ, Blumenthal JA, Andrea ED et al. Nucleic acid sequence and oncogenic properties of the HZ2 Feline sarcoma virus v–abl insert. J Virol 1987; 61:1265–1268.
9 Goddard JM, Weilard JJ, Capeichi MR. Isolation and characterization of *Caenorhabditis elegans* DNA sequences homologous to the *v–abl* oncogene. Proc Natl Acad Sci USA 1986; 83:2172–2176.
10 Henkermeyer MJ, Bennet RL, Gertler FB et al. DNA sequence, structure and tyrosine kinase activity of the Drosophila Abelson proto–oncogene homolog. Mol Cell Biol 1988; 8:843–853.
11 Schalken JA, Van den Ouweland AWM, Bloemers HPJ et al. Characterization of the feline *c–abl* protooncogene. Biochim Biophys Acta 1985; 824:104–112.
12 Durica DS, Restrepo MA, Thomas TL et al. Isolation and characterization of *abl* gene sequences in *Caliphora erythrocephala*. Gene 1987; 59:63–76.

13 Ben–Neriah Y, Bernards A, Paskind M et al. Alternative 5' exons in *c–abl* mRNA. Cell 1986; 44:577–586.
14 van Etten RA, Jackson P, Baltimore D. The mouse type IV *c–abl* gene product is a nuclear protein, and activation of transforming ability is associated with cytoplasmic localization. Cell 1989; 58:669–678.
15 McWhirter JR, Wang JY. An actin–binding function contributes to transformation by the Bcr–Abl oncoprotein of Philadelphia chromosome–positive human leukemias. Embo J 1993; 12:1533–46.
16 Shtivelman E, Lifshitz B, Gale RP et al. Alternative splicing of RNAs trancribed from the human *abl* gene and from the *bcr–abl* fused gene. Cell 1986; 47:277–284.
17 Renshaw MW, Capozza MA, Wang JYJ. Differential expression of type–specific *c–Abl* mRNAs in mouse tissues and cell lines. Mol Cell Biol 1988; 8:4547–4551.
18 Konopka JB, Watanabe SM, Witte ON. An alteration of the human c–abl protein in K562 leukemia cells unmasks associated tyrosine kinase activity. Cell 1984; 37:1035–42.
19 Shtivelman E, Lifshitz B, Gale RP et al. Fused transcript of *abl* and *bcr* genes in chronic myelogenous leukemia. Nature 1985; 315:550–1.
20 Shtivelman E, Lifshitz B, Gale RP et al. Alternative splicing of RNAs transcribed from the human *abl* gene and from the bcr–abl fused gene. Cell 1986; 47:277–84.
21 Clark SS, McLaughlin J, Timmons M et al. Expression of a distinctive *BCR–ABL* oncogene in Ph1–positive acute lymphocytic leukemia (ALL). Science 1988; 239:775–7.
22 Hesketh R. The Oncogene Facts Book. San Diego, CA: Academic Press Inc., 1995:370.
23 Kruh GD, King CR, Kraus MH et al. A novel human gene closely related to the *abl* proto–oncogene. Science 1986; 234:1545–8.
24 Kruh GD, Perego R, Miki T et al. The complete coding sequence of arg defines the Abelson subfamily of cytoplasmic tyrosine kinases. Proc Natl Acad Sci USA 1990; 87:5802–6.
25 Papadopoulos P, Ridge SA, Boucher CA et al. The novel activation of *ABL* by fusion to an *ets*–related gene, *TEL*. Cancer Res 1995; 55:34–38.
26 Golub TR, Goga A, Barker GF et al. Oligomerization of the ABL tyrosine kinase by the Ets protein TEL in human leukemia. Mol Cell Biol 1996; 16:4107–16.
27 Okunda K, Golub TR, Gilliland DG et al. p210BCR/ABL, p190BCR/ABL, and TEL/ABL activate similar signal transduction pathways in hematopoietic cell lines. Oncogene 1996; 13:1147–1152.
28 Lugo TG, Pendergast AM, Muller AJ et al. Tyrosine kinase activity and transformation potency of *bcr–abl* oncogene products. Science 1990; 247:1079–82.
29 Kipreos ET, Lee GJ, Wang JYJ. Isolation of temperature–sensitive tyrosine kinase mutants of *v–abl* oncogene by screening with antibodies for phophotyrosine. Proc Natl Acad Sci USA 1987; 84:1345–1349.
30 Kipreos ET, Wang JYJ. Reversible dependence on growth factor interleukin–3 in myeloid cells expressing temperature sensitive *v–abl* mutant. Oncogene Res 1988; 2:277–284.
31 Carlesso N, Griffin JD, Druker BJ. Use of a temperature–sensitive mutant to define the biological effects of the p210BCR–ABL tyrosine kinase on proliferation of a factor–dependent murine myeloid cell line. Oncogene 1994; 9:149–56.
32 Engelman A, Rosenberg N. Temperature–sensitive mutants of Abelson murine leukemia virus deficient in protein tyrosine kinase activity. J Virol 1990; 64:4242–51.
33 Engelman A, Rosenberg N. Isolation of temperature–sensitive Abelson virus mutants by site–directed mutagenesis. Proc Natl Acad Sci USA 1987; 84:8021–5.
34 Anderson SM, Mladenovic J. The BCR–ABL oncogene requires both kinase activity and *src*–homology 2 domain to induce cytokine secretion. Blood 1996; 87:238–244.
35 Franz WM, Berger P, Wang JYJ. Deletion of an N–terminal regulatory domain of the *c–abl* tyrosine kinase activates its oncogenic potential. EMBO J 1989; 8:137–147.
36 Jackson P, Baltimore D. N–terminal mutations activate the leukemogenic potential of the myristoilated form of *c–abl*. EMBO J 1989; 8:449–456.
37 Van Etten RA, Debnath J, Zhou H et al. Introduction of a loss–of–function point mutation from the SH3 region of the *Caenorhabditis elegans sem–5* gene activates the transformation ability of *c–abl* in vivo and abolishes binding of proline–rich ligands in vitro. Oncogene 1995; 10:1977–1988.

38 Pendergast AM, Muller AJ, Havlik MH et al. Evidence for regulation of the human ABL tyrosine kinase by a cellular inhibitor. Proc Natl Acad Sci USA 1991; 88:5927–31.
39 Cicchetti P, Mayer BJ, Thiel G et al. Identification of a protein that binds to the SH3 region of Abl and is similar to Bcr and GAP–rho. Science 1992; 257:803–6.
40 Dai Z, Pendergast AM. Abi–2, a novel SH3–containing protein interacts with the c–Abl tyrosine kinase and modulates c–Abl transforming activity. Genes Dev 1995; 9:2569–2582.
41 Shi Y, Ålin K, Goff SP. Abl–interactor–1, a novel SH3 protein binding to the carboxy–terminal portion of the Abl protein, suppresses *v–abl* transforming activity. Genes Dev 1995; 9:2583–2597.
42 Taniuchi I, Kitamura D, Maekawa Y et al. Antigen–receptor induced clonal expansion and deletion of lymphocytes are impaired in mice lacking HS1 protein, a substrate of the antigen–receptor–coupled tyrosine kinases. EMBO J 1995; 14:3664–3678.
43 Pendergast AM, Muller AJ, Havlik MH et al. BCR sequences essential for transformation by the *BCR–ABL* oncogene bind to the ABL SH2 regulatory domain in a non–phosphotyrosine–dependent manner. Cell 1991; 66:161–71.
44 Mayer BJ, Jackson PK, VanEtten RA et al. Point mutations in the *abl* SH2 domain coordinately impair phosphotyrosine binding in vitro and transforming activity in vivo. Mol Cell Biol 1992; 12:609–618.
45 Afar DE, Goga A, McLaughlin J et al. Differential complementation of Bcr–Abl point mutants with c–Myc. Science 1994; 264:424–6.
46 Ilaria RL Jr, Van Etten RA. The SH2 domain of $P210^{BCR/ABL}$ is not required for the transformation of hematopoietic factor–dependent cells. Blood 1995; 86:3897–3904.
47 McWhirter JR, Wang JY. Activation of tyrosinase kinase and microfilament–binding functions of c–abl by bcr sequences in bcr/abl fusion proteins. Mol Cell Biol 1991; 11:1553–65.
48 Sawyers CL, McLaughlin J, Witte ON. Genetic requirement for ras in the transformation of fibroblasts and hematopoietic cells by the Bcr–Abl oncogene. J Exp Med 1995; 181:307–313.
49 Pendergast AM, Quilliam LA, Cripe LD et al. BCR–ABL–induced oncogenesis is mediated by direct interaction with the SH2 domain of the GRB–2 adaptor protein. Cell 1993; 75:175–85.
50 Goga A, McLaughlin J, Afar DEH et al. Alternative signals to RAS for hematopoietic transformation by the BCR–ABL *oncogene*. Cell 1995; 82:981–8.
51 Daley GQ, Van Etten RA, Jackson PK et al. Nonmyristoylated Abl proteins transform a factor–dependent hematopoietic cell line. Mol Cell Biol 1992; 12:1864–71.
52 Tybulewicz VL, Crawford CE, Jackson PK et al. Neonatal lethality and lymphopenia in mice with a homozygous disruption of the *c–abl* proto–oncogene. Cell 1991; 65:1153–63.
53 Schwartzberg PL, Stall AM, Hardin JD et al. Mice homozygous for the ablm1 mutation show poor viability and depletion of selected B and T cell populations. Cell 1991; 65:1165–75.
54 Kharbanda S, Pandey P, Ren R et al. c–Abl acrivation regulates induction of the SEK1/stress–activated protein kinase pathway in the cellular response to 1–b–D–arabinofuranosylcytosine. J Biol Chem 1995; 270:30278–30281.
55 Kharbanda S, Ren R, Pandey P et al. Activation of the c–Abl tyrosine kinase in the stress response to DNA–damaging agents. Nature 1995; 376:785–788.
56 Kharbanda S, Bharti A, Pei D et al. The stress response to ionizing radiation involves c–Abl–dependent phosphorylation of SHPTP1. Proc Natl Acad Sci USA 1996; 93:6898–901.
57 Verheij M, Bose R, Lin XH et al. Requirement for ceramide–initiated SAPK/JNK signaling in stress–induced apoptosis. Nature 1996; 380:75–79.
58 Liu Z–G, Hsu H, Goeddel DV et al. Dissection of TNF receptor 1 effector functions: JNK activation is not linked to apoptosis while NF–kB activation prevents cell death. Cell 1996; 87:565–576.
59 Alnemri ES, Livingston DJ, Nicholson DW et al. Human ICE/CED–3 protease nomenclature. Cell 1996; 87:171.
60 Kharbanda S, Ren R, Pandey P et al. Activation of the c–Abl tyrosine kinase in the stress response to DNA–damaging agents. Nature 1995; 376:785–8.
61 Kipreos ET, Wang JY. Differential phosphorylation of c–Abl in cell cycle determined by cdc2 kinase and phosphatase activity. Science 1990; 248:217–20.
62 Kipreos ET, Wang JY. Cell cycle–regulated binding of c–Abl tyrosine kinase to DNA. Science 1992; 256:382–5.

63 Welch PJ, Wang JY. A C–terminal protein–binding domain in the retinoblastoma protein regulates nuclear c–Abl tyrosine kinase in the cell cycle. Cell 1993; 75:779–90.
64 Sawyers CL, McLaughlin J, Goga A et al. The nuclear tyrosine kinase c–Abl negatively regulates cell growth. Cell 1994; 77:121–31.
65 Renshaw MW, Kipreos ET, Albrecht MR et al. Oncogenic v–Abl tyrosine kinase can inhibit or stimulate growth, depending on the cell context. Embo J 1992; 11:3941–51.
66 Cortez D, Kadlec L, Pendergast AM. Structural and signaling requirements for BCR–ABL–mediated transformation and inhibition of apoptosis. Mol Cell Bio 1995; 15:5531–5541.
67 Varticovski L, Daley GQ, Jackson P et al. Activation of phosphatidylinositol 3–kinase in cells expressing *abl* oncogene variants. Mol Cell Bio 1991; 11:1107–1113.
68 Skorski T, Kanakaraj P, Nieborowska–Skorska M et al. Phosphatidylinositol–3 kinase activity is regulated by BCR/ABL and is required for the growth of philadelphia chromosome–positive cells. Blood 1995; 86:726–736.
69 Amarante–Mendes GP, Jascur T, Nishioka WK et al. Bcr–Abl–mediated resistance to apoptosis is independent of PI 3–kinase activity. Cell Death Diff. In Press.
70 Kurzrock R, Gutterman JU, Talpag M. The molecular genetics of Philadelphia chromosome–positive leukemias. New Eng J Med 1988; 319:1201–1208.
71 Foti A, Ahuja SL, Allen SL et al. Correlation between molecular and clinical events in the evolution of chronic myelocytic leukemia to blast crisis. Blood 1991; 77:2441–2444.
72 Stryckmans P, Debusscher L, Socquet M. Regulation of bone marrow myeloblast proliferation in chronic myeloid leukemia. Cancer Res 1976; 36:3034–8.
73 Strife A, Clarkson B. Biology of chronic myelogenous leukemia: is discordant maturation the primary defect? Semin–Hematol 1988; 25:1–19 issn: 0037–1963.
74 Koeffler PH, Golde D. Chronic Myelogenous Leukemia–New Concepts. N Eng J Med 1981; 304:1201–1209.
75 Renshaw MW, McWhirter JR, Wang JYJ. The human leukemia oncogene *bcr–abl* abrogates the anchorage requirement but not the growth factor requirement for proliferation. Mol Cell Biol 1995; 15:1286–1293.
76 Cook WD, de St. Groth BF, Miller JF et al. Abelson virus transformation of an interleukin 2–dependent antigen–specific T cell line. Mol–Cell–Biol 1987; 7:2631–5.
77 Pierce JH, DiFiore PP, Aaepmapm SA et al. Neoplastic transformation of mast cells by Abelson–MulV: Abrogation of IL–3 dependence by a non autocrine mechanism. Cell 1986; 41:685–693.
78 Mathey Prevot B, Nabel G, Palacios R et al. Abelson virus abrogation of interleukin–3 dependence in a lymphoid cell line. Mol Cell Biol 1986; 6:4133–5.
79 Rovera G, Valtieri M, Mavilio F et al. Effect of Abelson murine leukemia virus on granulocytic differentiation and interleukin–3 dependence of a murine progenitor cell line. Oncogene 1987; 1:29–35.
80 Daley GQ, Baltimore D. Transformation of an interleukin 3–dependent hematopoietic cell line by the chronic myelogenous leukemia–specific p210 $^{bcr-abl}$ protein. Proc Natl Acad Sci USA 1988; 88:506–512.
81 Hariharan IK, Adams JM, Cory S. bcr–abl oncogene renders myeloid cell line factor independent: potential autocrine mechanism in chronic myeloid leukemia. Oncogene Res 1988; 3:387–99.
82 Chapman RS, Whetton AD, Chresta CM et al. Characterization of drug resistance mediated via the suppression of apoptosis by Abelson protein tyrosine kinase. Mol Pharmacol 1995; 48:334–43.
83 Martin SJ, Lennon SV, Bonham AM et al. Induction of apoptosis (programmed cell death) in human leukemic cells by inhibition of RNA or protein synthesis. J Immunol 1990; 145:1859–1867.
84 McGahon A, Bissonnette R, Schmitt M et al. BCR–ABL maintains resistance of chronic myelogenous leukemia cells to apoptotic cell death. Blood 1994; 83:1179–87.
85 Bedi A, Zehnbauer BA, Barber JP et al. Inhibition of apoptosis by BCR–ABL in chronic myeloid leukemia. Blood 1994; 83:2038–44.
86 Vaerman JL, Lammineur C, Moureau P et al. BCR–ABL antisense oligodeoxyribonucleotides suppress the growth of leukemic and normal hematopoietic cells by a sequence–specific but nonantisense mechanism. Blood 1995; 86:3891–6.

87 Smetsers TF, van de Locht LT, Pennings AH et al. Phosphorothioate BCR–ABL antisense oligonucleotides induce cell death, but fail to reduce cellular bcr–abl protein levels. Leukemia 1995; 9:118–30.
88 Chapman RS, Whetton AD, Dive C. The suppression of drug–induced apoptosis by activation of v–ABL protein tyrosine kinase. Canc Res 1994; 54:5131–7.
89 McGahon AJ, Nishioka WK, Martin SJ et al. Regulation of the Fas apoptotic cell death pathway by Abl. J Biol Chem 1995; 270:22625–31.
90 Amarante–Mendes GP, McGahon Aj, Nishioka WK, Green DR. Manuscript in preparation.
91 McGahon AJ, Amarante–Mendes GP, Martin SJ et al. The v–Abl kinase induces a novel anti–apoptotic state dependent upon an associated downregulation of Bax. J Exp Med.
92 Sanchez Garcia I, Grutz G. Tumorigenic activity of the BCR–ABL oncogenes is mediated by BCL2. Proc Natl Acad Sci USA 1995; 92:5287–91.
93 Sakai N, Ogiso Y, Fujita H et al. Induction of apoptosis by a dominant negative H–RAS mutant (116Y) in K562 cells. Exp Cell Res 1994; 215:131–6.
94 Evans CA, Lord JM, Owen–Lynch PJ et al. Suppression of apoptosis by v–ABL protein tyrosine kinase is associated with nuclear translocation and activation of protein kinase C in an interleukin–3–dependent hemopoietic cell line. J Cell Sci 1995; 108:2591–8.
95 Darmon AJ, Nicholson DW, Bleackley RC. Activation of the apoptotic protease CPP32 by cytotoxic T cell–derived granzyme B. Nature 1995; 377:446–8.
96 Martin SJ, Amarante–Mendes GP, Shi L et al. The cytotoxic cell protease granzyme B initiates apoptosis in a cell–free system by proteolytic processing and activation of the ICE/CED–3 family protease, CPP32, via a novel two–step mechanism. EMBO J 1996; 15:2407–2416.
97 Fuchs EJ, Bedi A, Jones RJ et al. Cytotoxic T cells overcome BCR–ABL–mediated resistance to apoptosis. Canc Res 1995; 55:463–6.
98 Roger R, Issaad C, Pallardy M et al. BCR–ABL does not prevent apoptotic death induced by human natural killer or lymphokine–activated killer cells. Blood 1996; 87:1113–22.
99 Martin SJ, Green DR. Protease activation during apoptosis: death by a thousand cuts? Cell 1995; 82:349–352.
100 Thornberry NA, Bull HG, Calaycay JR et al. A novel heterodimeric cysteine protease is required for interleukin–1 beta processing in monocytes. Nature 1992; 356:768–74.
101 Cerretti DP, Kozlosky CJ, Mosley B et al. Molecular cloning of the interleukin–1 beta converting enzyme. Science 1992; 256:97–100.
102 Kaufmann SH, Desnoyers S, Ottaviano Y et al. Specific proteolytic cleavage of poly(ADP–ribose) polymerase: an early marker of chemotherapy–induced apoptosis. Canc Res 1993; 53:3976–3985.
103 Lazebnik YA, Kaufmann SH, Desnoyers S et al. Cleavage of poly(ADP–ribose) polymerase by a proteinase with properties like ICE. Nature 1994; 371:346–7.
104 Casciola Rosen LA, Miller DK, Anhalt GJ et al. Specific cleavage of the 70–kDa protein component of the U1 small nuclear ribonucleoprotein is a characteristic biochemical feature of apoptotic cell death. J Biol Chem 1994; 269:30757–60.
105 Martin SJ, O'Brien GA, Nishioka WK et al. Proteolysis of fodrin (non–erythroid spectrin) during apoptosis. J Biol Chem 1995; 270:6425–8.
106 Neamati N, Fernandez A, Wright S et al. Degradation of lamin B1 precedes oligonucleosomal DNA fragmentation in apoptotic thymocytes and isolated thymocyte nuclei. J Immunol 1995; 154:3788–95.
107 Voelkel Johnson C, Entingh AJ, Wold WS et al. Activation of intracellular proteases is an early event in TNF–induced apoptosis. J Immunol 1995; 154:1707–16.
108 Lazebnik YA, Takahashi A, Moir RD et al. Studies of the lamin proteinase reveal multiple parallel biochemical pathways during apoptotic execution. Proc Natl Acad Sci USA 1995; 92:9042–6.
109 Casciola Rosen LA, Anhalt GJ, Rosen A. DNA–dependent protein kinase is one of a subset of autoantigens specifically cleaved early during apoptosis. J Exp Med 1995; 182:1625–34.

Chapter 7

Apoptosis and Cancer, edited by Seamus J. Martin.

The Roles of p53 in Apoptosis and Cancer

Christopher O. C. Bellamy, Roger Malcomson and Andrew H. Wyllie

Sir Alister Currie Cancer Research Laboratories, University Department of Pathology, Medical School, Edinburgh, U.K.

Introduction

Fundamental to cancer biology is the concept that cancer pathogenesis is dependent upon disruption of a few key regulatory pathways. The recognition that relentless growth is a central property of malignant neoplasms led cancer biologists to focus upon the mechanisms that drive cell proliferation. However, recent attention has turned towards the cellular responses to DNA damage, examining particularly the roles of apoptosis and DNA repair in suppression of carcinogenesis. DNA repair restores genomic fidelity, whilst apoptosis is envisaged as an alternative outcome, eliminating cells that have sustained serious but not incapacitating DNA damage. Were such cells to replicate a defective genome, they could become the precursors of malignant clones. It follows therefore that defects on the signaling pathways from genotoxic damage to apoptosis could lead to the aberrant survival and subsequent propagation of mutated clones. However this concept also suggests that therapeutic restoration of the neoplastic cells' recognition and response to their genetic chaos would trigger selective cell suicide.

Until recently these processes have represented disparate fields of research. However evidence now suggests that a small set of common regulatory molecules are involved, and the integration of previously complementary fields is bringing a new level of insight into cancer biology. The *p53* tumor suppressor gene product has become central to this new focus. Directly or indirectly, roles for *p53* have been identified in aspects of DNA damage recognition, DNA repair, cell cycle regulation and most particularly in triggering apoptosis after genetic injury. However such wide involvement makes the *p53* gene a potential Achilles heel of cellular defenses against carcinogenesis, a defect of which may radically diminish the suppression of neoplasia. That *p53* is the most commonly mutated gene in human malignancy supports this hypothesis and makes *p53* a major potential target for gene–specific therapy of cancer.

In this article we focus upon the roles of p53 in the suppression and development of cancer, particularly the significance of p53–dependent apoptosis for genesis, progression and drug resistance of human cancers. The first part describes how wild type p53 mediates responses to DNA damage, a primary lesion in carcinogenesis. The second part discusses the evidence that dysfunction of p53 is important to human cancer. An attempt is made to discriminate evidence for a role of p53 in *carcinogenesis* from data concerning *neoplastic progression*, which is defined here as the stepwise acquisition by a neoplasm of increasingly aggressive growth features.

p53: Basic Biochemistry

p53 is a nuclear DNA–binding phosphoprotein that normally exists as a homotetramer or complex of tetramers (see ref.1 for review and references). It is a transcriptional activator of a specific set of target genes and can also exert a more general transcriptional repression, probably by interaction with other transcription factors or the general transcription machinery. The protein is present in vivo in a biochemically latent form and is normally rapidly degraded ($t_{1/2} \cong 30$ minutes), probably by ubiquitin–dependent proteolysis. p53 activity and stability are regulated post–transcriptionally and post–translationally by still incompletely understood mechanisms that include alternative splicing, conformational change, phosphorylation, protein–protein associations and regulation of nuclear localization. p53 also negatively regulates its own transcriptional activity through induction of the *mdm2* oncogene, forming an autoregulatory negative feedback loop. The normal activity and stability of the protein varies in a cell cycle dependent manner, although p53 is not required for normal mitotic or meiotic cycles.

The Cellular Response to DNA Damage

Upstream of p53

In many cell types DNA damage is followed by accumulation of p53 protein, due to an increase in its half–life. The precise events coupling damage to p53 response are still uncertain but DNA strand breaks are believed to be a primary stimulus [2]. p53 binds avidly to strand breaks and this stabilizes and activates the otherwise short–lived protein, perhaps facilitated by phosphorylation by DNA repair proteins such as DNA–PK [3, 4]. In patients with the inherited radiosensitivity syndrome ataxia telangiectasia and in Fanconi anemia, the p53 response to γ–irradiation is attenuated and delayed [5, 6], suggesting that other response–enabling pathways are important. γ–irradiation and some chemical agents produce strand breaks directly, but for many genotoxins strand breaks are only generated indirectly, during DNA repair, when short patches of DNA bearing the damaged nucleotides are excised. In this way p53 is made sensitive to a broad range of different DNA lesions. The short patches of single strand DNA can themselves also stimulate p53 [7].

At least in vitro the kinetics of p53 protein accumulation vary considerably after different genotoxic stimuli, for example γ– and UV–c irradiation [8]. The biological significance of this is uncertain but it relates at least in part to persistence of the damage. The kinetics also vary for different cell types. For example, within 3 hours after γ–irradiation in vivo, murine splenocytes, thymocytes and osteocytes show dramatic accumulation of immunoreactive p53, detectable up to 48–96 hours. By contrast hepatocytes show no accumulation [9], despite demonstrating p53–dependent growth arrest (Bellamy COC et al, submitted) and O^6–alkyl transferase induction after γ–irradiation [10]. Thus simple protein accumulation is perhaps not always necessary, and protein activation can be sufficient for a p53 response.

Downstream of p53: Choices

When normal proliferating cells sustain DNA damage they respond in one of two ways: cell cycle arrest or apoptosis, and p53 is implicated in both. It is unknown what determines the choice, although cell type is a factor in both outcome and the contribution of p53 to the response. For example, 5Gy γ–irradiation produces p53–dependent G_1 growth arrest in proliferating fibroblasts [11] but p53–dependent apoptosis in proliferating intestinal epithelium [12]. The latter does not show significant p53–dependent growth arrest in vivo. The pathways of p53–dependent cycle arrest and apoptosis are distinct, as demonstrated by p21–deficient mice, in whom embryonic fibroblasts have defective G_1 arrest responses to γ–irradiation, but thymocytes and small intestinal crypt epithelium have intact apoptotic responses [13, 14]. Nevertheless in culture growth arrest will switch to apoptosis if growth arrest signals (e.g., serum depletion or activated p53) are challenged by forced growth activation signals (e.g., from deregulated *c–myc* or E2F oncogenes) [15–18]. This probably represents a defense against autonomous oncogene–driven proliferation, deleting the offending cell by apoptosis unless other permissive factors are present. Such experiments are clearly relevant to understanding tumor suppression, but it is not certain whether they reproduce mechanisms of choice in normal cells. One influential hypothesis suggests that cells have the capacity to recognize when DNA damage is too profound to be repaired completely or sufficiently rapidly, whereupon p53–dependent apoptosis is triggered, preventing replication of a damaged genome [19]. This remains unproven, but new data that p53–dependent apoptosis and DNA repair involve common molecules demonstrate potential for such a decision mechanism (see below). Other data that the Wilms tumor suppressor gene product, WT1, binds p53 and inhibits p53–dependent apoptosis without affecting p53–dependent growth arrest [20], also suggests a determining mechanism for p53 downstream events.

The following subsections consider in more detail the individual cellular responses engaged or regulated by p53, and attempt to highlight how interactive and interdependent these apparently distinct responses can be.

Apoptosis

Analysis of cells and tissues of mice homozygous–deficient for p53 has definitively shown that p53 is required for DNA damage–induced apoptosis of cortical thymocytes [21], myeloid progenitor cells [22], marrow pre–B cells [23], quiescent peripheral B and T lymphocytes [23], keratinocytes [24] and crypt epithelium of large and small intestine [12]. Interestingly and significantly for carcinogenesis, cells with only one functioning copy of the *p53* gene have intermediate rates of apoptosis between null and wild type cells, a gene dose effect that demonstrates the relevance of physiological levels of expression of p53 to the decision to respond to injury in this way.

The biochemical steps by which p53 triggers apoptosis are still incompletely defined, but there is evidence for distinct mechanisms that predominate according to cell type [25], and may be interactive [16, 26–29]. One mechanism requires specific transcriptional transactivation by p53, perhaps of Bax [30–32], a member of the *bcl–2* family that in relative excess to *bcl–2* permits apoptosis (see chapter 5). Transcriptional repression by p53 may also contribute to apoptosis, for example of *bcl–2* [31, 32]. Modifying influences by other members of the bcl–2 family are probably relevant, for example bcl_{xl} can inhibit p53–mediated apoptosis in vitro [33]. p53 also interacts directly with other cellular proteins and there is exciting new evidence that binding of the TFIIH protein complex is critical for a pathway of p53–dependent apoptosis that does not involve specific transcriptional transactivation [27]. p53 inhibits the helicase activity of this complex by binding to the XPB (ERCC3) and XPD (ERCC2) subunits, both of which are required in this apoptotic pathway [27, 34]. Thus a single protein defect can compromise both repair and apoptosis responses to DNA damage. The data are particularly provocative since TFIIH participates in basal transcription, nucleotide excision repair, and probably also in cell cycle control [35, 36]. Thus a core element is identified, through which dynamic regulation and coupling of these critical cellular processes can be achieved and the balance shifted according to circumstance. Moreover there is a basis for understanding how defects in one pathway can have effects on the others and the balance between them.

Once p53–dependent apoptosis is triggered there is no evidence that it differs in any way from apoptosis induced by other means, and it is not doubted that downstream events feed into a common effector pathway of apoptosis (reviewed elsewhere) [37, 38].

Growth Arrest

DNA damage can prevent entry to and progression through the cell cycle at multiple points (late G_1, S phase, G_2) [39]. These "checkpoint" delays may allow time for DNA repair, thereby preventing replication or segregation of a damaged genome. p53 is best characterized as triggering the G_1 checkpoint arrest, through specific transcriptional activation of the broad spectrum cyclin–dependent kinase inhibitor, p21 (WAF1, sdi1, cip1) [5, 40, 41]. p53–null cells lose this arrest [5], although no gene dose effect has been reported as for apoptosis. In fibroblasts it has been estimated that a single unrepaired DNA strand break may be sufficient to prevent progression beyond the G_1 checkpoint indefinitely, through persistent activation of p53/p21 [11].

p21 expression can be upregulated by several p53–independent pathways, and at least one, involving the transcription factor Interferon Regulatory Factor–1 (IRF–1), is also stimulated by DNA damaging agents [42]. Intriguingly, new evidence using p53–null and IRF–1–null mouse embryo fibroblasts suggests that alone each pathway is insufficient to trigger growth arrest and that they must act in concert [42]. This suggests that a threshold of p21 activation exists that is not easily achievable by a single damage response pathway. In this way, paradoxically, p53–dependent growth arrest is mutually dependent upon co–activation of one or more p53–independent pathways. Thus, once again, the context in which an individual signal acts is seen to be critical to outcome. The concept provides a basis for understanding how p53–dependent G_1 arrest can be overridden by p53–independent mechanisms, for example after transformation by activated H–ras

[43]. The cytokine TGFβ_1 provides a further example of a distinct pathway of p21 activation that could cooperate with p53 to effect growth arrest [44]. Indeed in hepatocytes p53 deficiency compromises TGFβ_1–induced growth arrest (Bellamy COC et al, submitted).

Involvement of p53 in triggering cell cycle arrest at other points is less certain, and is probably more variable between cell types. However there is evidence for regulation of S phase progression [45], G_2 arrest [46, 47] and a mitotic spindle checkpoint [48]. Moreover, although p53–deficiency completely abolishes G_1 arrest in γ–irradiated mouse embryo fibroblasts, p21 deficiency only partially abolishes it [13, 14, 26], suggesting that other p53–responsive pathways to G_1 arrest exist, perhaps involving the p53 target gene GADD45 [49],

DNA Repair

p53 was suggested to facilitate DNA repair simply through establishing growth arrest and so providing extra time for completion of repair before DNA replication or mitosis. However a more active contribution has become evident, although many details are still incomplete. The evidence can be summarized as follows. First, p53 may act as a DNA damage recognition protein, recruiting repair proteins to DNA strand breaks [50] and insertion/deletion mismatches [51]. Second, p53 itself reanneals DNA strand breaks [52]. Third, induction of the DNA repair enzyme O^6 alkyl guanyl transferase by γ–irradiation is p53–dependent [10]. Fourth, fibroblasts heterozygous for functional p53 are defective in nucleotide excision repair (NER) [34, 53]. A colorectal carcinoma cell line also showed decreased NER when p53 function was blocked by a mutant p53 transgene or HPV E6 viral protein [54]. However although global NER is deficient, transcription–coupled NER is intact in human p53 mutant fibroblasts [53]. Finally, p53 directly interacts with DNA repair proteins, including RP–A [55], TFIIH subunits (XPB,XPD) [34, 56] and ERCC6 (CSB) [34].

Many of these proteins participate in both replication and repair, but these activities are separately regulable. Hence during growth arrest the p53 target gene p21 selectively downregulates the DNA replication activity of PCNA but spares its repair activity [57–59]. Thus growth arrest after DNA damage constitutes a switch of cellular priorities from DNA replication to DNA repair.

A detailed consideration of this aspect of p53 function is beyond the scope of this article, but it is clearly of great importance to clarify whether p53–dependent apoptosis is coupled to its repair activities, as hinted by the interaction with TFIIH. For example, do repair and apoptosis after DNA damage share damage sensing mechanisms? Is apoptosis triggered by repair rather than damage?

p53–Independent Pathways

Although p53 is essential in many circumstances for a normal response to DNA damaging agents, p53–independent pathways also determine apoptosis and growth arrest. These are less well understood but are easily demonstrated in p53 homozygous knockout mice. For example mitogenically activated peripheral T lymphocytes do not show p53–dependence for apoptosis after γ–irradiation, but instead undergo apoptosis by a pathway dependent upon IRF–1 [60]. IRF–1 null thymocytes retain p53–dependent apoptosi [60]. Therefore in contrast to growth arrest responses in embryo fibroblasts (see earlier), p53 and IRF–1 pathways to apoptosis are not mutually dependent and appear to operate in distinct cell populations.

Even cell types such as intestinal crypt epithelium that do show p53–dependent apoptosis—generally manifested as early phase apoptosis, induced within 24 hours of genotoxic insult—have additional late phase p53–independent apoptosis (Clarke AR, unpublished results). Moreover, genotoxins can affect cellular components other than DNA and can trigger apoptosis through pathways quite unrelated to the genetic injury, in a p53–independent manner. For example activation of membrane sphingolipase by γ–irradiation is not related to the DNA damage [61], but triggers apoptosis in some tissue types through a ceramide pathway. This pathway is probably p53–independent and may be responsible for most of the pulmonary endothelial apoptosis observed after γ–radiation injury [62]. Nevertheless there is clear evidence that loss of the contribution of p53 to DNA damage responses is often critical to carcinogenesis, at least in some cell types, as will be discussed in the next section.

p53 in Cancer

The Spectrum of Defects

Abnormalities associated with loss of p53 function are prevalent in human cancers of many primary sites and histological types [63]. They include lesions in the p53 gene itself (point mutation or allele loss) or binding of normal p53 by amplified gene products such as *mdm–2* or viral proteins that quench its normal function. Whilst allele loss clearly abrogates p53 function, more subtle defects in only some aspects of function can be sufficient to provide selective advantage. Mutation analysis shows that the function of p53 most commonly disrupted in neoplasms is specific transcriptional activation. Typically, a missense mutation at or near the DNA binding surface of p53 either alters the binding affinity of critical contact sites for consensus sites on DNA (so–called "contact mutants") or disrupts a structural element holding the protein in the conformation necessary to position residues for DNA binding ("structural mutants") [64]. Such mutants are probably unfolded or partially folded forms of the wild type. Mutant p53 can form heterotetramers with wild type protein, compromising the wild type allele's function. However since most cancers with p53 mutation show loss of the other allele, implying selective advantage, this *dominant negative* effect is at least incomplete in vivo. Certain p53 mutants also show *gain of function*. This may be demonstrated by introduction of the mutant gene into cells null for p53; for example the Ser249 mutant of p53 conferred a growth advantage to p53 null Hep3B hepatoblastoma cells [65].

The gene dose effect, dominant negative and gain of function properties of various p53 mutants are important areas of departure from the classical recessive tumor suppressor gene hypothesis and allow only a single abnormal allele to alter phenotype. When combined with the vulnerability of p53 function to the effects of a single base change, this goes some way to explain why this particular tumor suppressor is so often altered in human cancer.

Since different p53 mutants differ in phenotype, mutations prevalent in one type of neoplasm may not provide equivalent selective advantage to different neoplasms. Viewed alternatively it may be stated that cellular context can influence the pathobiological effects of specific p53 mutants. In support of this, Dumenco et al showed that transfection into hepatocytes of a val135 p53 mutant did not confer reduced serum dependence for growth and gave fewer, smaller colonies than a Ser246 mutant; however, this difference was not found in transfected fibroblasts [66]. Thus the nature of the mutation and the cellular context of expression (including cell type [67], stage of carcinogenesis) are important considerations.

Although the situation just described for p53 in human cancers of missense mutation associated with loss of transcriptional activation is typical, for certain tumor types this pattern is violated. For example hepatocellular and bladder carcinomas frequently demonstrate p53 non–missense mutation (frameshift, stop codon), located outside exons 5–8 that are typical for missense mutation [63]. Furthermore, although loss of specific transactivation activity by p53 mutants correlates well with loss of growth suppression in vitro, there is not a complete correlation with loss of tumor suppression, as assessed by the ability to suppress transformation of primary fibroblasts by oncogene combinations [68, 69]. This is due to the contribution of other p53 activities to tumor suppression, particularly apoptosis which as discussed earlier can be independent of transactivation activity.

Contributions of Defective p53 to Human Cancers

The evidence that p53 dysfunction is relevant to human cancer is compelling. The prevalence data on p53 abnormalities in human cancers is complemented by the observation that humans with critical germline mutations in one p53 allele suffer the Li–Fraumeni syndrome of cancer susceptibility [70]. Affected individuals may develop multiple neoplasms and are predisposed to specific types of malignancy, most frequently breast carcinomas, soft tissue and osteosarcomas, acute lymphocytic leukemia, astrocytomas and rarely, adrenocortical carcinomas. Interestingly, certain sporadic cancers that frequently show p53 dysfunction, such as colorectal carcinoma, are not part of the syndrome. The developing neoplasms show loss of the wild type p53 allele and retain the mutant allele. It is probable that the group of patients with malignancy who carry germline p53 mutation is more diverse than Li–Fraumeni syndrome [71] and that germline mutation of p53 may be relatively common for selected malignancies, for example, rhabdomyosarcoma in young children [72]. In vitro observations complement the clinical evidence, showing that mutant p53 proteins derived from human tumors have lost the transformation and/or growth suppression properties of wild type protein (see ref. 63).

It is difficult or impossible to understand the contribution to tumor suppression of different p53 functions without the use of model systems, of which transgenic and gene–targeted mice continue to provide critical data. Even so the overlapping contributions to genomic stability and tumor suppression of growth arrest, DNA repair, apoptosis and other functions are probably not completely separable, and it may be the interaction and coupling between these activities which will prove critical in cancer development. Animal models and the ability to generate double or triple mutant mice carrying defects in multiple oncogenes or oncosuppressor genes have also demonstrated unexpected genetic interactions relevant to carcinogenesis. For example the *min* mutant mouse crossed to the p53 null mouse does not show accelerated intestinal carcinogenesis, as could have been expected, but produces a new phenotype for susceptibility towards pancreatic carcinoma, to which neither genotype alone is predisposed [73].

Nevertheless, and despite the caveats just discussed, the contribution of defective p53 to human cancer development can be usefully examined by considering separately the individual cellular responses regulated by p53.

A) Loss of p53–dependent apoptosis

p53 can limit the oncogenic potential of aberrant oncogene activation, by triggering apoptosis. This is an effective protection against not only endogenous oncogene activation but also viral oncogene expression in host cells. Not surprisingly however, oncogenic viruses such as EBV, adenovirus and HPV have evolved counterstrategies to evade or block the induction of apoptosis. For example, the adenovirus E1A oncogene is unable alone to transform primary cells since the affected cells die by p53–dependent apoptosis. However adenovirus produces a second protein, E1B, that inhibits p53–dependent apoptosis, and this allows sustained proliferation and transformation to occur [74].

The significance of oncogene–activated, p53–dependent apoptosis for suppression of carcinogenesis has been directly tested in vivo. In a transgenic mouse model of experimental choroid plexus tumors, a variant of the SV40 T antigen that functionally disrupts only the Retinoblastoma family of proteins (leading to aberrant E2F oncogene activation) but not p53, produced atypical hyperplasia associated with increased apoptosis [75, 76]. The added effect of p53 inactivation was shown to reduce the excess apoptosis, without affecting proliferation rates, leading to the rapid development of highly malignant tumors. A similar demonstration was provided by mice bearing an HPV16 E7 transgene that was expressed in photoreceptor cells [77]. The transgene caused functional Rb–1 inactivation, leading to abnormal activation of the E2F oncogene. In these transgenic mice the photoreceptor cells failed to terminally differentiate and instead underwent apoptosis. However when the analysis was repeated on a p53–deficient genetic background, a similar pattern of apoptosis was not observed and the transgenic animals developed neoplasms arising from the photoreceptor cell layer.

These elegant experimental systems show that p53–dependent apoptosis in response to inappropriately activated proliferation signals inhibits carcinogenesis at an early stage by deleting the potentially neoplastic cells. They do not however directly test the hypothesis that DNA damage–induced p53–dependent apoptosis suppresses carcinogenesis and indeed such data as is available is not conclusive. This is because of difficulty in isolating the contribution of p53–dependent apoptosis from its repair or growth arrest functions. For example, p53 deficient keratinocytes from gene–targeted mice were reported to show a gene dose–dependent reduction in the normal apoptotic response to uv–irradiation, the prime etiological agent of squamous carcinoma of the skin [24]. This leads to survival of clones which, over successive exposures to uv, should predominate over wild type in the epidermis and acquire further mutations. However p53 dysfunction also renders keratinocytes deficient in repair of uv–induced DNA damage [78] and so the absolute contribution of an apoptotic effect to carcinogenesis is uncertain. Moreover, in mice carrying a mutant p53 transgene the transgenic keratinocytes showed only decreased DNA repair but no alteration in apoptosis following uv irradiation, despite increased susceptibility to uv–induced squamous skin carcinomas [78]. Thus the role of p53–dependent apoptosis in uv–induced skin carcinogenesis remains uncertain. The intestine provides an alternative tissue to explore this question since p53–dependent apoptosis following genotoxic damage of crypt stem cells has been independently documented by different groups [12, 79]. However in γ–irradiated murine small intestine at doses adequate to trigger p53–dependent stem cell apoptosis, p53 deficiency was not associated with increased numbers of mutated stem cells compared with wild type (as assessed in an endogenous indicator gene), suggesting that inappropriate survival of genetically damaged cells does not necessarily equate to increased tissue mutability in vivo (Clarke AR et al, in preparation). A different model, p21–deficient mice, have defective p53–dependent G_1 growth arrest but do not show increased susceptibility to spontaneous cancers [14]. Whilst p53–dependent apoptosis is preserved in these mice and might explain the preservation of tumor suppression, p53–dependent repair and other cell cycle checkpoints may also be intact. Susceptibility of the mice to DNA damage–induced carcinogenesis has not yet been reported.

Thus at the present time, and somewhat surprisingly, the true significance of DNA damage–induced p53–dependent apoptosis for suppression of carcinogenesis in normal cells is not certain. Perhaps other mechanisms provide redundancy in vivo.

Independently of apoptosis triggered by DNA damage or oncogene activation, p53 can regulate the dependence of some cells on hormone or cytokine survival factors, as demonstrated for hemopoietic cells [22, 80], prostate [81] and hepatocytes (Bellamy COC et al, submitted). A potential contribution of p53 dysfunction to carcinogenesis and tumor progression is therefore through increased survival (decreased apoptosis) in competitive or unfavorable environments, for example within solid neoplasms or during neoplastic spread to other tissues. One aspect of this has been investigated recently in an important study. Hypoxia and associated necrosis are prevalent in solid malignant neoplasms, and the latter has often been correlated with a poor prognosis. Although normal cells were relatively resistant to hypoxia, transformation was found to make them sensitive to apoptosis under conditions of extreme hypoxia [82]. However, p53 deficiency (or Bcl–2 overexpression) protected against hypoxia–induced cell death both in vitro and in vivo, and over successive exposures to hypoxia resulted in overgrowth of cultures by an initially small fraction of p53–deficient cells [82]. Hypoxia was thus shown to exert a selective pressure for loss of p53–dependent apoptosis from neoplastic cells.

Wilms' tumor provides an authentic human example in which p53–dependent apoptosis may be relevant to tumor progression. p53 mutations are rare in this pediatric malignancy except in the poor prognosis anaplastic variant, characterized by focal areas of anaplastic morphology within the neoplasm. In these tumors the p53 mutations are restricted to the histologically anaplastic tissue, which was also shown to display much reduced apoptosis compared with surrounding nonanaplastic tumor [83]. These studies therefore imply a pathogenetic role for p53 inactivation in Wilms' tumor progression and suggest that loss of p53–dependent apoptosis may be the critical event, analogous to the mouse models described above.

B) Proliferative advantage

Wild type p53 can inhibit progression through the cell cycle through activation of p21 and may also influence entry into the cell cycle from G_0, perhaps through mediating the effects of proteins such as the growth arrest specific protein Gas–1 [84]. Therefore cells defective in p53 function could obtain a proliferative advantage independently of enhanced survival. Thus, early passage p53–deficient mouse embryo fibroblasts display greater proliferative indices and grow faster than wild type in vitro [85]. p53 null hepatocytes (Bellamy COC et al, submitted) and cortical astrocytes [86] also have increased proliferation indices in vitro compared with wild type.

Loss of normal p53 function may also reduce the effect of mitosuppressive cytokines such as TGFβ [87]. Moreover p53 mutants can increase proliferation through a gain of function mechanism, demonstrated by increased growth rates when added to p53 null cell lines. This may be due to direct effects on growth regulatory genes or secretion of paracrine mitogens such as bFGF.

C) Genomic instability and DNA repair

Aneuploidy and gross chromosomal aberrations are characteristic of neoplastic progression and are a cardinal manifestation of genetic instability. Chromosomal abnormalities in human and experimental malignancies are often nonrandom, consistent with a process of clonal selection, and implying that genetic instability can accelerate tumor progression by providing increased opportunity for favorable mutations to occur. Evidence from several sources indicates that loss of p53 precedes and is causally associated with chromosomal instability. Cultured fibroblasts from Li–Fraumeni patients lose or mutate the remaining wild type allele and become aneuploid [88], a feature also shown by cultured p53–deficient mouse embryo fibroblasts [85]. In vivo, p53 deficient mouse bone marrow cells display a gene dose–dependent increase in spontaneous stable chromosomal aberrations [89], and in a variety of human malignancies p53 mutations have been associated with or shown to precede aneuploid clonal divergence [90]. Loss of p53 is probably not sufficient to cause genetic instability in normal cells since the majority of p53 deficient mice survive normally to adulthood, although they are subject to an increase in developmental abnormalities [91]. However, the possibility of early intrauterine selection in these mice for some compensatory mechanism of genomic stability cannot be completely excluded.

The mechanisms by which p53 affects genetic stability are still under evaluation. Evidence suggests that p53 dysfunction may only contribute to certain types of genetic instability, and again perhaps only in specific cell types or situations. For example, p53 deficiency is associated with abnormal gene amplification in vitro and karyotypic abnormalities [85, 92], but has been shown not to affect spontaneous rates of point mutation in vivo [93]. Moreover lack of p53 does not increase the spontaneous rate of intestinal adenoma formation in heterozygous APC mutated (*Min*) mice [73], suggesting that there is no increased genetic instability conferred to the preneoplastic *Min* cells, since adenoma formation in these mice is associated with loss of the remaining wild type APC allele. In contrast however, in a transgenic (*Wnt–1*) mouse model of breast carcinoma, germline p53 deficiency resulted in more rapid development of mammary adenocarcinomas, with an increased prevalence of aneuploidy compared with carcinomas arising on a p53 wild type background [94]. Therefore, some stage of carcinogenesis or tumor progression was sensitive to p53 function in this model, with a consequence of greater genetic instability in the incident neoplasms. Interestingly, p53 abnormality was not detected in the carcinomas arising on a p53 wild type background in this model [94], raising the possibility that the imposition of germline p53 deficiency might have caused carcinogenesis to proceed in a completely different fashion to usual, rather than simply to accelerate events through increased genetic instability, a conclusion supported by the different histology of the carcinomas in mice from different p53 backgrounds. These models are not necessarily conflicting, if it is supposed that the effect of p53 on genetic stability only usually becomes apparent in already neoplastic cells, bearing other permissive genetic lesions.

At a molecular level, the combination of checkpoint arrest and facilitation of DNA repair activities should contribute to maintenance by p53 of genomic integrity, as should apoptotic deletion of DNA damaged cells. However the role of apoptosis after DNA damage in maintaining genomic stability is uncertain, as discussed above. Furthermore, in contrast to p53–deficient mice, p21–deficient mice show no predisposition to spontaneous carcinogenesis despite defective G_1 checkpoint arrest after DNA damage [13, 14], showing this particular function of p53 to be at least redundant if not irrelevant to its tumor suppressor activity in vivo. It is possible therefore that the influence of p53 on DNA repair activities may be more important to genetic stability than so far appreciated. Another only recently recognized activity of p53 is regulation of centrosome duplication [95]. This function is yet to be explored but, in contrast to the damage response activities of p53 discussed so far, suggests a more fundamental contribution of p53 to the normal cell cycle.

D) Angiogenesis

Neoplastic cells, like all others, are completely dependent on an adequate local blood supply to provide gas and metabolite exchange. Therefore the ability to recruit neovascularization is fundamental to the survival of a growing tumor. Angiogenesis is regulated in part by the relative levels of cytokine inhibitors such as thrombospondin, and angiogenic factors such as vascular endothelial cell growth factor (VEGF). By regulating secretion of these and other cytokines, wild type p53 has been shown to promote an antiangiogenic phenotype in fibroblasts, breast carcinoma cells and glioblastoma cells in vitro (see ref. 96 for review). This effect is cell–type specific and does not for example occur in keratinocytes. Moreover, there are differences between cell types in the specific factors regulated by p53. Nevertheless, loss of p53 function could directly assist primary tumors and metastases to establish their own microcirculation. In addition to the loss of angiogenesis suppression by wild type p53, a specific gain of function p53 mutant has been described that upregulates expression of pro–angiogenic βFGF.

Taken together with the data discussed earlier that p53 dysfunction reduces susceptibility to hypoxia–induced apoptosis, p53–deficient neoplastic cells could be expected to be better equipped than p53 wild type to withstand periods of metabolic stress, such as related to local vascular stasis or intratumoral thrombosis, but also to resolve the situation through neovascularization.

The Timing of p53 Dysfunction in Human Cancers

For Li–Fraumeni patients or p53 heterozygous deficient mice, loss of p53 function is an early event in genesis of the neoplasms that commonly arise. However, to some extent depending on the tissue type and etiology of the cancer, there is evidence to show that p53 dysfunction can occur early or late in the neoplastic process and this will now be discussed, beginning with an evaluation of the assumptions inherent in most of such work.

Of necessity human tissue datasets for study of the evolution of carcinogenesis are usually cross–sectional (i.e., single time point) rather than longitudinal (i.e., sequential), and reflect an advanced stage in cancer development (a clinically apparent lesion). However for most epithelial malignancies there are morphologically defined noninvasive lesions that are recognized as reflecting "stages" of carcinogenesis preceding the invasive carcinoma and which are therefore suitable for study. Generally these lesions take the form of graded degrees of cytological and architectural disorder, the identification of which has been shown to correlate with increased risk of subsequent malignancy at that site. It is important to recognize that there is no evidence that these stages are a prerequisite for the development of malignancy, which may well appear without previously manifesting any of these changes (e.g., papillary carcinoma of thyroid or aggressive "interval" cancers in breast screening patients). Nor does their presence necessarily imply subsequent development of a malignant lesion, and a proportion may regress (e.g., cervical intraepithelial neoplasia) or never progress (perhaps most adenomas of any epithelium) in the lifetime of the individual. For example uterine leiomyosarcomas do not arise from uterine leiomyomas (fibroids), which are a stable endpoint of neoplastic progression in almost all patients. There can thus be alternate routes of neoplastic evolution within the same tissue, and the final single designation of carcinoma and relative accessibility for study of selected subsets of "premalignant" lesions can conceal pathogenetic and etiological diversity. This is important since the lesions most susceptible to develop into malignancy may be less likely to be recognized than more indolent lesions which, even in multiplicity, have little likelihood of progression, but are more likely to be sampled and studied. Thus in the colorectal mucosa, unifocal high grade flat dysplasia/carcinoma in situ may be ultimately more significant to carcinogenesis than adenomatous polyps. The so–called adenoma–carcinoma sequence is therefore attractive and useful, but has limitations. Likewise it may be unnecessarily restrictive to conceive of p53 dysfunction as occurring and contributing only to particular "stages of carcinogenesis". It seems not unlikely that p53 dysfunction may contribute differently to carcinogenesis depending on the particular mutation, as discussed earlier (which may differ according to the prevalent mutagen), and also depending on the particular mutational load already accrued by the afflicted cells, i.e., the context in which the mutation occurs.

In this respect it is interesting to consider a study in which a classic 2 step initiation–promotion protocol for skin carcinogenesis was tested on p53 deficient mice. In wild type mice papillomas arise after an interval, and after a further interval a proportion of papillomas progress to carcinomas. When compared with wild type, p53 heterozygotes displayed increased probability of development of carcinoma (which showed loss of the wild type allele), without a shortening of the minimum interval before carcinomas appeared or change in the incidence of papillomas [97]. This suggests a role for p53 late in carcinogenesis, close to the transition to carcinoma. However, there was a qualitatively different effect in p53 null mice. The mice showed a reduction in papilloma development, shortening of the latency interval for carcinomas to almost that for papillomas, a greatly increased probability of development of carcinoma and a different histology of the carcinomas (high grade, spindle cell, deeply invasive). The experiment suggests that differences in the timing of loss of p53 function caused carcinogenesis, once initiated, to proceed in a qualitatively different fashion. It is of note that p53 deficiency could not replace the requirement for chemical initiation to generate the skin lesions.

Nevertheless, the ability to identify high–risk lesions for subsequent malignancy has prompted attempts to identify the timing of p53 dysfunction and mutation in various human cancers. In the colorectum, p53 mutation is rare in adenomatous polyps but frequent in carcinomas and moreover precedes aneuploid clonal divergence, suggesting a role in a tightly defined window of adenoma/carcinoma progression [90]. In contrast to these findings, p53 mutation and allelic deletion are already present in the dysplastic colorectal mucosa of chronic ulcerative colitis, suggesting an alternative pathway in which p53 dysfunction occurs earlier to promote colorectal carcinogenesis [98–100]. p53 mutations, or immunohistochemical accumulation suggestive of defective p53, have been similarly found in bronchial squamous dysplasia [101–103], esophageal dysplasia [104–107] and oropharyngeal dysplastic mucosa [108]. In other tissues p53 mutations are generally felt to arise late, perhaps after carcinogenesis, and may be associated with the outgrowth of aggressive tumor nodules as for example in some hepatocellular carcinomas [109, 110]. Even where such nodules cannot be identified p53 deficient HCCs are often of high histological grade, and correlate with a clinically aggressive phenotype [111, 112]. Hepatocellular carcinogenesis is a good example of variability in timing of p53 dysfunction. p53 mutations may be identified in non–neoplastic liver, associated with aflatoxin exposure [113], and there is also evidence that functional inactivation of p53 by hepatitis B viral protein X may precede development of carcinoma [114]. However most clinical studies do not report p53 mutation except after carcinogenesis. Thus epigenetic inactivation may sufficiently inactivate p53 pathways to affect carcinogenesis, and in the absence of exposure to dietary mutagens such as aflatoxin, p53 mutation and allelic deletion typically occur only later, perhaps sealing leaky functional suppression or altering the phenotype through gain of function. Clearly the issues involved are complex and remain largely unresolved.

Clinical Exploitation of p53

Despite the large body of experimental data that p53 dysfunction can be critical to the genesis, progression and responsiveness to therapy of malignant neoplasms, a utility for the detection and management of the common human cancers is unproven. Nevertheless, it seems plausible that evaluation and manipulation of p53 function will find a place in clinical oncology, as will now be discussed.

A) Prognostication

A plethora of studies have evaluated the role of p53 as a prognostic marker in malignancy; however, there are significant criticisms for many. Study groups are often too small to draw firm conclusions. The majority of data available at present is retrospective, and is often correlated with *markers* of poor prognosis such as stage or histological grade, rather than final clinical outcome. It is therefore impossible to test whether p53 is an independent prognostic variable using such data. In particular, data from prospective, controlled clinical trials with multivariate analyses of outcome are lacking. Moreover there are no standardized protocols for the detection of gene mutation and protein accumulation. These techniques have significant limitations in sensitivity and specificity, and variations in methodology can confound interpretation and comparison between studies [63, 115].

Nonetheless, there are several well designed large studies of p53 dysfunction as a prognostic indicator in human cancers, usually primarily by immunohistochemistry. In general p53 is found to be prognostically significant and independent of other variables, but this is not unanimous and any effect may not be strong enough to be of clinical utility for the common human cancers [115]. Indeed overall, the assessment of p53 status has proved somewhat disappointing as a tool in diagnosis or prognosis, when compared with existing indices, although with some exceptions. For example, p53 immunopositivity was found to be an insensitive but highly specific indicator of neoplasia in equivocal cervical smears [116]. Another interesting application has been to use the ability to detect mutant p53 by sensitive molecular genetic techniques such as PCR to enhance the evaluation of completeness of surgical excision of malignancy, a major predictive factor for disease recurrence [117].

p53 status might help determine therapy for certain neoplasms, by predicting therapeutic response and thus, indirectly, prognosis. For example, in one small study p53 wild type status strongly correlated with induction of remission of chronic B lymphocytic leukemia, whilst p53 mutant neoplasms generally did not enter remission with the same protocol [118]. However the relationship of p53 status to treatment sensitivity is complex and this will now be considered in more detail.

b) p53 and Responses to Cancer Therapy: Sensitivity or Resistance?

Irradiation and many drugs used to treat cancers are primarily genotoxic, either directly or by disrupting DNA metabolism, and at therapeutic doses trigger apoptosis in the target cells. If the apoptosis is due to DNA damage then the p53 status of the neoplastic cells might be expected to modify the drug effect. Two contrasting scenarios are envisaged. Firstly, in cells that readily undergo p53–dependent apoptosis as the preferred response to DNA damage, p53 dysfunction could allow survival, and therefore resistance to treatment. Indeed the surviving cells may have acquired further mutations as a result of exposure to the treatment agent and behave more aggressively than before. However if p53–dependent apoptosis is not a readily invoked consequence of DNA damage, and instead growth arrest and repair activities are compromised by loss of p53 pathways, then neoplastic cells are more likely to enter S phase and mitosis bearing high levels of unrepaired damage and viability could be decreased. Moreover the cycling fraction of p53 deficient cell populations is often high, and if the drug is one that preferentially acts during S phase or mitosis, lethality would be increased simply because a greater proportion of neoplastic cells pass into susceptible cell cycle phases during exposure of the patient to the drug. The consequences of altered p53 status for drug efficacy in cancer therapy are therefore difficult to predict. Indeed, in vitro evidence exists for sensitization, resistance and no effect conferred by loss of wild type p53 on the effects of chemo– or radiotherapy [119–122]. Given these caveats to broad generalizations, some of the experimental evidence that p53 is relevant to cancer therapy will now be discussed.

As discussed earlier transformation of normal cells by oncogenes sensitizes them to triggering of apoptosis that is often p53–dependent. This lowers the threshold at which irradiation and many drugs used in cancer chemotherapy induce apoptosis, as demonstrated in vitro for fibroblasts transformed by E1A+ras oncogenes. In that system the p53 genetic background of the transformed fibroblasts was a critical determinant of drug–induced apoptosis; p53 null transformed cells were resistant to doses of γ–irradiation or adriamycin that efficiently killed p53 wild type transformed fibroblasts by apoptosis [120]. Significantly, identical results were found in vivo when tumors derived from the transformed cells were grown in mice and γ–irradiated or adriamycin treated [123]. In other systems, oncogenes such as c–myc [124] and HPV E7 + ras [125] have been similarly shown to induce p53–dependent sensitivity to irradiation or chemotherapeutic drugs.

The elegant model systems just outlined are informative, but the role of p53 in the efficacy of human cancer therapy is likely to be less clear–cut. One important prediction of these observations however is that where cytotoxic agents commonly induce p53–dependent apoptosis in transformed cells, loss of p53 pathways can produce a *multiresistant* phenotype. This concept has implications for both *de novo* resistance to treatment and the development of acquired resistance in recurrent or relapsing malignancy. Indeed in the murine E1A/ras–transformed fibroblast tumor model, described in the previous paragraph, Lowe et al found that over 50% of the initially treatment–resistant or recurrent tumors derived from transformed fibroblasts on a p53 wild type background had acquired p53 gene mutations [123]. Thus the cytotoxic treatment (γ–irradiation) had selected for apoptosis resistance, and consequently enriched the tumor population in cells with defective p53, which would be predicted to show resistance to other cytotoxic agents. It also follows that reintroduction of wild type p53 function to such cancers should restore sensitivity to therapy, and the prospects for achieving this will be discussed in the next section.

As well as determining responses to genotoxic therapies for malignancy, p53 status may affect responses to hormone ablation therapy, such as anti–androgen therapy of prostate carcinoma [81]. The principle of this type of treatment is that some element of hormone responsiveness or dependence is retained by the neoplasm and is therefore a target for slowing tumor growth or induction of regression by shifting the balance between cell proliferation and apoptosis. Since p53 can regulate dependence of some cell types on survival factors, loss of wild type p53 may confer resistance to hormone ablation therapy.

c) Gene Specific Therapy

As can readily be appreciated from the discussion so far, the ability to reactivate p53–dependent pathways in neoplasms is a potentially powerful therapeutic tool, by either directly provoking apoptosis or by returning sensitivity to cytotoxic cancer chemotherapeutic drugs. In vivo testing of this hypothesis is clearly a priority, but it is too early yet for substantial data to have accumulated. The availability of inducible p53 gene constructs, cloned into tumorigenic cell lines or introduced into the germline will allow p53 expression to be suddenly switched on within established neoplasms by exposure of the cells or tissue to the pharmacological inducing agent.

If benefit is shown for p53–specific therapy, the options for clinical intervention are manifold [126–128]. They include gene delivery systems by use of lipid vehicles or viral vectors, or alternatively peptides designed to mimic or activate particular aspects of p53 function and tagged for target cell specificity might be used. Finally, structural mutants of p53 that are unable to maintain a stable wild type conformation might be stabilized by specific pharmacological agents, or perhaps even vaccinated against, using mutant specific epitopes as immunogens.

Conclusion

The position of p53 at the head of key cellular pathways, without actually being essential for life, and its susceptibility to dysfunction through loss or mutation of a single allele, make it at once both a powerful and fragile instrument of tumor suppression. The downstream links to central molecular regulators like TFIIH provide mechanisms for the coregulation of apoptosis, cell cycle and DNA repair after DNA damage, and also illustrate how defects in one pathway could impinge upon the others. p53 is pleiotropic but there are important differences between cell types in both the upstream induction of p53 and the downstream responses evoked, and the consequences of p53 dysfunction for carcinogenesis must be read in the context of the specific lesion, the cell type, differentiation state, genetic background and cellular environment.

p53–dependent apoptosis is important for tumor suppression in some tissues, such as thymus, or in particular biochemical situations, such as forced oncogene activation. However apoptosis may not be the important tumor suppressor function of p53 in other tissues, perhaps such as liver, and in some cell types p53 may be partially or completely redundant to pathways dependent on other genes such as IRF–1. Moreover, the particular balance of p53 downstream pathways that operates in normal cells may be distorted in neoplasia, perhaps for example giving the potential to therapeutically trigger p53–dependent apoptosis in a cell type that would not normally engage this response. Achieving better understanding of these issues is essential to more fully comprehend the contribution of p53 to tumor suppression in different tissues and is a major goal of cancer biology.

Despite these complexities, evidence for a clinical utility of gene–specific therapy of human cancer in specific situations is accumulating, and the practicalities of implementing such therapy are already being addressed.

Acknowledgments

C.O.C.B. is a CRC Gordon Hamilton Fairley Clinical Research Fellow. The authors work is supported by the Cancer Research Campaign.

References

1 Gottlieb TM, Oren M. p53 in growth control and neoplasia. Biochim Biophys Acta Rev Cancer 1996; 1287:77–102.
2 Nelson WG, Kastan MB. DNA strand breaks: the DNA template alterations that trigger p53–dependent DNA damage response pathways. Mol Cell Biol 1994; 14:1815–1823.
3 Lees–Miller SP, Sakaguchi K, Ullrich S et al. The human DNA–activated protein kinase phosphorylates serines 15 and 37 in the amino–terminal transactivation domain of human p53. Mol Cell Biol 1992; 11:5041–5049.
4 Molinari M, Milner J. p53 in complex with DNA is resistant to ubiquitin–dependent proteolysis in the presence of HPV–16 E6. Oncogene 1995; 10:1849–1854.
5 Kastan MB, Zhan Q, El–Deiry WS et al. A mammalian cell cycle checkpoint pathway utilizing p53 and GADD45 is defective in Ataxia–telangiectasia. Cell 1992; 71:587–597.
6 Rosselli F, Ridet A, Soussi T et al. p53–dependent pathway of radio–induced apoptosis is altered in Fanconi anemia. Oncogene 1995; 10:9–17.
7 Jayaraman J, Prives C. Activation of p53 sequence–specific DNA binding by short single strands of DNA requires the p53 C–terminus. Cell 1995; 81:1021–1029.

8 Lu X, Lane DP. Differential induction of transcriptionally active p53 following UV or ionizing radiation: defects in chromosome instability syndromes? Cell 1993; 75:765–778.
9 Midgley CA, Owens B, Briscoe CV et al. Coupling between gamma irradiation, p53 induction and the apoptotic response depends upon cell type in vivo. J Cell Sci 1995; 108:1843–1848.
10 Rafferty JA, Clarke AR, Sellappan D et al. Induction of murine O^6–alkylguanine–DNA–alkyltransferase in response to ionizing radiation is p53 gene dose dependent. Oncogene 1996; 12:693–697.
11 Di Leonardo A, Linke SP, Clarkin K et al. DNA damage triggers a prolonged p53–dependent G_1 arrest and long–term induction of Cip1 in normal human fibroblasts. Genes Dev 1994; 8:2540–2551.
12 Clarke AR, Gledhill S, Hooper ML et al. p53 dependence of early apoptotic and proliferative responses within the mouse intestinal epithelium following gamma–irradiation. Oncogene 1994; 9:1767–1773.
13 Brugarolas J, Chandrasekaran C, Gordon JI et al. Radiation–induced cell cycle arrest compromised by p21 deficiency. Nature 1995; 377:552–557.
14 Deng C, Zhang P, Harper JW et al. Mice lacking p21CIP1/WAF1 undergo normal development, but are defective in G_1 checkpoint control. Cell 1995; 82:675–684.
15 Evan GI, Wyllie AH, Gilbert CS et al. Induction of apoptosis in fibroblasts by c–myc protein. Cell 1992; 69:119–128.
16 Wagner AJ, Kokontis JM, Hay N. Myc–mediated apoptosis requires wild–type p53 in a manner independent of cell cycle arrest and the ability of p53 to induce p21waf1/cip1. Genes Dev 1994; 8:2817–2830.
17 Qin XQ, Livingston DM, Kaelin WG, Jr. et al. Deregulated transcription factor E2F–1 expression leads to S–phase entry and p53–mediated apoptosis. Proc Natl Acad Sci USA 1994; 91:10918–10922.
18 Wu X, Levine AJ. p53 and E2F–1 cooperate to mediate apoptosis. Proc Natl Acad Sci USA 1994; 91:3602–3606.
19 Lane DP. p53, guardian of the genome. Nature 1992; 358:15–16.
20 Maheswaran S, Englert C, Bennett P et al. The WT1 gene product stabilizes p53 and inhibits p53–mediated apoptosis. Genes Dev 1995; 9:2143–2156.
21 Clarke AR, Purdie CA, Harrison DJ et al. Thymocyte apoptosis induced by p53–dependent and independent pathways. Nature 1993; 362:849–852.
22 Lotem J, Sachs L. Hematopoietic cells from mice deficient in wild–type p53 are more resistant to induction of apoptosis by some agents. Blood 1993; 82:1092–1096.
23 Strasser A, Harris AW, Jacks T et al. DNA damage can induce apoptosis in proliferating lymphoid cells via p53–independent mechanisms inhibitable by Bcl–2. Cell 1994; 79:329–339.
24 Ziegler A, Jonason AS, Leffell DJ et al. Sunburn and p53 in the onset of skin cancer. Nature 1994; 372:773–776.
25 Haupt Y, Barak Y, Oren M. Cell type–specific inhibition of p53–mediated apoptosis by mdm2. EMBO J 1996; 15:1596–1606.
26 Attardi LD, Lowe SW, Brugarolas J et al. Transcriptional activation by p53, but not induction of the p21 gene, is essential for oncogene–mediated apoptosis. EMBO J 1996; 15:3693–3701.
27 Wang XW, Vermeulen W, Coursen JD et al. The XPB and XPD DNA helicases are components of the p53–mediated apoptosis pathway. Genes Dev 1996; 10:1219–1232.
28 Caelles C, Helmberg A, Karin M. p53–dependent apoptosis in the absence of transcriptional activation of p53–target genes. Nature 1994; 370:220–223.
29 Haupt Y, Rowan S, Shaulian E et al. Induction of apoptosis in HeLa cells by transactivation–deficient p53. Genes Dev 1995; 9:2170–2183.
30 Miyashita T, Reed JC. Tumor suppressor p53 is a direct transcriptional activator of the human *bax* gene. Cell 1995; 80:293–299.
31 Miyashita T, Krajewski S, Krajewska M et al. Tumor suppressor p53 is a regulator of bcl–2 and bax gene expression in vitro and in vivo. Oncogene 1994; 9:1799–1805.
32 Selvakumaran M, Lin HK, Miyashita T et al. Immediate early up–regulation of bax expression by p53 but not TGF beta 1: a paradigm for distinct apoptotic pathways. Oncogene 1994; 9:1791–1798.

33 Schott AF, Apel IJ, Nuñez G et al. Bcl–x_L protects cancer cells from p53–mediated apoptosis. Oncogene 1995; 11:1389–1394.
34 Wang XW, Yeh H, Schaeffer L et al. p53 modulation of TFIIH–associated nucleotide excision repair activity. Nature Genetics 1995; 10:188–195.
35 Friedberg EC. Relationships between DNA repair and transcription. Annu Rev Biochem 1996; 65:15–42.
36 Roy R, Adamcsewski JP, Seroz T et al. The MO15 cell cycle kinase is associated with the TFIIH transcription–DNA repair factor. Cell 1994; 79:1093–1101.
37 Patel T, Gores GJ, Kaufmann SH. The role of proteases during apoptosis. FASEB J 1996; 10:587–597.
38 Whyte M. ICE/CED–3 proteases in apoptosis. Trends Cell Biol 1996; 6:245–248.
39 Kaufmann WK, Paules RS. DNA damage and cell cycle checkpoints. FASEB J 1996; 10:238–247.
40 Harper JW, Adami GR, Wei N et al. The p21 Cdk–interacting protein Cip1 is a potent inhibitor of G_1 cyclin–dependent kinases. Cell 1993; 75:805–816.
41 El–Deiry WS, Harper JW, O'Connor PM et al. WAF1/CIP1 is induced in p53–mediated G_1 arrest and apoptosis. Cancer Res 1994; 54:1169–1174.
42 Tanaka N, Ishihara M, Lamphier MS et al. Cooperation of the tumor suppressors IRF–1 and p53 in response to DNA damage. Nature 1996; 382:816–818.
43 Kadohama T, Tsuji K, Ogawa K. Indistinct cell cycle checkpoint after u.v. damage in H–*ras*–transformed mouse liver cells despite normal p53 gene expression. Oncogene 1994; 9:2845–2852.
44 Datto MB, Li Y, Panus JF et al. Transforming growth factor beta induces the cyclin–dependent kinase inhibitor p21 through a p53–independent mechanism. Proc Natl Acad Sci USA 1995; 92:5545–5549.
45 Cox LS, Lane DP. Tumor suppressors, kinases and clamps: how p53 regulates the cell cycle in response to DNA damage. BioEssays 1995; 17:501–508.
46 Stewart N, Hicks GG, Paraskevas F et al. Evidence for a second cell cycle block at G_2/M by p53. Oncogene 1995; 10:109–115.
47 Guillouf C, Rosselli F, Krishnaraju K et al. p53 involvement in control of G_2 exit of the cell cycle: Role in DNA damage–induced apoptosis. Oncogene 1995; 10:2263–2270.
48 Cross SM, Sanchez CA, Morgan CA et al. A p53–dependent mouse spindle checkpoint. Science 1995; 267:1353–1356.
49 Smith ML, Chen IT, Zhan Q et al. Interaction of the p53–regulated protein Gadd45 with proliferating cell nuclear antigen. Science 1994; 266:1376–1380.
50 Reed M, Woelker B, Wang P et al. The C–terminal domain of p53 recognizes DNA damaged by ionizing radiation. Proc Natl Acad Sci USA 1995; 92:9455–9459.
51 Lee S, Elenbaas B, Levine A et al. p53 and its 14 kDa C–terminal domain recognize primary DNA damage in the form of insertion/deletion mismatches. Cell 1995; 81:1013–1020.
52 Bakalkin G, Yakovleva T, Selivanova G et al. p53 binds single–stranded DNA ends and catalyzes DNA renaturation and strand transfer. Proc Natl Acad Sci USA 1994; 91:413–417.
53 Ford JM, Hanawalt PC. Li–Fraumeni syndrome fibroblasts homozygous for p53 mutations are deficient in global DNA repair but exhibit normal transcription–coupled repair and enhanced UV resistance. Proc Natl Acad Sci USA 1995; 92:8876–8880.
54 Smith ML, Chen IT, Zhan Q et al. Involvement of the p53 tumor suppressor in repair of u.v.–type DNA damage. Oncogene 1995; 10:1053–1059.
55 Dutta A, Ruppert SM, Aster JC et al. Inhibition of DNA replication factor RPA by p53. Nature 1993; 365:79–82.
56 Leveillard T, Andera L, Bissonnette N et al. Functional interactions between p53 and the TFIIH complex are affected by tumor–associated mutations. EMBO J 1996; 15:1615–1624.
57 Pan ZQ, Reardon JT, Li L et al. Inhibition of nucleotide excision repair by the cyclin–dependent kinase inhibitor p21. J Biol Chem 1995; 270:22008–22016.
58 Li R, Waga S, Hannon GJ et al. Differential effects by the p21 CDK inhibitor on PCNA–dependent DNA replication and repair. Nature 1994; 371:534–537.
59 Shivji MK, Grey SJ, Strausfeld UP et al. Cip1 inhibits DNA replication but not PCNA–dependent nucleotide excision–repair. Curr Biol 1994; 4:1062–1068.

60 Tamura T, Ishihara M, Lamphier MS et al. An IRF–1–dependent pathway of DNA damage–induced apoptosis in mitogen–activated T lymphocytes. Nature 1995; 376:596–599.
61 Haimovitz–Friedman A, Kan C–C, Ehleiter D et al. Ionizing radiation acts on cellular membranes to generate ceramide and initiate apoptosis. J Exp Med 1994; 180:525–535.
62 Santana P, Pena LA, Haimovitz–Friedman A et al. Acid Sphingomyelinase–deficient human lymhoblasts and mice are defective in radiation–induced apoptosis. Cell 1996; 86:189–199.
63 Greenblatt MS, Bennett WP, Hollstein M et al. Mutations in the p53 tumor suppressor gene: clues to cancer etiology and molecular pathogenesis. Cancer Res 1994; 54:4855–4878.
64 Arrowsmith CH, Morin P. New insights into p53 function from structural studies. Oncogene 1996; 12:1379–1385.
65 Ponchel F, Puisieux A, Tabone E et al. Hepatocarcinoma–specific mutant p53–249ser induces mitotic activity but has no effect on transforming growth factor beta 1–mediated apoptosis. Cancer Res 1994; 54:2064–2068.
66 Dumenco L, Oguey D, Wu J et al. Introduction of a murine p53 mutation corresponding to human codon 249 into a murine hepatocyte cell line results in growth advantage, but not in transformation. Hepatology 1995; 22:1279–1288.
67 Forrester K, Lupold SE, Ott VL et al. Effects of p53 mutants on wild–type p53–mediated transactivation are cell type dependent. Oncogene 1995; 10:2103–2111.
68 Rowan S, Ludwig RL, Haupt Y et al. Specific loss of apoptotic but not cell–cycle arrest function in a human tumor derived p53 mutant. EMBO J 1996; 15:827–838.
69 Zhang W, Guo XY, Hu GY et al. A temperature–sensitive mutant of human p53. EMBO J 1994; 13:2535–2544.
70 Vogelstein B. A deadly inheritance. Nature 1990; 348:681–682.
71 Toguchida J, Yamaguchi T, Dayton SH et al. Prevalence and spectrum of germline mutations of the p53 gene among patients with sarcoma. New England Journal of Medicine 1992; 326:1301–1308.
72 Diller L, Sexsmith E, Gottlieb A et al. Germline p53 mutations are frequently detected in young children with rhabdomyosarcoma. J Clin Invest 1995; 95:1606–1611.
73 Clarke AR, Cummings MC, Harrison DJ. Interaction between murine germline mutations in p53 and APC predisposes to pancreatic neoplasia but not to increased intestinal malignancy. Oncogene 1995; 11:1913–1920.
74 Han JH, Sabbatini P, Perez D et al. The E1B 19K protein blocks apoptosis by interacting with and inhibiting the p53–inducible and death–promoting Bax protein. Genes Dev 1996; 10:461–477.
75 Lowe SW, Bodis S, Bardeesy N et al. Apoptosis and the prognostic significance of p53 mutation. Cold Spring Harbor Symposia on Quantitative Biology 1994; 59:419–426.
76 Symonds H, Krall L, Remington L et al. p53–dependent apoptosis suppresses tumor growth and progression in vivo. Cell 1994; 78:703–711.
77 Howes KA, Ransom N, Papermaster DS et al. Apoptosis or retinoblastoma: alternative fates of photoreceptors expressing the HPV–16 E7 gene in the presence or absence of p53 [published erratum appears in Genes Dev 1994 Jul 15; 8(14):1738]. Genes Dev 1994; 8:1300–1310.
78 Li G, Mitchell DL, Ho VC et al. Decreased DNA repair but normal apoptosis in ultraviolet–irradiated skin of p53–transgenic mice. Am J Pathol 1996; 148:1113–1123.
79 Merritt AJ, Potten CS, Kemp CJ et al. The role of p53 in spontaneous and radiation–induced apoptosis in the gastrointestinal tract of normal and p53–deficient mice. Cancer Res 1994; 54:614–617.
80 Yonish–Rouach E, Resnitzky D, Lotem J et al. Wild–type p53 induces apoptosis of myeloid leukemic cells that is inhibited by interleukin–6. Nature 1991; 352:345–347.
81 Colombel M, Radvanyi F, Blanche M et al. Androgen suppressed apoptosis is modified in p53 deficient mice. Oncogene 1995; 10:1269–1274.
82 Graeber TG, Osmanian C, Jacks T et al. Hypoxia–mediated selection of cells with diminished apoptotic potential in solid tumors. Nature 1996; 379:88–91.
83 Bardeesy N, Beckwith JB, Pelletier J. Clonal expansion and attenuated apoptosis in Wilms' tumors are associated with p53 gene mutations. Cancer Res 1995; 55:215–219.
84 Del Sal G, Ruaro EM, Utrera R et al. Gas1–induced growth suppression requires a transactivation–independent p53 function. Mol Cell Biol 1995; 15:7152–7160.

85 Harvey M, Sands AT, Weiss RS et al. In vitro growth characteristics of embryo fibroblasts isolated from p53–deficient mice. Oncogene 1993; 8:2457–2467.
86 Bogler O, Huang HJ, Cavenee WK. Loss of wild–type p53 bestows a growth advantage on primary cortical astrocytes and facilitates their in vitro transformation. Cancer Res 1995; 55:2746–2751.
87 Ewen ME, Oliver CJ, Sluss HK et al. p53–dependent repression of CDK4 translation in TGF–beta–induced G_1 cell–cycle arrest. Genes Dev 1995; 9:204–217.
88 Rogan EM, Bryan TM, Hukku B et al. Alterations in p53 and p16INK4 expression and telomere length during spontaneous immortalization of Li–Fraumeni syndrome fibroblasts. Mol Cell Biol 1995; 15:4745–4753.
89 Bouffler SD, Kemp CJ, Balmain A et al. Spontaneous and ionizing radiation–induced chromosomal abnormalities in p53–deficient mice. Cancer Res 1995; 55:3883–3889.
90 Carder PJ, Wyllie AH, Purdie CA et al. Stabilized p53 facilitates aneuploid clonal divergence in colorectal cancer. Oncogene 1993; 8:1397–1401.
91 Armstrong JF, Kaufman MH, Harrison DJ et al. High–frequency developmental abnormalities in *p53*–deficient mice. Curr Biol 1995; 5:931–936.
92 Livingstone LR, White A, Sprouse J et al. Altered cell cycle arrest and gene amplification potential accompany loss of wild–type p53. Cell 1992; 70:923–935.
93 Sands AT, Suraokar MB, Sanchez A et al. p53 deficiency does not affect the accumulation of point mutations in a transgene target. Proc Natl Acad Sci USA 1995; 92:8517–8521.
94 Donehower LA, Godley LA, Aldaz CM et al. Deficiency of p53 accelerates mammary tumorigenesis in Wnt–1 transgenic mice and promotes chromosomal instability. Genes Dev 1995; 9:882–895.
95 Fukasawa K, Choi T, Kuriyama R et al. Abnormal centrosome amplification in the absence of p53. Science 1996; 271:1744–1747.
96 Bouck N. P53 and angiogenesis. Biochimica et Biophysica Acta: Reviews on Cancer 1996; 1287:63–66.
97 Kemp CJ, Donehower LA, Bradley A et al. Reduction of p53 gene dosage does not increase initiation or promotion but enhances malignant progression of chemically induced skin tumors. Cell 1993; 74:813–822.
98 Burmer GC, Rabinovitch PS, Haggitt RC et al. Neoplastic progression in ulcerative colitis: histology, DNA content, and loss of a p53 allele. Gastroenterology 1992; 103:1602–1610.
99 Yin J, Harpaz N, Tong Y et al. p53 point mutations in dysplastic and cancerous ulcerative colitis lesions. Gastroenterology 1993; 104:1633–1639.
100 Brentnall TA, Crispin DA, Rabinovitch PS et al. Mutations in the p53 gene: an early marker of neoplastic progression in ulcerative colitis. Gastroenterology 1994; 107:369–378.
101 Bennett WP, Colby TV, Travis WD et al. p53 protein accumulates frequently in early bronchial neoplasia. Cancer Res 1993; 53:4817–4822.
102 Nuorva K, Soini Y, Kamel D et al. Concurrent p53 expression in bronchial dysplasias and squamous cell lung carcinomas. Am J Pathol 1993; 142:725–732.
103 Boers JE, ten Velde GP, Thunnissen FB. P53 in squamous metaplasia: a marker for risk of respiratory tract carcinoma. American Journal of Respiratory & Critical Care Medicine 1996; 153:411–416.
104 Bennett WP, Hollstein MC, Hsu IC et al. Mutational spectra and immunohistochemical analyses of p53 in human cancers. Chest 1992; 101:19S–20S.
105 Wang LD, Hong JY, Qiu SL et al. Accumulation of p53 protein in human esophageal precancerous lesions: a possible early biomarker for carcinogenesis. Cancer Res 1993; 53:1783–1787.
106 Wang X, Matsumoto H, Takahashi A et al. p53 accumulation in the organs of low–dose X–ray–irradiated mice. Cancer Lett 1996; 104:79–84.
107 Kitamura K, Kuwano H, Yasuda M et al. What is the earliest malignant lesion in the esophagus?. Cancer 1996; 77:1614–1619.
108 Raybaud–Diogène H, Tétu B, Morency R et al. p53 overexpression in head and neck squamous cell carcinoma: Review of the literature. Eur J Cancer [B] 1996; 32B:143–149.
109 Tanaka S, Toh Y, Adachi E et al. Tumor progression in hepatocellular carcinoma may be mediated by p53 mutation. Cancer Res 1993; 53:2884–2887.

110 Oda T, Tsuda H, Sakamoto M et al. Different mutations of the p53 gene in nodule–in–nodule hepatocellular carcinoma as a evidence for multistage progression. Cancer Lett 1994; 83:197–200.
111 Oda T, Tsuda H, Scarpa A et al. p53 gene mutation spectrum in hepatocellular carcinoma. Cancer Res 1992; 52:6358–6364.
112 Hayashi H, Sugio K, Matsumata T et al. The clinical significance of *p53* gene mutation in hepatocellular carcinomas from Japan. Hepatology 1995; 22:1702–1707.
113 Aguilar F, Harris CC, Sun T et al. Geographic variation of p53 mutational profile in nonmalignant human liver. Science 1994; 264:1317–1319.
114 Ueda H, Ullrich SJ, Gangemi JD et al. Functional inactivation but not structural mutation of p53 causes liver cancer. Nature Genetics 1995; 9:41–47.
115 Dowell SP, Hall PA. The p53 tumor suppressor gene and tumor prognosis: is there a relationship? J Pathol 1995; 177:221–224.
116 Dowell SP, Wilson PO, Derias NW et al. Clinical utility of the immunocytochemical detection of p53 protein in cytological specimens. Cancer Res 1994; 54:2914–2918.
117 Brennan JA, Mao L, Hruban RH et al. Molecular assessment of histopathological staging in squamous–cell carcinoma of the head and neck. New England Journal of Medicine 1995; 332:429–435.
118 El Rouby S, Thomas A, Costin D et al. p53 gene mutation in B cell chronic lymphocytic leukemia is associated with drug resistance and is independent of MDR1/MDR3 gene expression. Blood 1993; 82:3452–3459.
119 Brachman DG, Beckett M, Graves D et al. p53 mutation does not correlate with radiosensitivity in 24 head and neck cancer cell lines. Cancer Res 1993; 53:3667–3669.
120 Lowe SW, Ruley HE, Jacks T et al. p53–dependent apoptosis modulates the cytotoxicity of anticancer agents. Cell 1993; 74:957–967.
121 Slichenmyer WJ, Nelson WG, Slebos RJ et al. Loss of a p53–associated G_1 checkpoint does not decrease cell survival following DNA damage. Cancer Res 1993; 53:4164–4168.
122 Malcomson RDG, Oren M, Wyllie AH et al. p53–independent death and p53–induced protection against apoptosis in fibroblasts treated with chemotherapeutic drugs. Br J Cancer 1995; 72:952–957.
123 Lowe SW, Bodis S, McClatchey A et al. *p53* status and the efficacy of cancer therapy in vivo. Science 1994; 266:807–810.
124 Lotem J, Sachs L. Regulation by bcl–2, c–myc, and p53 of susceptibility to induction of apoptosis by heat shock and cancer chemotherapy compounds in differentiation–competent and –defective myeloid leukemic cells. Cell Growth and Differentiation 1993; 4:41–47.
125 Bristow RG, Jang A, Peacock J et al. Mutant p53 increases radioresistance in rat embryo fibroblasts simultaneously transfected with HPV16–E7 and/or activated H–ras. Oncogene 1994; 9:1527–1536.
126 Roemer K, Friedmann T. Mechanisms of action of the p53 tumor suppressor and prospects for cancer gene therapy by reconstitution of p53 function. Annals of the New York Academy of Sciences 1994; 716:265–282.
127 Lesoon–Wood LA, Kim WH, Kleinman HK et al. Systemic gene therapy with p53 reduces growth and metastases of a malignant human breast cancer in nude mice. Human Gene Therapy 1995; 6:395–405.
128 Mujoo K, Maneval DC, Anderson SC et al. Adenoviral–mediated p53 tumor suppressor gene therapy of human ovarian carcinoma. Oncogene 1996; 12:1617–1623.

Apoptosis and Cancer, edited by Seamus J. Martin.

The Retinoblastoma Gene and The Control of Apoptosis

Kay Macleod and Tyler Jacks

Howard Hughes Medical Institute, Research Laboratories, Massachusetts Institute of Technology, Center for Cancer Research, Cambridge, Massacheusetts, U.S.A.

Introduction

The existence of the retinoblastoma tumor suppressor gene (RB) was predicted before it was genetically mapped and molecularly cloned. Based on the pattern of inheritance of familial retinoblastoma and the latency of hereditary and nonhereditary forms of retinoblastoma (a childhood tumor of the eye), Knudson postulated the existence of a gene for which both alleles need to be mutated in order for tumors to develop [1]. It is now known that individuals with familial retinoblastoma inherit a mutated copy of the RB gene through the germline and suffer a somatic mutation in the second allele at a high frequency, usually the consequence of gene conversion or chromosomal nondisjunction. Individuals with the nonhereditary form of retinoblastoma develop malignant tumors following somatic mutation of both RB alleles. This "two–hit" mechanism of tumor suppressor gene mutation was confirmed following mapping of the RB gene to human chromosome 13q14.2, careful "loss of heterozygosity" studies, and ultimately through cloning of the gene [2, 3]. Mutation of RB also occurs at a significant frequency during the development of sporadically occurring osteosarcoma [3, 4], breast carcinoma [5], acute lymphoblastic leukemia [6], prostate carcinoma [7] and small cell lung carcinoma [8–10]. The status of RB as a tumor suppressor gene was further established when its overexpression in cultured tumor cells was shown to inhibit their tumorigenic properties [11, 12].

pRb and Cell Cycle Control

In addition to helping to establish the tumor suppressor gene concept, dissecting the function of the protein encoded by RB (pRB) has been central to our understanding of cell cycle control [13]. Passage of the cell through the G_1 phase of the cell cycle is dependent on the presence of growth factors until the cell passes the so–called "restriction point" in late G_1 [14, 15], after which mitogens are no longer required for continued cell cycle progression. Accumulating evidence suggests that the restriction point may reflect the function of pRB, which controls passage of the cell from G_1 to S phase in response to the growth factor environment [16]. The role of pRB in regulating G_1 progression was first demonstrated when its overexpression was shown to induce a reversible G_1 arrest [17]. pRb is thought to regulate cell cycle progression by down–modulating the activity of members of the E2F/DP family of transcription factors [18–21]. E2F/DP complexes regulate the expression of genes required for S–phase entry and completion, such as *dhfr* and *cyclin E* [22–27]. The repression of E2F/DP by pRB is in turn regulated by phosphorylation [19, 28–30]. In early G_1, pRB is in a hypophosphorylated state and able to bind to a subset of E2F/DP complexes (those containing E2F–1, –2 or –3), thereby repressing expression of E2F target genes and blocking cell cycle progression. When pRB becomes hyperphosphorylated in mid– to late G_1, transactivation by E2F/DP is derepressed and cells progress into S–phase [18]. Thus, by regulating the phosphorylation of pRB, the cell determines, at least in part, the timing of S–phase entry.

Phosphorylation of pRB is mediated by cyclin dependent kinases (CDKs) with their associated cyclin subunits [13]. Activation of cyclin D/CDK4(6) in mid–G_1 is thought to be the critical event in initiating pRB phosphorylation. It is the first cyclin/CDK to become active in G_1 and will phosphorylate pRB in vitro on most of the sites on which the protein becomes phosphorylated in vivo.[31–33] Expression of D–type cyclins is growth factor inducible, and this may link the phosphorylation of pRB and cell cycle progression with the growth factor environment of the cell [34, 35]. Cyclin E/CDK2, which is active in late G_1/early S–phase, also plays an important role in pRB phosphorylation [17]. Other cyclin/CDK complexes may ensure hyperphosphorylation of pRB through S phase and into G_2 and M. The protein is actively dephosphorylated upon exit from mitosis [43]. Finally, as a specific inhibitor of cyclin D/CDK4(6) kinases, $p16^{INK4a}$ is an important regulator of pRB function as well [36].

Abrogation of pRb Function by Viral Oncoproteins

Much of the early insight into pRB function has come from studies on its interaction with the viral oncoproteins SV40 T antigen, adenovirus E1A and human papilloma virus E7 [37–41]. In particular, the observation that SV40 T antigen binds dephosphorylated pRB but not the hyperphosphorylated form led to an understanding of the importance of the state of pRB phosphorylation in protein:protein interactions and cell cycle regulation [42, 43]. The viral oncoproteins are thought to displace cellular factors such as E2F that are negatively regulated in a pRB complex, releasing them from their normal control and promoting cell cycle entry [21]. pRB contains a protein domain referred to as the "pocket," mapping from amino acids 379 to 792, which is essential for oncoprotein binding [44]. A larger region (amino acids 379–928) is required for full growth suppression by pRB, defining a larger "pocket" domain required for binding to E2F [44]. The study of the interaction of viral oncoproteins with pRB has led not only to a functional dissection of the domains of pRB but also to an improved understanding of how loss of function of RB can lead to growth deregulation and tumorigenesis in the whole animal (see below).

pRb–Related Proteins, p107 and p130

RB is now known to be a member of a larger gene family, including the genes p107 and p130, which are more closely related to each other than either is to RB [39, 45–49]. All three proteins contain a "pocket" domain that mediates their binding to known interacting proteins, including adenovirus E1a, SV40 T antigen, papilloma virus E7 and E2F/DP complexes [46–49]. pRB, p107 and p130 have different specificities for E2F/DP complexes. Of the five different E2F proteins known (E2F–1, –2, –3, –4 and –5), pRB binds E2F–1, –2 and –3 complexes preferentially, while p107 and p130 are more commonly associated with E2F–4 and –5 complexes [50–54]. Importantly, p107 and p130 both differ from RB in that neither gene has been shown to be mutated in human cancer. Moreover, in the mouse, loss of function of p107 or p130 has a less significant effect on embryological or adult development than Rb (see below). Finally, the expression and protein–protein interactions of the three family members differ through the cell cycle [47–49, 55–57].

The importance of maintaining the integrity of the pRB pathway for normal growth control is illustrated by the tumorigenic effects of mutation of genes involved in its regulation [58]. For example, the cyclin D/CDK4(6) inhibitor p16^{INK4a} is a potent tumor suppressor gene, which is frequently in sporadic cancers and responsible for certain forms of familial melanoma [36, 59–63]. Also, mice lacking p16^{INK4a} function are predisposed to formation of lymphoma and fibrosarcoma [64]. In addition, cyclin D1 is overexpressed in a proportion of parathyroid adenomas as a consequence of chromosomal inversion involving 11q13 [65]. Furthermore, amplification of 11q13 in some breast and esophageal carcinomas results in overexpression of cyclin D1 [66, 67], and CDK4 is mutated in some human melanomas such that it can no longer be bound and repressed by p16^{INK4a} [68, 69]. Finally, targeted disruption of the E2F–1 gene in the mouse leads to lymphoproliferation and tumor development, suggesting that

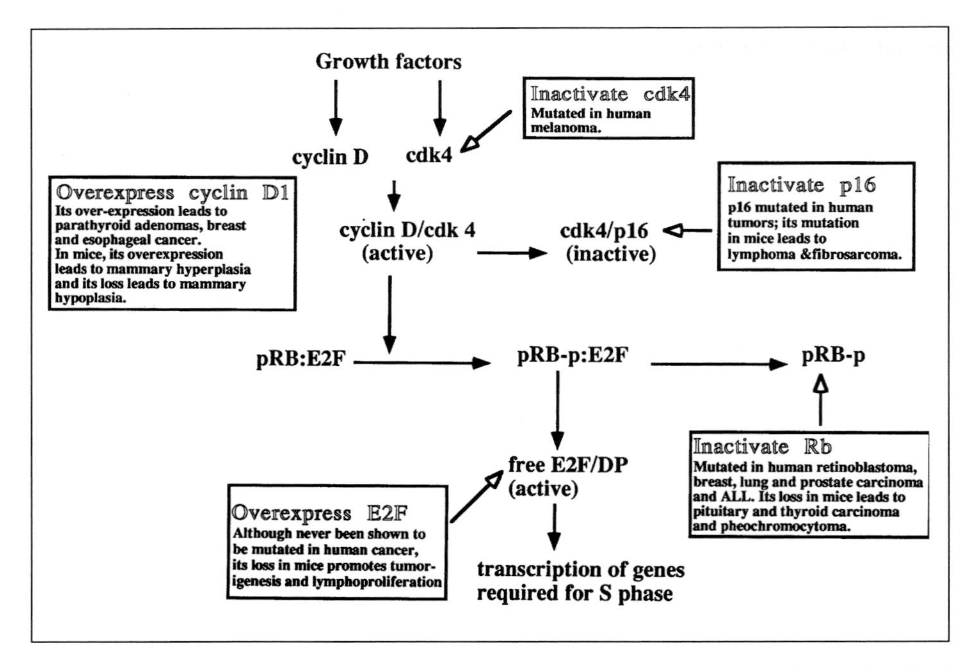

Fig. 8.1. Mutation of the RB pathway in cancer. Mutation of RB or of genes involved in regulating pRB activity and its downstream effectors can all lead to tumor formation, in both the mouse and in humans. The pRB pathway is shown along with the points at which mutations (shown in boxes) have been associated with cancer.

E2F–1, as part of the pRB complex, is also involved in negative growth control.[70,71] Thus, as illustrated in Figure 8.1, tumor development can be promoted by mutation of the RB gene itself or of genes encoding modulators of pRB activity or its downstream effectors.

Rb and Apoptosis

As discussed above, there is considerable evidence that mutation of RB (or other genes in this regulatory pathway) can lead to loss of proper growth control and tumorigenesis. Therefore, it was at first quite surprising that in several contexts, abrogation of RB function was not associated with increased proliferation but rather cell death. As reviewed below, in the context of the Rb mutant mouse as well as in tissue culture and in vivo experiments using viral oncoproteins, the fate of many cells lacking pRB control is apoptosis. This observation, along with a determination of the underlying mechanism, has clear and important implications for the understanding of the connection between deregulated cell growth, the induction of cell death, and the tumorigenic process. Moreover, these studies have laid the groundwork for establishing the relationship between the escape from apoptosis, tumor progression, and the resistance of tumor cells to therapy.

Rb and Apoptosis During Development

The first evidence that loss of Rb function could lead to programmed cell death came with the observation of apoptosis in the nervous system and fetal liver of Rb homozygous mutant mouse embryos [72–74]. Cell death was also subsequently observed in the developing lens of the eye of Rb–/– embryos [75]. These embryos die between days 12.5 and 14.5 of gestation exhibiting anemia and edema, probably due to failed fetal hematopoiesis. Examination of cell death in the affected tissues in the Rb mutant mouse has led to new insights into the role of Rb in regulating cell cycle exit during differentiation, of the mechanisms by which deregulated proliferation induces apoptosis and of how the microenvironment can affect the outcome of aberrant growth control.

Loss of Rb Function in the Lens and Nervous System Leads to Failure of Cell Cycle Withdrawal During Differentiation and Apoptosis

The lens develops from the lens vesicle which pinches off from embryonic ectoderm around day 11.5 of mouse gestation. Within this spherical ball of lens epithelia, the cells closest to the retina start to synthesize crystallins, elongate and differentiate into lens fibers. As they differentiate and elongate, lens fibers cease proliferating and become enucleate. The cells at the front of the lens continue to proliferate and can be recruited to differentiate during the life of the organism as required. Lens development is essentially complete by day 14.5 of mouse development. However, in the Rb mutant embryo, this process is severely disrupted. Rb–/– lens fibers fail to exit the cell cycle as they begin the differentiation program and consequently undergo apoptosis. Cell death in the lens of the Rb–/– embryo is evident by standard histological analysis and by TUNEL staining [75].

The apoptotic death of cells in the central nervous system (CNS) of the Rb mutant embryo shows many similarities with that occurring in the lens. Lee and co–workers have shown that neuroblasts in Rb–/– embryos fail to exit the cell cycle as they migrate away from the lumen of the neural tube during differentiation [76]. This is probably due to the continued expression of genes involved in S–phase entry [77]. Normally, differentiating neurons downregulate E2F–controlled genes, such as *cyclin E* and *dhfr*. In the Rb mutant CNS, however, these genes continue to be expressed, and cells aberrantly enter S–phase [77]. As in the case of the inappropriately cycling lens fiber cells, the ultimate fate of these abnormal neuronal cells is to die by apoptosis. In addition to overexpressing certain S–phase–specific genes, Rb–deficient neurons exhibit reduced levels of some markers of neuronal differentiation, such as βII tubulin, TrkA, TrkB and $p75^{NGFR}$ [76]. The underexpression of these neurotrophin receptors may help explain why Rb mutant neurons die.

In related studies, adenovirus E1A expression in differentiating neuroectoderm–derived cells causes apoptosis in a manner dependent on the ability of E1A to inactivate pRB [78]. Furthermore, overexpression of cyclin D1 (which inactivates pRB by phosphorylation) in differentiating neurons, induces cell death; this effect can be blocked by $p16^{INK4a}$ [79]. Thus, control of cell cycle exit and prevention of cell death in nervous tissue appears to be critically dependent on Rb function and regulation. This is depicted in Figure 8.2, which represents schematically the effect of loss of Rb on proliferation and differentiation, in both the lens and the CNS.

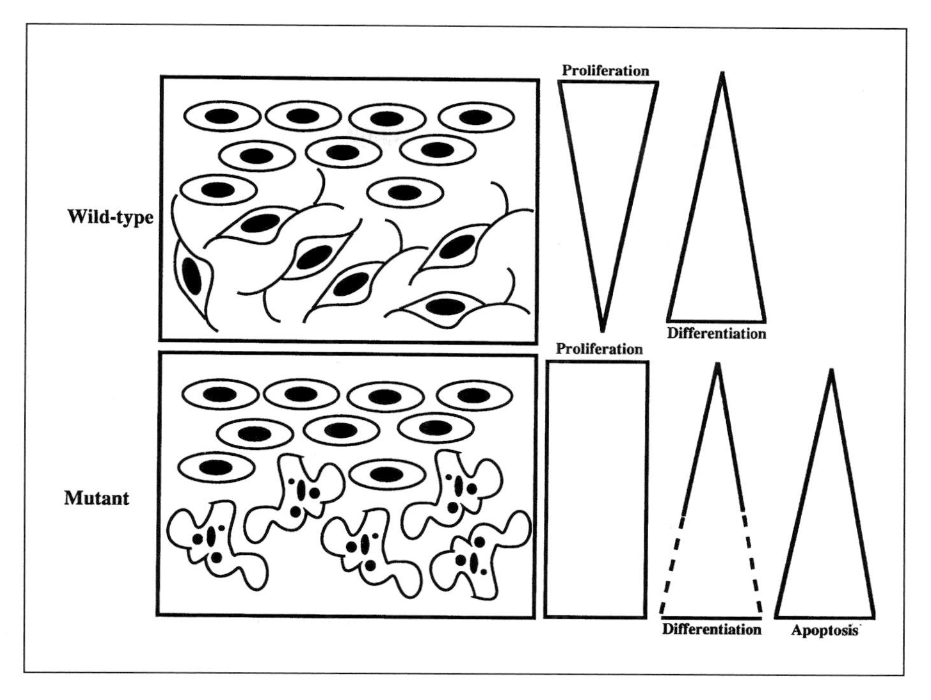

Fig. 8.2. Discoordination of cell cycle withdrawal and differentiation leads to cell death. This model depicts how loss of Rb function can lead to deregulated proliferation, failure to exit the cell cycle during differentiation and, ultimately, programmed cell death. This is most obvious in the developing lens and central nervous system of Rb mutant embryos where proliferating and differentiating cells are spatially organized.

Loss of Rb Can Lead to Both p53–Dependent and –Independent Apoptosis

The p53 tumor suppressor gene is required for apoptosis in response to a variety of different stresses as has been demonstrated in cycling thymocytes and intestinal cells following irradiation and after chemotherapeutic treatment of some tumor cells [80–83]. Similarly, cell death in the lens and CNS of Rb mutant mice requires p53 function [75, 77]. This has been determined by creating embryos mutant for both p53 and Rb; these embryos still exhibit abnormal proliferation in the lens and CNS, but there is a marked suppression of apoptosis [75, 77]. It is unclear what the long–term fate of the Rb–/–;p53–/– cells would be, because the doubly mutant embryos still die at approximately day 13.5 of gestation (probably due to continuing erythropoietic defects).

Macleod et al [77] have also demonstrated that p53 is induced in Rb–/– CNS, and that there is p53–dependent upregulation of the $p21^{WAF-1}$ gene in this tissue [77]. The mechanism by which Rb loss leads to p53 induction is still unclear. p53 may be activated by the products generated by aberrant S–phase entry, such as the accumulation of DNA replication intermediates, or depletion of nucleotides by premature DNA replication [84]. Alternatively, p53 may be induced by the absence of the appropriate survival factors in the microenvironment or by a more general failure of cell cycle regulation [85–88].

It is equally unclear how p53 acts to bring about the death of Rb deficient cells. Evidence from different systems is conflicting regarding the requirement for p53–dependent transcriptional activation in this process [89–93]. Moreover, p53 target genes important in cell death pathways have not been identified, although the Bax gene remains an attractive target [94]. One critical p53 target gene, $p21^{WAF-1}$, has been shown to be dispensable for cell death in a number of systems [89, 95, 96].

Although cell death in the lens and CNS of the Rb mutant embryo occurs by a p53–dependent mechanism(s), the situation in the peripheral nervous system (PNS) is distinct [77]. The neurons of the developing PNS respond similarly to the loss of Rb, including failure to exit the cell cycle upon differentiation and overexpression of S–phase genes. However, unlike the CNS, apoptosis in the PNS is not dependent on p53 function [77]. There is as much, if not more, programmed cell death in the sensory ganglia of the Rb–/–;p53–/– embryos than in those lacking Rb alone [77].

Beyond the finding that it is not dependent on p53, the mechanism of apoptosis in the Rb–deficient PNS has not been defined. However, there are some intriguing connections to the Bcl–2 gene family. Bcl–2 is highly expressed in the developing nervous system and has been shown to inhibit the cell death of sensory neurons deprived of nerve growth factors but not ciliary neurons deprived of CNTF [97, 98]. Bcl–X, another member of the family, is also expressed in the CNS and PNS of developing mouse embryos, and loss of Bcl–X function by targeted disruption results in embryonic lethality at the same gestational age as Rb deficient embryos [99]. Furthermore, these Bcl–X deficient embryos show extensive cell death in the nervous system and fetal liver, suggesting that loss of Bcl–X expression may play a role in inducing the cell death observed in the Rb mutant. In fact, we have observed p53–independent downregulation of Bcl–X expression in the Rb mutant embryo in a manner which correlates with the programmed cell death of cells in the fetal liver and peripheral nervous system (Macleod and Jacks, unpublished). How loss of Rb might affect the expression of *Bcl–2*–related proteins remains to be determined.

Apoptosis in Interacting Cell Populations: The Fetal Liver

The fetal liver is the major site of hematopoiesis in the mid–gestational embryo [100]. At day 13.5 of gestation, when apoptosis is evident in the Rb mutant fetal liver, the organ is composed of both hepatocytes and hematopoietic cells. Histological and TUNEL analysis has revealed that both cell types undergo cell death (Macleod and Jacks, unpublished). It is presently unclear whether both hematopoietic cells and hepatocytes require Rb function directly for their survival, or whether the death of one cell type is a consequence of the death of the other. This issue is particularly relevant in the fetal liver, where hepatocytes comprise the microenvironment that normally supports hematopoietic differentiation. Hepatocytes produce the necessary cytokines, survival factors, cell–cell contacts and matrix components that are necessary for the hematopoietic cells to survive and mature. For example, erythropoietin, which is required for red cell maturation and survival, is expressed in the liver during embryogenesis [101, 102]. Also, stem cell factor (SCF), which is necessary for the migration and survival of hematopoietic progenitors, is expressed in fetal liver hepatocytes. Interestingly, SCF expression is downregulated in the Rb mutant fetal liver (Macleod, unpublished).

The possibility that the failed differentiation and apoptosis of Rb–/– erythroid progenitors is a secondary consequence of the death of hepatocytes is further supported by the observation that erythropoiesis from Rb mutant cells is restored in chimeric embryos and adult animals composed in part of Rb–/– and in part of wild–type cells (see below) [103, 104]. It appears that some transacting factor produced by the wild–type cells in these chimeras allows the Rb mutant erythroid progenitors to survive and reach end–stage differentiation. The specific transacting factor(s) has not been defined, but SCF would appear to be a good candidate.

In embryos mutant for both Rb and p53, there is significantly less apoptosis in the liver, although apoptotic cells are still detectable (Macleod and Jacks, unpublished). This result may indicate that the death of only one of the two major cell types in this tissue is dependent on p53 function. Interestingly, these double mutant embryos exhibit signs of anemia and inefficient erythropoiesis, suggesting that erythroid development is not restored in this context.

Rescue of Rb Mutant Cells from Death in Chimeric Mice

As discussed above with respect to the fetal liver, the fate of certain Rb mutant cells in vivo can be altered by the presence of wild–type cells. Williams et al [104] and Maandag et al [103] have constructed chimeric animals by injection of Rb–/– ES cells into wild–type blastocysts. Surprisingly, chimeras with

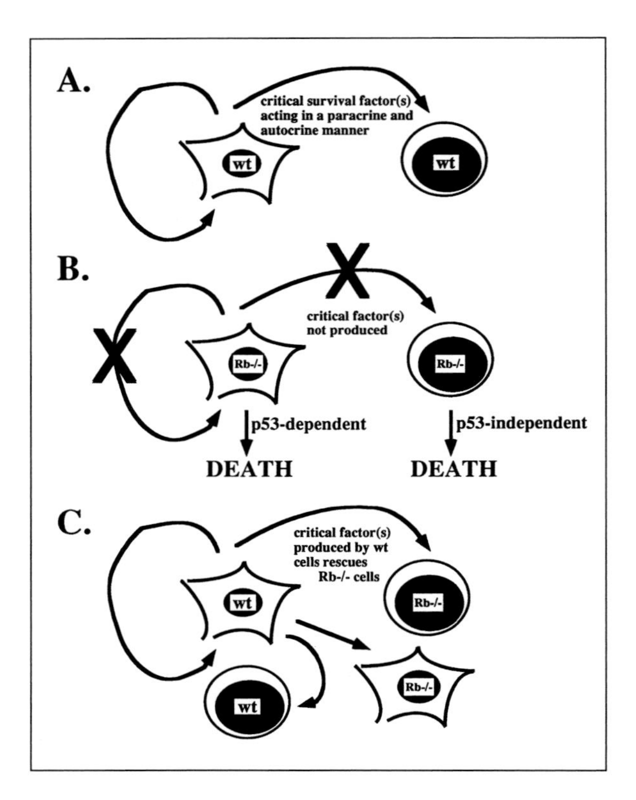

Fig. 8.3. Rescue of Rb mutant cells by surrounding normal cells. Wild–type cells can rescue some Rb mutant cells from apoptosis in the context of a chimeric mouse embryo. The figure suggests a model to explain this rescue. (*a*) Wild–type cells of one type produce a factor(s) required for their own survival and the survival of other cell types. (*b*) In the Rb mutant mouse, these cells fail to produce this factor and both they and cells dependent on this factor(s) undergo programmed cell death. (*c*) In the chimeric mouse, wild–type cells produce enough factor(s) to maintain their own survival and that of the Rb mutant cells.

extensive contribution of Rb–/– cells survived to adulthood [103, 104]. Moreover, in both the CNS and erythroid compartments of these chimeras, there was a high percentage of Rb–/– cells. Thus, it appears that the presence of wild–type cells can "rescue" the Rb mutant cells from incomplete differentiation and death (Fig. 8.3). In the case of erythroid differentiation, the production of SCF by wild–type hepatocytes might account for this effect, but the nature of the relevant factors in the rescue of Rb–/– neuronal cells is less clear. Recently, the analysis of chimeras generated with β–galactosidase–expressing Rb–/– cells has allowed a more precise determination of the point at which wild–type cells exert their effect. In chimeric embryos at 13.5 days of gestation, Rb–/– cells in the PNS and CNS overexpress cyclin E and enter S phase inappropriately, but they fail to undergo apoptosis (Macleod, Williams and Jacks, unpublished). These results suggest that the effect of wild–type cells is either to affect the decision of these abnormal cells to undergo apoptosis or to inhibit directly the apoptotic program. Once again, the secretion of one or more survival factors by neighboring wild–type cells may be the mechanism.

Conclusions from Developmental Analyses of Rb Function

Lens fiber cells, hepatocytes, hematopoietic cells and neurons of developing mouse embryos undergo programmed cell death when Rb function is absent. The requirement for Rb function is apparently at the moment when they start to differentiate and attempt to exit the cell cycle. Failure of cell cycle exit at this time seems to be incompatible with survival of these cell types. Continued proliferation induces cell death by intrinsic mechanisms and is clearly influenced by changes in the microenvironment occurring as part of the normal differentiation process. Apoptosis induced by loss of Rb in the lens and CNS occurs by p53–dependent mechanisms, while in the PNS (and hematopoietic cells of the fetal liver) death is p53–independent.

It is still unclear why other tissues which are also undergoing differentiation at this time in development do not submit to the same fate of programmed cell death in the Rb–/– embryo. One possibility is that the related genes p107 and p130 functionally compensate for Rb mutation in a tissue–specific fashion. In fact, p107 has been shown to be upregulated in cultured Rb–deficient myotubes and to be downregulated again when these cells are induced to re–enter the cell cycle [105]. Importantly, however, muscle differentiation appears normal in the Rb–/–;p107–/– embryo, perhaps suggesting that p130 can compensate for loss of both Rb and p107 in this tissue. Nor is it clear why it takes until at least day 12.5 of gestation for cell death to become apparent in the Rb–/– embryo. Again, p107 or p130 may be compensating in the nervous system and fetal liver earlier in development. This possibility is supported by the observation that embryos mutant for both Rb and p107 die earlier and manifest more extensive cell death in both of these tissues compared to the Rb mutant embryos [106]. Part of the timing and tissue specificity of apoptosis in the Rb–/– embryos is also likely to be a consequence of the pattern of terminal differentiation and cell cycle withdrawal in different tissues relative to the stage at which the embryos die.

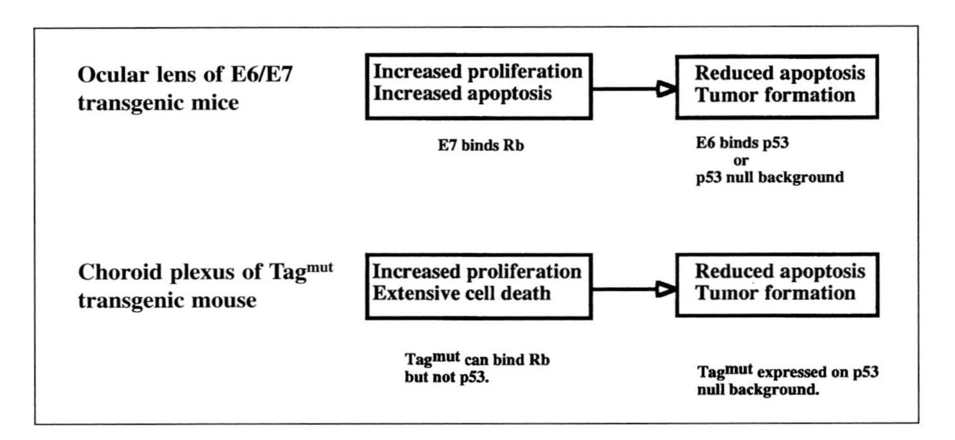

Fig. 8.4. Relationship between escape from apoptosis and tumorigenesis. Viral oncogenesis involves subversion of Rb family function (leading to hyperproliferation and increased apoptosis) plus subversion of p53 function (leading to reduced apoptosis and tumor progression). This is exemplified by studies using HPV E7 and E6 expression in the lens (ref. 112) or SV40 Tag in the choroid plexus (ref. 115).

Rb and Apoptosis During Tumorigenesis

Some of the interest in understanding the role of Rb in embryogenesis has come from the expectation that information gleaned from a developmental system is relevant to our understanding of cancer. As a tumor suppressor gene, RB is thought to function in the adult to negatively regulate the growth of many different cell types. In addition to its defining role in retinoblastoma, loss of RB function has been observed in approximately 30% of human tumor types, including breast, lung and prostate carcinoma, acute lymphoblastic leukemia and osteosarcoma. RB is unlikely to be the only genetic lesion in these tumors, and determining how its mutation contributes mechanistically to the tumorigenic phenotype remains to be determined in detail. Of particular interest is the question of why RB is rarely found mutated in some tumor types (e.g., malignant melanoma), while being frequently mutated in others (e.g., retinoblastoma). As mentioned above, this may be due to the mutation of other genes upstream or downstream of RB (for example, cyclin D1 or $p16^{INK4a}$) in growth control pathways that makes RB mutation unnecessary. It will also be important to understand how loss of RB cooperates with other mutations in the course of tumorigenesis, and, relevant to the topic at hand, how the induction of apoptosis and escape from apoptosis factor into the genetic progression of the disease.

Mouse models of human cancer have been used extensively, especially since the advent of transgenic and gene knockout technology [107, 108]. These models have several useful features: (1) the effect of tumor suppressor gene/oncogene mutation can be readily monitored over the lifetime of the animal and on subsequent generations; (2) the effect of diet or environmental stress can be assessed; (3) strains can be interbred to determine the extent of cooperativity between mutations in different genes; and (4) therapeutic protocols can be tested in vivo. One major disadvantage of using the mouse as a model for human cancer is the failure of some genetic lesions to induce the same disease in mice as they do in humans. Nevertheless, these mice have been central to much of our current understanding of tumorigenesis.

Viral Oncoproteins in Tumorigenesis

Viral oncoproteins that target the pRB family of proteins (sometimes along with other cellular proteins) have been invaluable tools in demonstrating the connection between changes in growth control and the induction of apoptosis. Indeed, the initial demonstration that loss of pRB family function could induce apoptosis was made in tissue culture using the adenovirus E1A oncogene [109]. As illustrated in Figure 8.4, this system has also been used to show the importance of p53 in the death response [110, 111].

In vivo, these viral oncoproteins have been tested in a number of systems. The need for functional Rb family function to ensure proper lens development in vivo has been examined in mice transgenic for the human papilloma virus (HPV–16) E7 gene expressed under the control of the α–crystallin promoter [112]. These E7 transgenic mice develop cataracts and show both ectopic proliferation and increased apoptosis compared to nontransgenic littermates. Furthermore, neonatal mice transgenic for both E7 and E6 (which inactivates p53) show reduced apoptosis, and the adults develop lens tumors. This study is a nice complement to the developmental studies of the lens in Rb–/– and Rb–/–;p53–/– embryos [75], and makes the additional important connection between the inhibition of apoptosis and tumorigenesis.

Interestingly, a more detailed examination of the timing of death in the E7–expressing lens has demonstrated that p53 was necessary for the apoptosis occurring at 13.5 day but by day 17.5, lens fiber cell death was p53–independent mechanisms [113]. Surprisingly, E6 expression blocks both forms of death, indicating that E6 has anti–apoptotic properties beyond its effects on p53.

Transgenic expression of HPV E7 expression has also been used to probe the importance of the pRB family in the regulation of cellular proliferation and death in the retina [114]. In mice expressing an E7 transgene under the control of the retinal–specific IRBP gene promoter, photoreceptor cells die by apoptosis at a time when they should be undergoing terminal differentiation. Again, this cell death is predominantly p53–dependent, as judged by the reduced levels of apoptosis in IRBP–E7 p53–/– animals. As with the lens studies, the abrogation of the cell death pathway promotes transformation of retinal cells, and these animals go on to develop retinoblastoma. Interestingly, this effect may be specific to the retinal cells of the mouse, since p53 mutations have not been reported in human retinoblastoma.

Because HPV E7 binds and inactivates pRB, p107, and p130, it is not possible to determine from the transgenic experiments which of these proteins is required for normal growth in murine retinal development. However, analysis of mice carrying germline mutations in both Rb and p107 suggest that both of these genes are important in this tissue [106]. Specifically, animals with the genotype Rb+/–;p107–/– develop bi–lateral, multifocal lesions in the retina characterized as retinal dysplasia. Such lesions have not been observed in either Rb+/– or p107–/– animals. Although not yet determined experimentally, it is has been suggested that retinal dysplasia in the Rb+/–;p107–/– mice develops following somatic loss of the remaining wild–type Rb allele, transient hyperproliferation of Rb–/–;p107–/– mutant cells, and, ultimately death of these cells by apoptosis [106]. Based on the results from the IRBP–E7 transgenic mice, it is likely that death of doubly mutant cells would be p53–dependent. Therefore, Rb+/–; p107–/–;p53–/– animals would be expected to develop retinoblastoma, and this possibility can be readily tested. These experiments underscore the importance of functional overlap between members of the Rb gene family in determining the phenotypic consequences of mutations of individual members of the family.

Studies using mice expressing wild–type and mutant alleles of SV40 T antigen (T–ag) in the choroid plexus region of the brain have more firmly established the link between the loss of p53–dependent apoptosis and tumor progression. Wild–type T–ag expression in the choroid plexus has been shown to cause rapid tumor development, with 50% of transgenic animals dying within 6 weeks of birth [115]. However, when the p53 binding domain of T–ag was removed, transgenic mice succumbed to tumors more slowly (50% dead by 26 weeks) [116]. The T–ag fragment used in these studies was still capable of binding to the pRB family, and the longer tumor latency was correlated with increased levels of apoptosis in pretumorigenic tissue [117]. Importantly, on a p53–deficient genetic background, levels of apoptosis caused by expression of the T–ag mutant were reduced, and the kinetics of tumor development increased. More informative still, results from transgenic p53 heterozygous mice demonstrated a clear transition from slow–growing tumors with high levels of apoptosis and functional p53 to more–rapidly growing tumor nodules lacking p53 in which the frequency of apoptosis was significantly lower.

Transgenic mice expressing wild–type T–ag in the islet cells of the pancreas develop pancreatic carcinoma by approximately 5 months of age [118]. The progression of these tumors has been analyzed at its different stages from prevascular hyperplasia through angiogenic hyperplasia to the solid tumor. The prevascular tumor is apoptotic and loss of apoptosis during tumor progression is associated with the focal induction of the survival factor IGF–II [119]. Targeted disruption of IGF–II in this model leads to increased apoptosis and increased malignancy. IGF–II becomes activated by loss of imprinting and transcriptional activation of both IGF–II alleles [120]. Because T–ag inactivates p53, apoptosis in the premalignant tumors is thought to be p53 independent. Thus, in both tumorigenesis and embryonic development, loss of Rb function can lead to both p53–dependent and–independent death.

Tumor Formation in Rb Heterozygous Mice: Cooperativity with Loss of p53

In mice heterozygous for a disrupted Rb allele, three tumor types have been detected: intermediate lobe pituitary adenocarcinoma forms in 100% of Rb+/– mice, while thyroid carcinoma and pheochromocytoma develop in 70% and 5%, respectively [121].

It was initially surprising that loss of Rb did not lead to a broader spectrum of tumors given its key role in growth control. However, based on the analysis of Rb–/– embryo and viral oncogene transgenic mouse strains, it is quite possible that the majority of cells in Rb heterozygous mice that lose the wild–type allele of Rb are eliminated by apoptosis.

Intermediate lobe pituitary tumor development in the Rb+/– mouse takes approximately nine months, suggesting additional mutational events are necessary for full tumor formation. One of these is the inactivation of the wild–type Rb allele [122]. Indeed, this event has been shown to occur early in tumor formation and is associated with increased proliferation and programmed cell death [122]. One assumes that additional mutations that abrogate these cell death pathways (whether they be p53–dependent or –independent) would also be selected for during tumor progression.

Support for this notion has come from analysis of tumor spectrum in mice mutant for both Rb and p53 [121, 123]. p53 homozygous mutant mice develop a large spectrum of different tumor types but die predominantly with lymphomas and sarcomas by 4–6 months of age [124, 125]. Mice deficient for p53 and also heterozygous for Rb develop a large series of tumors, including those associated with the individual mutations plus a set of novel tumor types [121]. Although not yet demonstrated experimentally, it is likely that the elimination of p53–dependent apoptosis is a major contributing factor in the tumorigenic cooperativity caused by germline mutations in these two tumor suppressor genes.

Why Does Loss of Rb Function Lead to Apoptosis?

This chapter has outlined several examples in which loss of Rb function results in cell death, both in embryogenesis and tumorigenesis. The question of why this occurs can be asked at two levels. First, why should the response to loss of Rb function be cell death instead of proliferation? And second, what is the mechanism through which Rb deficiency causes cell death? The answers to these questions are not yet known, but there is enough available information to allow for some speculation. Regarding the more teleological question, it is interesting that in most cases, cell death associated with Rb deficiency is preceded by inappropriate cell cycle activity. This raises the possibility that it is the abnormal activity itself that triggers the apoptotic response. Indeed, perhaps proper passage through the cell cycle is in some way monitored, and when events are disorderly, the cell death machinery is in some way accessed and engaged. This hypothetical fail–safe mechanism would act to guard against the development of abnormally proliferating cells that if left unchecked could go on to compromise the further development or health of the organism. In this scenario, Rb mutation would not directly activate cell death; rather, the consequences of Rb deficiency would indirectly do so.

On the more mechanistic level, there is accumulating evidence that the Rb–regulated transcription factor E2F may be an important cofactor in the death of Rb deficient cells. For example, E2F overexpression in tissue culture can cause premature S–phase entry and apoptosis, at least some of which has been shown to be p53–dependent [126–128]. More recently, transgenic expression of E2F in the eye imaginal disc in Drosophila was shown to cause inappropriate cell cycle activity and cell death [129]. The specific mechanism by which deregulated E2F activity might lead to apoptosis is not understood. It is possible that E2F directly regulates the expression of some component of the cell death machinery. Alternatively, inappropriate S–phase entry may indirectly trigger cell death, perhaps through the induction of DNA strand breaks. Finally, cellular proliferation driven by deregulated E2F may exhaust the local supply of survival factors.

Conclusions

Loss of Rb Can Lead to the Failure of Differentiating Cells to Exit the Cell Cycle and to Their Apoptosis

Rb function is critically required during the development of certain tissues (the nervous system, the lens, the retina and the fetal liver) as cells undergo terminal differentiation and exit the cell cycle. In the absence of Rb, cells in some these tissues fail to exit the cell cycle, enter S phase aberrantly and undergo programmed cell death. Not all tissues are equally affected by Rb loss. This may reflect timing in development, the ability to induce growth arrest by other means, and/or the expression and role of the pRb related proteins, p107 and p130, in these tissues.

Rb Loss Induces Apoptosis by p53–Dependent and –Independent Mechanisms

Some tissues respond to loss of pRb by undergoing p53–dependent cell death. The precise mechanism by which Rb loss induces p53 DNA–binding activity and transactivation potential is unclear but it appears to be post–translational. Whether apoptosis induced by Rb loss is p53–dependent or –independent is specified in part by cell type. This is true for both developmental and tumorigenic processes. For example, loss of Rb induces p53–dependent death in the central nervous system but p53–independent death in the peripheral nervous system of developing embryos. Similarly, T antigen induced transformation of the choroid plexus involves circumvention of p53–dependent apoptosis while cell death in premalignant pancreatic tumors of T antigen transgenic mice is p53–independent. Neither of these cell death pathways induced by Rb loss is well understood.

The Fate of a Cell Lacking Rb is Affected by its Microenvironment

The presence of wild–type cells can rescue Rb deficient cells from apoptosis in the nervous system and fetal liver of chimeric mouse embryos. Wild–type cells may be providing a survival factor which mutant cells are defective in producing. Survival factor activity is equally critical during the malignant progression of T antigen–induced pancreatic tumors, where activation of IGF–II expression results in reduced cell death. Thus, in a similar manner to that described for rescue of c–Myc–induced apoptosis following serum starvation, production of the appropriate survival signal can prevent apoptosis of Rb–deficient cells.

Implications for Tumor Therapy

Loss of Rb function leads to apoptosis by p53–dependent and p53–independent means. p53 is the most frequently mutated gene in human cancers, and its role in inducing the apoptosis of premalignant cells has made the gene the focus of intense study. In particular, p53 mutation is linked to the development of drug resistance in many different human tumor types, and the ability to restore p53 function to these late–stage tumors may lead to their successful treatment. However, an alternative way to treat these tumors may be to induce programmed cell death by p53–independent means. An understanding of the mechanisms by which Rb induces death in a p53–independent manner may lead to alternative therapies for tumors lacking p53.

References

1 Knudson AG. Mutation and cancer: statistical study of retinoblastoma. Proc Natl Acad Sci USA 1971; 68:820–823.
2 Cavenee WK, Dryja TP, Phillips RA et al. Expression of recessive alleles by chromosomal mechanisms in retinoblastoma. Nature 1983; 305:779–784.
3 Friend SJ, Bernards R, Rogelj S et al. A human DNA segment with properties of the gene that predisposes to retinoblastoma and osteosarcoma. Nature 1986; 323:643–646.
4 Toguchida J, Ishizaki K, Sasaki MS et al. Chromosomal reorganisation for the expression of recessive mutation of retinoblastoma susceptibility gene in the development of osteosarcoma. Cancer Res 1988; 48:3939–3943.
5 Lee EY, To J, Shew JY et al. Inactivation of the retinoblastoma susceptibility gene in human breast cancers. Science 1988; 241:218–221.
6 Cheng J, Scully P, Shew JY et al. Homozygous deletion of the retinoblastoma gene in acute lymphoblastic leukemia. Blood 1990; 75:730–735.
7 Bookstein R, Shew JY, Chen PL et al. Suppression of tumorigenicity of human prostate carcinoma cells be replacing a mutated RB gene. Science 1990; 247:712–715.
8 Harbor JW, Lai SL, Whang–Pend J et al. Abnormalities in structure and expression of the human retinoblastoma gene in SCLC. Science 1988; 241:353–356.
9 Hensel CH, Hsieh CL, Gazdar AF et al. Altered structure and expression of the human retinoblastoma susceptibility gene in small cell lung cancer. Cancer Res 1990; 50:3067–3072.
10 Horowitz JM, Park SH, Begenmann E et al. Frequent inactivation of the retinoblastoma anti–oncogene is restricted to a subset of human tumor cells. Proc Natl Acad Sci USA 1990; 87:2775–2779.
11 Huang HJ, Yee JK, Shew J et al. Suppression of the neoplastic pheotype by replacement of the Rb gene in human cancer cells. Science 1988; 242:1563–1566.
12 Goodrich DW, Wang NP, Qian YW et al. The retinoblastoma gene product regulates progression through the G_1 phase of the cell cycle. Cell 1991; 67:293–302.
13 Weinberg RA. The Retinoblastoma Protein and Cell Cycle Control. Cell 1995; 81:323–330.
14 Pardee AB. G_1 events and regulation of cell proliferation. Science 1989; 246:603–608.
15 Zetterberg A, Larsson O, Wiman KG et al. What is the restriction point? Curr Opinion in Cell Biol 1995; 7:835–842.
16 Laiho M, DeCaprio JA, Ludlow JW et al. Growth inhibition by TGF–B linked to suppression of retinoblastoma protein phosphorylation. Cell 1990; 62:175–189.
17 Hinds PW, Mittnacht S, Dyulic V et al. Regulation of retinoblastoma protein functions by ectopic expression of human cyclins. Cell 1992; 70:993–1006.
18 Helin K, Harlow E, Fattaey AR et al. Inhibition of E2F–1 transactivation by direct binding of the retinoblastoma protein. Mol Cell Biol 1993; 13:6501–6508.
19 Hiebert SW, Chellappan SP, Horowitz JM et al. The interaction of RB with E2F coincides with an inhibition of the transcriptional activity of E2F. Genes & Dev 1992; 6:177–185.
20 Lees JA, Saito M, Vidal M *et al.* The retinoblastoma protein binds to a family of E2F transcription factors. Mol Cell Biol 1993; 13:7813–7825.

21 Nevins J. E2F: A link between the Rb tumor suppressor protein and viral oncoproteins. Science 1992; 258:424–429.

22 Hsiao KM, McMahon SL, Farnham PJ. Multiple DNA elements are required for the growth regulation of the mouse E2F1 promoter. Genes & Dev 1994; 8:1526–1537.

23 Lam EF, Watson RJ. An E2F–binding site mediates cell–cycle regulated repression of mouse B–myb transcription. The EMBO J 1993; 12:2705–2713.

24 Johnson DG, Ohtani K, Nevins JR. Autoregulatory control of E2F1 expression in response to positive and negative regulators of the cell cycle. Genes & Dev 1994; 8:1514–1525.

25 Means AL, Slansky JE, McMahon SL et al. The HIP1 binding site is required for growth regulation of the dehydrofolate reductase promoter. Mol Cell Biol 1992; 12:1054–1063.

26 Muller R. Transcriptional regulation during the mammalian cell cycle. Trends in Genetics 1995; 11:173–178.

27 Slansky JE, Li Y, Kaelin WG et al. A protein synthesis –dependent increase in E2F1 mRNA correlates with growth regulation of the dihydrofolate reductase promoter. Mol Cell Biol 1993; 13:1610–1618.

28 Kaelin WG, Krek W, Sellers WR *et al.* Expression cloning of a cDNA encoding a retinoblastoma binding protein with E2F–like properties. Cell 1992; 70:351–364.

29 Helin K, Lees JA, Vidal M *et al.* A cDNA encoding a pRB–binding protein with properties of the transcription factor E2F. Cell 1992; 70:337–350.

30 Chellappan SP, Hiebert S, Mudryj M et al. The E2F transcription factor is a cellular target for the RB protein. Cell 1991; 65:1053–1061.

31 Dowdy SF, Hinds PW, Louie K et al. Physical interaction of the retinbolastoma protein with human D cyclins. Cell 1993; 73:499–511.

32 Ewen ME, Sluss HK, Sherr CJ et al. Functional interactions of the retinoblastoma protein with mammalian D–type cyclins. Cell 1993; 73:487–497.

33 Kato J–Y HM, Hiebert SW et al. Direct binding of cyclin D to the retinoblastoma gene product and pRb phosphorylation by the cyclin D–dependent kinase, cdk4. Genes Dev 1993; 7:331–342.

34 Sherr CJ. Mammalian G_1 cyclins. Cell 1993; 73:1059–1065.

35 Sherr CJ. D–type cyclins. TIBS 1995; 20:187–190.

36 Serrano M., Hannon GJ, Beach D. A new regulatory motif in cell cycle control causing specific inhibition of cyclin D/cdk4. Nature 1993; 366:704–707.

37 DeCaprio JA, Ludlow JW, Figge J et al. SV40 large tumor antigen forms a specific complex with the product of the retinblastoma susceptibility gene. Cell 1988; 54:275–283.

38 DeCaprio JA, Ludlow JW, Lynch Y et al. The product of the retinblastoma susceptibility gene has properties of a cell cycle regulatory element. Cell 1989; 58:1085–1095.

39 Ewen ME, Ludlow JW, Marsilio E et al. An N–terminal transformation–governing sequence of SV40 large T antigen contributes to the binding of both p110Rb and a second cellular protein, p120. Cell 1989; 58:257–267.

40 Vousden K. Interactions of human papillomavirus transforming proteins with the products of tumor suppressor genes. FASEB J 1993; 7:872–879.

41 Whyte P, Buchkovich KJ, Horowitz JM et al. Association between an oncogene and an anti–oncogene: the adenovirus E1A proteins bind to the retinoblastoma gene product. Nature (London) 1988; 334:124–129.

42 Ludlow JW, De Caprio JA, Huang CM et al. SV40 large T antigen binds preferentially to an under–phosphorylated member of the retinoblastoma susceptibility gene product family. Cell 1989; 56:57–65.

43 Ludlow JW, Shon J, Pipas JM et al. The retinoblastoma susceptibility gene product undergoes cell cycle–dependent dephosphorylation and binding to and release from SV40 large T. Cell 1990; 60:387–396.

44 Qin X, Chittenden T, Livingston DM et al. Identification of a growth suppression domain within the retinoblastoma gene product. Genes & Dev 1992; 6:953–964.

45 Whyte P, Williamson NM, Harlow E. Cellular targets for transformation by the adenovirus E1A proteins. Cell 1989; 56:67–75.

46 Ewen ME, Xing Y, Lawrence JB et al. Molecular cloning, chromosomal mapping and expression of the cDNA for p107, a retinoblastoma gene product–related protein. Cell 1991; 66:1155–1164.

47 Cobrinik D, Whyte P, Peeper DS et al. Cell cycle–specific association of E2F with the p130 E1A–binding protein. Genes & Dev 1993; 7:2392–2404.
48 Hannon GJ, Demetrick D, Beach D. Isolation of the Rb–related p130 through its interaction with CDK2 and cyclins. Genes & Dev 1993; 7:2378–2391.
49 Li Y, Graham C, Lacy S et al. The adenovirus E1A–associated 130–kD protein is encoded by a member of the retinoblastoma gene family and physically interacts with cyclins A and E. Genes & Dev 1993; 7:2366–2377.
50 Vairo G, Livingston D, Ginsberg D. Functional interaction between E2F–4 and p130: evidence for distinct mechanisms underlying growth suppression by different retinoblastoma protein family members. Genes & Dev 1995; 9:869–881.
51 Ginsberg D, Vairo G, Chittenden T et al. E2F–4, a new member of the E2F transcription factor family, interacts with p107. Genes & Dev 1994; 8:2665–2679.
52 Beijersbergen RL, Kerkhoaven RM, Zhu L et al. E2F–4, a new member of the E2F family, has oncogenic activity and associates with p107 in vivo. Genes & Dev 1994; 8:2680–2690.
53 Sardet C, Vidal M, Cobrinik D et al. E2F–4 and E2F–5, two members of the E2F family, are expressed in the early phases of the cell cycle. Proc Natl Acad Sci USA 1995; 92:2403–2407.
54 Hijmans EM, Voorhoeve PM, Beijersbergen RL et al. E2F–5, a new E2F family member that interacts with p130 in vivo. Mol Cell Biol 1995; 15:3082–3089.
55 Beijersbergen RL, Carlee L, Kerkhoven RM et al. Regulation of the retinoblastoma protein–related p107 by G_1 cyclin complexes. Genes & Dev 1995; 9:1340–1353.
56 Zhu L, van den Heuvel S, Helin K et al. Inhibition of cell proliferation by p107, a relative of the retinoblastoma protein. Genes & Dev 1993; 7:1111–1125.
57 Lees E, Faha B, Dulic V et al. CyclinE/cdk2 and cyclin A/cdk2 kinases associate with p107 and E2F in a temporally distinct manner. Genes & Dev 1992; 6:1874–1885.
58 Hunter T, Pines J. Cyclins and Cancer II: Cyclin D and CDK inhibitors come of age. Cell 1994; 79:573–582.
59 Kamb A, Shattuck–Eidens D, Eeles R et al. Analysis of the p16 gene as a candidate for the chromosome 9p melanoma susceptibility locus. Nature Genetics 1994; 8:23–26.
60 Hussussian CJ, Struewing JP, Goldstein AM et al. Germline p16 mutations in familial melanoma. Nature Genetics 1994; 8:15–21.
61 Koh J, Enders GH, Dynlacht BD et al. Tumor derived p16 alleles encodong proteins defective in cell cycle inhibition. Nature 1995; 506–5100.
62 Okamoto A, Demetrick DJ, Spillare EA et al. Mutations and altered expression of p16INK4a in human cancer. Proc Natl Acad Sci USA 1994; 91:11045–11049.
63 Serrano M, Gomez–Lahoz E, DePinho RA et al. Inhibition of Ras–induced proliferation and cellular transformation by p16INK4a. Science 1995; 267:249–252.
64 Serrano M, Lee HW, Chin L et al. Role of the INK4a locus in tumor suppression and cell mortality. Cell 1996; 85:27–37.
65 Motokura T, Bloom T, Kim HG et al. A novel cyclin encoded by a bcl–linked candidate oncogene. Nature 1991; 350:512–515.
66 Buckley MF, Sweeney KJ, Hamilton JA et al. Expression and amplification of cyclin genes in human breast cancer. Oncogene 1993; 2127.
67 Jiang W, Kahn SM, Tomita N et al. Amplification and expression of the human cyclin D gene in esophageal cancer. Cancer Res 1992; 52:2980–2983.
68 Wolfel T, Hauer M, Schneider J et al. A p16INK4a insensitive cdk4 mutant targeted by cytolytic lymphocytes in a human melanoma. Science 1995; 269:1281–1284.
69 Zuo L, Weger J, Yang Q et al. Germline mutations in the p16INK4a binding domain of CDK4 in familial melanoma. Nature Genetics 1996; 12:97–99.
70 Yamasaki L, Jacks T, Bronson R et al. Tumor induction and tissue atrophy in mice lacking E2F–1. Cell 1996; 85:537–548.
71 Field SJ, Tsai FY, Kuo F et al. E2F–1 functions in mice to promote apoptosis and suppress proliferation. Cell 1996; 85:549–561.
72 Clarke AR, Maandag ER, van Roon M et al. Requirement for a functional *Rb–1* gene in murine development. Nature 1992; 359:328–330.

73 Jacks T, Fazeli A, Schmitt E et al. Effects of an Rb mutation in the mouse. Nature 1992; 359:295–300.
74 Lee EY, Chang CY, Hu N et al. Mice deficient for Rb are nonviable and show defects in neurogenesis and hematopoiesis. Nature 1992; 359:288–295.
75 Morgenbesser SD, Williams BO, Jacks T et al. p53–dependent apoptosis produced by Rb–deficiency in the developing mouse lens. Nature 1994; 371:72–74.
76 Lee EY, Hu N, Yuan SSF et al. Dual Roles of the retinoblastoma protein in cell cycle regulation and neuron differentiation. Genes & Dev 1994; 8:2008–2021.
77 Macleod K, Hu Y, Jacks T. Loss of Rb activates both p53–dependent and –independent cell death pathways in the developing mouse nervous system. EMBO J 1996; (in press).
78 Slack RS, Skerjanc IS, Lach B et al. Cells Differentiating into Neuroectoderm undergo apoptosis in the absence of functional retinoblastoma family proteins. The Journal of Cell Biology 1995; 129:779–788.
79 Kranenburg O, van der Eb AJ, Zantema A. Cyclin D1 is an essential mediator of apoptotic neuronal cell death. EMBO J. 1996; 15:46–54.
80 Merrit AJ, Potten CS, Kemp CJ et al. The role of p53 in spontaneous and radiation –induced apoptosis in the gastrointestinal tract of normal and p53–deficient mice. Cancer Res 1994; 54:614–617.
81 Lowe SW, Schmitt ES, Smith SW et al. p53 is required for radiation–induced apoptosis in mouse thymocytes. Nature 1993; 362:847–849.
82 Lowe SW, Bodis S, McClatchey A et al. p53 status and the efficacy of cancer therapy in vivo. Science 1994; 266:807–810.
83 Clarke AR, Purdie CA, Harrison DJ et al. Thymocyte apoptosis induced by p53–dependent and independent pathways. Nature 1993; 362:849–852.
84 Linke SP, Clarkin KC, Di Leonardo A et al. A reversible, p53–dependent G_0/G_1 cell cycle arrest induced by ribonucleotide depletion in the absence of detectable DNA damage. Genes & Dev 1996; 10:934–947.
85 Evan GI, Brown L, Whyte M et al. Apoptosis and the cell cycle. Curr Op Cell Biol 1995; 7:825–834.
86 Canman CE, Gilmer TM, Coutts SB et al. Growth factor modulation of p53–mediated growth arrest versus apoptosis. Genes & Dev 1995; 9:600–611.
87 Harrington EA, Bennett MR, Fanidi A et al. c–Myc induced apoptosis in fibroblasts is inhibited by specific cytokines. EMBO J 1994; 13:3286–3295.
88 Raff MC, Barres BA, Burne JF et al. Programmed Cell Death and the Control of cell Survival: Lessons from the Nervous System. Science 1993; 262:695–700.
89 Attardi LD, Lowe SD, Brugarolas J et al. Transcriptional activation by p53 but not by induction of the p21 gene is essential for oncogene–mediated apoptosis. EMBO J 1996; 15:3693–3701.
90 Caelles C, Helmberg A, Karin M. p53–dependent apoptosis in the absence of transcriptional activation of p53–target genes. Nature 1994; 370:220–223.
91 Haupt Y, Rowan S, Shaulian E et al. Induction of apoptosis in HeLa cells by transactivation–deficient p53. Genes & Dev 1995; 9:2170–2183.
92 Sabbatini P, Lin J, Levine AJ et al. Essential role for p53–mediated transcription in E1A–induced apoptosis. Genes & Dev. 1995; 9:2184–2192.
93 Wagner AJ, Kokontis JM, Hay N. Myc–mediated apoptosis requires wild–type p53 in a manner independent of cell cycle arrest and the ability of p53 to induce p21 waf1/cip1. Genes & Dev 1994; 8:2817–2830.
94 Miyashita T, Reed JC. Tumor suppressor p53 is a direct transcriptional activator of the human bax gene. Cell 1995; 80:293–299.
95 Brugarolas J, Chandrasekaran C, Gordon JI et al. Radiation–induced cell cycle arrest compromised by p21 deficiency. Nature 1995; 377:552–557.
96 Deng C, Zhang P, Harper JW et al. Mice lacking p21CIP1/WAF1 undergo normal development but are defective in G_1 checkpoint control. Cell 1995; 82:675–684.
97 Allsop TE, Wyatt S, Paterson HF et al. The proto–oncogene bcl–2 can selectively rescue neurotrophic factor–dependent neurons from apoptosis. Cell 1993; 73:295–307.

98 Merry DE, Veis DJ, Hickey WF et al. bcl–2 protein expression is widespread in the developing nervous system and retained in the adult PNS. Development 1994; 120:301–311.
99 Motoyama N, Wang F, Roth KA et al. Massive Cell Death of Immature Hematopoietic Cells and Neurons in Bcl–X deficient Mice. Science 1995; 267:1506–1510.
100 Dzierzak E, Medvinsky A. Mouse embryonic hematopoiesis. TIGS 1995; 11:359–366.
101 Krantz SB. Erythropoietin. Blood 1991; 77:419–434.
102 Koury ST, Bondurant MC, Semenza GL. Localization of cells producing erythropoietin in murine liver by in situ hybridisation. Blood 1991; 77:2497–2503.
103 Maandag ECR, van der Valk M, Vlaar M et al. Developmental rescue of an embryonic–lethal mutation in the retinoblastoma gene in chimeric mice. EMBO J 1994; 13:4260–4268.
104 Williams BO, Schmitt EM, Remington L et al. Extensive contribution of Rb–deficient cells to adult chimeric mice with limited histopathological consequences. EMBO J 1994; 13:4251–4259.
105 Schneider JW, Gu W, Zhu L et al. Reversal of terminal differentiation mediated by p107 in Rb–/– muscle cells. Science 1994; 264:1467–1471.
106 Lee MH, Williams BO, Mulligan G et al. Targeted disruption of p107: functional overlap between p107 and Rb. Genes & Dev 1996; 10:1621–1632.
107 Hanahan D. Transgenic mice as probes into complex systems. Science 1989; 246:1265–1275.
108 Adams JM, Cory S. Transgenic models of tumor development. Science 1991; 254:1161–1167.
109 Rao L, Debbas M, Sabbatini P et al. The adenovirus E1A proteins induce apoptosis which is inhibited by the E1B 19K and Bcl–2 proteins. Proc Natl Acad Sci USA 1992; (in press).
110 Debbas M, White E. Wild–type p53 mediates apoptosis by E1A, which is inhibited by E1B. Genes and Dev 1993; 7:546–554.
111 Lowe SW, Jacks T, Housman DE et al. Abrogation of oncogene–associated apoptosis allows transformation of p53–deficient cells. Proc Natl Acad Sci USA 1994; 91:2026–2030.
112 Pan H, Griep A. Altered cell cycle regulation in the lens of HPV–16 E6 or E7 transgenic mice: implications for tumor suppressor gene function in development. Genes & Dev 1994; 8:1285–1299.
113 Pan H, Griep, AE. Temporally distinct patterns of p53–dependent and p53–independent apoptosis during mouse lens development. Genes & Dev 1995; 9:2157–2169.
114 Howes KA, Ransom N, Papermaster DS et al. Apoptosis or retinoblastoma: alternative fates of photoreceptors expressing the HPV–16 E7 gene in the presence or absence of p53. Genes and Dev 1994; 8:1300–1310.
115 Chen J, van Dyke T. Uniform cell–autonomous tumorigenesis of the choroid plexus by papovavirus large T antigens.. Mol Cell Biol 1991; 11:5968–5976.
116 Saenz Robles MT, Symonds H, Chen J et al. Induction versus progression of brain tumor development: differential functins for the pRb and p53–targeting domains of Simian virus 40 T antigen. Mol Cell Biol 1994; 14:2686–2698.
117 Symonds H, Krall L, Remington L et al. p53–dependent apoptosis suppresses tumor growth and progression in vivo. Cell 1994; 78:703–711.
118 Hanahan D. Heritable formation of pancreatic B cell tumors in transgenic mice expressing recombinant insulin/simian virus 40 genes. Nature 1985; 315:115–122.
119 Christofori G, Naik P, Hanahan D. A second signal supplied by insulin–like growth factor II in oncogene–induced tumorigenesis. Nature 1994; 369:414–418.
120 Christofori G, Naik P, Hanahan D. Deregulation of both imprinted and expressed alleles of the insulin–like growth factor 2 gene during b–cell tumorigenesis. Nature Genetics 1995; 10:196–201.
121 Williams BO, Remington L, Bronson RT et al. Cooperative tumorigenic effects of germline mutations in Rb and p53. Nat Genet 1994; 7:480–484.
122 Nikitin AY, Lee WH. Early loss of the retinoblastoma gene is associated with impaired growth inhibitory innervation during melanotroph carcinogenesis in Rb+/– mice. Genes & Dev 1996; 10:1870–1879.
123 Harvey M, Vogel H, Lee EY et al. Mice deficient in both p53 and rb develop tumors primarily of endocrine origin. Cancer Res. 1995; 55:1146–1151.
124 Donehower LA, Harvey M, Slagle BL et al. Mice deficient for p53 are developmentally normal but susceptible to spontaneous tumors. Nature 1992; 356:215–221.

125 Jacks T, Remington L, Williams BO et al. Tumor spectrum analysis in p53–mutant mice. Curr Biol 1994; 4:1–7.
126 Kowalik TF, DeGregori J, Schwarz JK et al. E2F1 overexpression in quiescent fibroblasts leads to induction of cellular DNA synthesis and apoptosis. J Virol 1995; 69:2491–2500.
127 Qin XQ, Livingston DM, Kaelin WG et al. De–regulated transcription factor E2F1 expression leads to S–phase entry and p53–mediated apoptosis. Proc Natl Acad Sci USA 1994; 91:10918–10922.
128 Wu X, Levine AJ. p53 and E2F–1 cooperate to mediate apoptosis. Proc. Natl. Acad. Sci. USA 1994; 91:3802–3806.
129 Asano M, Nevins JR, Wharton R. Ectopic E2F expression induces S phase and apoptosis in Drosophila imaginal discs. Genes & Dev 1996; 10:1422–1432.

Chapter 9

Apoptosis and Cancer, edited by Seamus J. Martin.

Programmed (Apoptotic) Cell Death and Prostate Cancer

Samuel R. Denmeade[a] *and John T. Isaacs*[b]

[a] Johns Hopkins Oncology Center, Johns Hopkins School of Medicine, Baltimore, Maryland, U.S.A.

[b] Johns Hopkins Oncology Center and James Buchanan Brady Urological Institute, Johns Hopkins School of Medicine, Baltimore, Maryland, U.S.A.

Abstract

The programmed cell death (apoptotic) process involves an epigenetic reprogramming of a cell that results in an energy–dependent cascade of biochemical and morphologic changes (also termed apoptosis) within the cell, resulting in its death and elimination. Although the final steps (i.e., DNA and cellular fragmentation) are common to cells undergoing programmed cell death, the activation of this death process is initiated either by sufficient injury to the cell induced by various exogenous damaging agents (e.g., radiation, chemicals, viruses, etc.) or by changes in the levels of a series of endogenous signals (e.g., hormones and growth/survival factors). Within the prostate, androgens are capable of both stimulating cell proliferation as well as inhibiting the rate of cell death. Androgen withdrawal triggers the programmed cell death pathway in both normal prostate glandular epithelia and androgen dependent prostate cancer cells. Androgen independent prostate cancer cells do not initiate the programmed cell death pathway upon androgen ablation; however, they do retain the cellular machinery necessary to activate the programmed cell death cascade when sufficiently damaged by exogenous agents. In the normal prostate epithelium, cell proliferation is balanced by an equal rate of programmed cell death, such that neither involution nor overgrowth occurs. In prostatic cancer, however, this balance is lost, such that there is greater proliferation than death producing continuous net growth. Thus, an imbalance in programmed cell death must occur during prostatic cancer progression. The goal of effective therapy of prostate cancer, therefore, is to correct this imbalance. Unfortunately, this has not been achieved and metastatic prostate cancer is still a lethal disease for which no curative therapy is currently available. In order to develop such effective therapy, an understanding of the programmed death pathway, and what controls it, is critical. Presented here is a review of the current state of knowledge concerning programmed cell death of normal and malignant prostatic cells.

Introduction

The study of programmed cell death/apoptosis in both the normal prostate gland and in prostate cancer has become a major area of prostate research. The prostate gland affords a unique opportunity to study programmed cell death during the normal process of glandular self renewal. At the same time, the prostate represents an unparalleled system for studying the mechanisms of programmed cell death in neoplasia, both in terms of response to an initially effective therapy, androgen ablation, and in subsequent resistance to programmed cell death with progression to an androgen independent state. Currently, nearly all men with metastatic prostate cancer treated with surgical or medical castration have an initial beneficial response to androgen withdrawal. While this initial response has substantial palliative value, almost all treated patients relapse to an androgen insensitive state. Unfortunately, once prostate cancer progresses to become androgen independent it is uniformly fatal because presently no effective systemic therapy exists. In 1995, an estimated 41,000 American men died from prostate cancer [1]. This number represents 15% of all cancer deaths for men in 1995, making prostate cancer the second leading cause of cancer deaths in this group [1]. Even more startling is the fact that, in 1995, approximately one out of every three newly diagnosed cancers in men were due to prostate cancer [1]. Therefore, understanding the mechanisms of PCD could prove critical to developing new, effective therapies for prostate cancer.

Apoptosis was originally defined by Kerr et al [2] as the orderly and characteristic sequence of structural changes resulting in the programmed death of the cell. Morphologically, apoptosis is characterized by a temporal sequence of events consisting of chromatin aggregation, nuclear and cytoplasmic condensation, and eventual fragmentation of the dying cell into a cluster of membrane–bound segments (apoptotic bodies) which often contain morphologically intact organelles. These apoptotic bodies are rapidly recognized, phagocytized and digested by either macrophages or adjacent epithelial cells. In PCD, the cell also progresses through an orderly series of biochemical and molecular changes, similar to the sequential changes involved in progression through the proliferative cell cycle (Fig. 9.1). The hallmark of the PCD process is the fragmentation of genomic DNA, an irreversible event that commits the cell to die and occurs before changes in plasma and internal membrane permeability [3–6]. This period

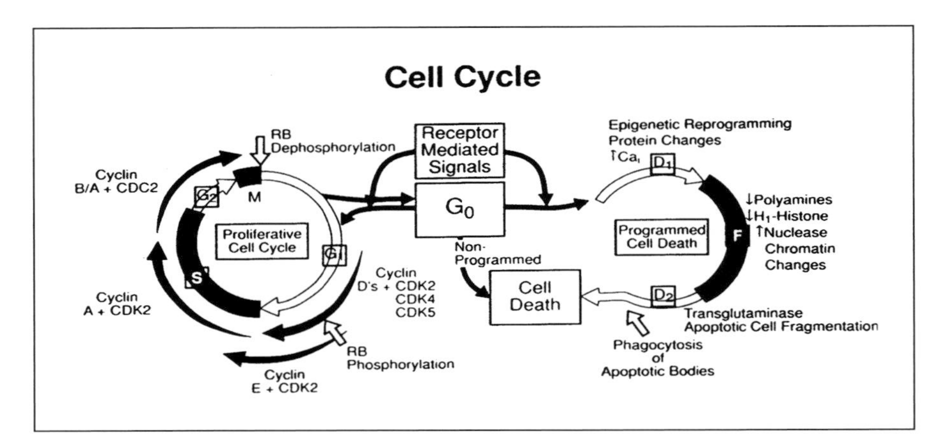

Fig. 9.1. Revised cell cycle denoting the options of a G_0 prostatic glandular cell. D1 denotes the period during which new gene and protein expression required for induction of the DNA fragmentation period (denoted F phase) occurs as part of the programmed cell death pathway. D2 denotes the period during which the cell itself fragments into apoptotic bodies as part of its programmed death.

of DNA fragmentation (the F phase) (Fig. 9.1) can be used to divide the temporal series of events involved in PCD, much as the period of DNA synthesis (the S phase) is used to divide the proliferative cell cycle. The overall cell cycle controlling cell number is thus composed of a multi–compartment system in which the cell has at least three possible options (Fig. 9.1). The cell can be: (a) metabolically active but not undergoing either proliferation or death (G_0 cell); (b) undergoing cell proliferation (G_0 to mitosis); or (c) undergoing cell death by either the programmed pathway (G_0–D1–F–D2 apoptotic cellular fragmentation) or the nonprogrammed (necrotic) pathway [7]. The endogenous systemic and local growth factor signals that regulate the progression within this cell cycle are cell type specific and are uniquely determined as part of the differentiated phenotype of the particular cell. Thus the same growth factor (e.g., $TGF\beta_1$) can have either agonistic or antagonistic effects within the cell cycle for different cell types. Therefore the specific details of the regulatory pathway for the cell cycle vary between different cell types.

Programmed Cell Death of Normal Prostatic Glandular Cells Following Androgen Ablation

In the normal adult prostate, the epithelial cells are continuously turning over with time [8, 9]. In this self–renewing condition, the rate of prostatic cell death is balanced by an equal rate of prostatic cell proliferation such that neither involution nor overgrowth of the gland normally occurs [8, 9]. If an adult male is castrated, the serum testosterone level rapidly decreases to below a critical value [5, 10]. As a result, the prostate rapidly involutes due to a major loss in the glandular epithelial cells but not the basal epithelial or stomal cells of the prostate [11]. Only the glandular epithelial cells are androgen dependent and undergo PCD following castration [11]. The chronic requirements for androgen by the glandular epithelia is due to the fact that androgens can act as agonists and antagonists by simultaneously stimulating the rate of cell proliferation while inhibiting the rate of cell death [5, 8].

In the ventral prostate of an intact adult rat, glandular cells contain androgen receptor [12] and constitute approximately 80% of the total cells [12]. Approximately 70% of these glandular cells die by 7 days postcastration [11]. Using the ventral prostate of the rat as a model system, the temporal sequence of events involved in the programmed cell death pathway induced by androgen ablation has begun to be defined. In the androgen–maintained ventral prostate of an intact adult male rat, the rate of cell death is very low, approximately 2% per day, and this low rate is balanced by an equally low rate of cell proliferation, also 2% per day [8, 9]. If animals are castrated, the serum testosterone level drops to less than 10% of the intact control value within 2 hours [5]. By 6 hours postcastration the serum testosterone level is only 1.2% of intact control [5]. By 12 to 24 hours following castration, the prostatic dihydrotestosterone (DHT) levels (i.e., the active intracellular androgen in prostatic cells) are only 5% of intact control values. This lowering of prostatic DHT leads to changes in nuclear androgen receptor function (i.e., by 12 hours after castration, androgen receptors are no longer retained in biochemically isolated ventral prostatic nuclei) [5]. While the lowering of prostatic DHT and resultant androgen receptor changes are maximal by 24 h postcastration, the programmed death of the prostatic glandular cells occurs continuously during the first two weeks postcastration.

These observations demonstrate that the reduction of occupancy of the androgen receptor by DHT is not sufficient alone to activate PCD of the glandular cells. Likewise, the temporally asynchronous nature of the death demonstrates that activation of programmed death of glandular cells is initiated when some other cellular survival factor besides DHT, whose level is regulated by DHT, decreases to a critical level. An excellent candidate for such a DHT dependent survival factor is the andromedin peptide factor, keratinocyte growth factor, normally produced and secreted by prostatic stromal cells under the stimulation of androgen [13]. Once the level of such peptide survival factors decreases to below a critical level within a particular glandular cell, a major epigenetic reprogramming of this cell occurs, resulting in the activation phase (D1) of the programmed death pathway (Fig. 9.1).

During this D1–activation phase, (D1a phase) (Fig. 9.2), the earliest events that can be seen upon androgen withdrawal are inhibition of glandular cell proliferation [14] coupled with a generalized atrophy of these secretory cells in individual acini [15, 16] (Fig. 9.2). Universally, tall columnar secretory cells rapidly shrink and become cuboidal in shape within 24 hours of androgen deprivation. Concurrent with these global morphological changes is the initial downregulation of a series of proteins (to be described later). At this stage the process is completely reversible simply by replacement of exogenous androgen [17, 18]. After this point, individual cells stochastically enter the D1b phase (Fig. 9.2) during which the activated cells morphologically round up and undergoes changes in its nuclear chromatin structure. During this phase, a series of proteins become upregulated and polyamine levels decrease [19]. An increase in intracellular calcium levels also occurs that appears to be derived from extracellular pools [20, 21]. The mechanism for the induced change in intracellular calcium is not fully known; however, there are indications that enhanced expression of TGF–β1 mRNA and protein [22] as well as the receptor for TGF–β1 [22] following castration are somehow involved.

With continued androgen deprivation, prostatic glandular epithelial cells undergo a further series of changes that result in an irreversible progression through the programmed cell death pathway. During the D1b phase (Fig. 9.2), Ca^{2+}/Mg^{2+} dependent endonuclease present within the nuclei of the prostatic glandular cells are enzymatically activated [20]. Levels of both histone H_1 and polyamines are decreased during this D1b phase [23, 24]. Both are involved in maintaining DNA compaction [25, 26] and decrease in their respective levels allows for opening of the genomic DNA conformation in the ladder region between nucleosomes. Once this occurs, the cell progresses through the F phase (Fig. 9.2) and DNA fragmentation begins at sites located between nucleosomal units and cell death is no longer reversible. Recent unpublished studies using inverted pulse–gel electrophoresis have demonstrated the initial DNA fragmentation produces 300–50 kb size DNA pieces. Once formed, these 300–50 kb size pieces are further degraded into nucleosomal size pieces (i.e., > 1Kb). During the F–phase the plasma and lysosomal membranes are still intact and mitochondria are still functional [11].

Subsequent to the F phase, the laminins in the nuclear membrane are degraded and the nucleus itself undergoes fragmentation. During this D2a phase, proteases are activated, including the ICE–like protease that hydrolyzes

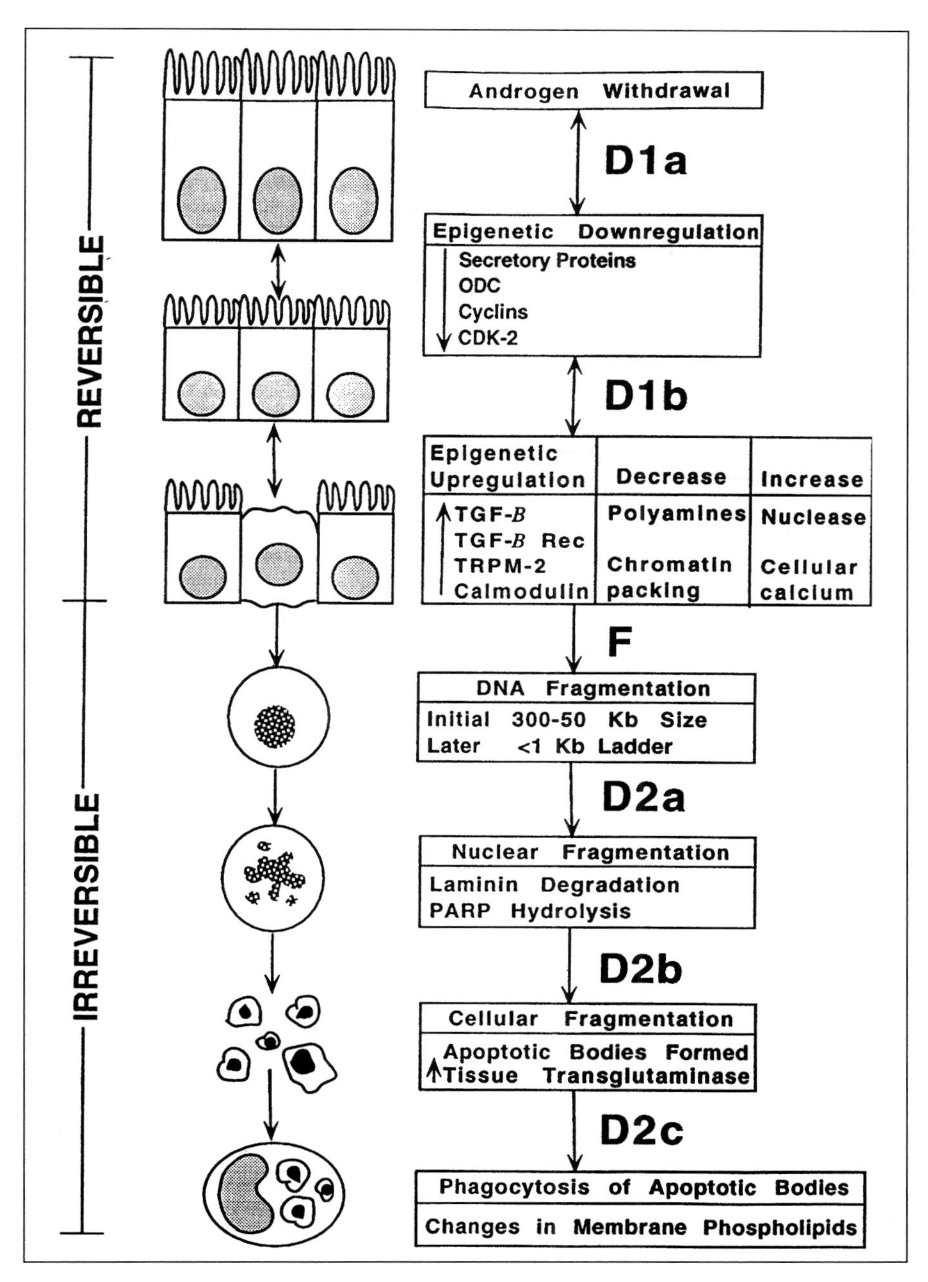

Fig. 9.2. Schematic diagram of the biochemical and morphological events occurring during the different phases of programmed death of normal prostatic glandular epithelium and prostatic cancer cells (see text for specific details).

Table 9.1. Epigenetic response in the rat ventral prostate during glandular cell proliferation/death induced by androgen manipulation [31–38]

Genes	Changes in mRNA Expression During	
	Proliferation	Programmed Cell Death
Ornithine decarboxylase	Induced	Decreased
Thymidine kinase	Induced	Decreased
H_4–histone	Induced	Decreased
c–*fos*	Induced	Decreased
Glucose–reg protein 78 kDa.	Induced	Decreased
Cyclin C	Induced	Decreased
Cyclin D_1	Induced	Decreased
Cyclin E	Induced	Decreased
DNA polymerase α	Induced	Decreased
C_3–prostatein	Restored	Decreased
TRPM–2	Repressed	Induced
$TGF\beta_1$	Repressed	Induced
Calmodulin	Decreased	Induced
α–prothymosin	Decreased	induced
c–*myc*	Induced	Induced
H–*ras*	Induced	Induced
Tissue transglutaminase	Induced	Induced

poly(ADP–ribose) polymerase (PARP) [27, 28]. Subsequent to the nuclear fragmentation, plasma membrane blebbing and cellular fragmentation into clusters of membrane bound apoptotic bodies occur. This D2b phase involves an upregulation in the Ca^{2+}–dependent tissue transglutaminase activity which cross-links various membrane proteins [29]. Once formed, these apoptotic bodies are rapidly phagocytized, during the D2c phase, by macrophages and/or neighboring epithelial cells [11, 15]. This phagocytosis is induced by changes in the membrane phospholipids in the apoptotic cell and cell bodies that is recognized by the phagocytic cells [30]. Thus, within 7–10 days post castration 80% of the glandular epithelial cells die and are eliminated from the rat prostate [11].

Prostate Gene Expression During PCD Pathway Induced by Androgen Ablation

The expression of a series of genes are upregulated during the D1b phase of programmed death by prostatic glandular cells induced by castration (Table 9.1) [31–36]. TRPM–2 [37], calmodulin [38] and tissue transglutaminase [29] have also previously been demonstrated to be induced in a variety of other cell types undergoing programmed cell death. At the same time, several of the genes (i.e., c–*myc*, H–*ras)* previously have been demonstrated to be involved in cell proliferation. Thus, as a comparison, the relative level of expression of these same genes was determined during the androgen–induced proliferation regrowth of the involved prostate in animals previously castrated 1 week before beginning androgen replacement. These comparative results demonstrate that the expression of *c–myc*, H–*ras, and* tissue transglutaminase are enhanced in both prostatic cell death and proliferation [36]. In contrast, the expression of calmodulin, TRPM–2, TGFβ_1 [36], glutathione S–transferase subunit Yb_1 [33] and α–prothymosin [36] are enhanced only during prostatic cell death and not prostatic cell proliferation.

Additional analysis demonstrated that the expression of a series of genes are decreased during the D1a phase following castration (Table 9.1). For example, the C_3 subunit of the prostatein gene (i.e., the major secretory protein of the glandular cells), ornithine decarboxylase (ODC), histone–H_4, p53, and glucose regulated protein 78 all decrease following castration [36]. In contrast to the decrease in the mRNA expression of these latter genes during programmed cell death in the prostate following castration, the expression of each of these genes is enhanced during the androgen induced prostatic cell proliferation [36].

Cell Kinetics During Progression of Prostate Cancer

Growth of a cancer is determined by the relationship between the rate of cell proliferation and the rate of cell death. Only when the rate of cell proliferation is greater than cell death does tumor growth continue. If the rate of cell proliferation is lower than the rate of cell death, then regression of the cancer occurs. Metastatic prostate cancers, like the normal prostates from which they arise, are sensitive to androgenic stimulation of their growth. This is due to the presence of androgen dependent prostatic cancer cells within such metastatic patients. These cells are androgen dependent since androgen stimulates their daily rate of cell proliferation (i.e., Kp) while inhibiting their daily rate of death (i.e., Kd) [19]. In the presence of adequate androgen, continuous net growth of these dependent cells occurs since their rate of proliferation exceeds their rate of death. In contrast, following androgen ablation, androgen dependent prostatic cancer cells stop proliferating and activate programmed cell death [19]. This activation results in the elimination of these androgen dependent prostatic cancer cells from the patient since under these conditions their death rate value now exceeds their rate of proliferation. Due to this elimination, eighty to ninety percent of all men with metastatic prostatic cancer treated with androgen ablation therapy have an initial positive response. Eventually, all of these patients relapse to a state unresponsive to further anti–androgen therapy, no matter how completely given [39]. This is due to the heterogeneous presence of androgen independent prostatic cancer cells within such metastatic patients. These latter cells are androgen independent since their rate of proliferation exceeds their rate of cell death even after complete androgen blockage is performed [40].

Attempts to use nonandrogen ablative chemotherapeutic agents to adjust the kinetic parameters of these androgen independent prostatic cancer cells so that their rate of death exceeds their rate of proliferation have been remarkable in their lack of success [41]. The agents tested in patients failing androgen ablation have been targeted at inducing DNA damage directly or indirectly via inhibition of DNA metabolism or repair. These agents are thus critically dependent upon an adequate rate of proliferation to be cytotoxic [42]. In vitro cell culture studies have demonstrated that when androgen independent, metastatic, prostatic cancer cells are rapidly proliferating (i.e., high Kp value), these cells are highly sensitive to the induction of programmed cell death via exposure to the same anti–proliferative chemotherapeutic agent which are of limited value when used in vivo in prostatic cancer patients [43]. The paradox between the in vitro and in vivo responsiveness to the same chemotherapeutic agents by androgen independent prostatic cancer cells is due to major differences in the rate of proliferation occurring in the two states. Likewise, for chemotherapeutic agents to be effective, not only must the cancer cells have a critical rate of proliferation but also a critical sensitivity to induction of cell death [44]. The sensitivity to induction of cell death is reflected in the magnitude of the rate of cell death in the untreated condition.

The daily rates of cell proliferation (i.e., Kp) and cell death (i.e., Kd) were determined for normal, premalignant, and cancerous prostatic cells within the prostate as well as for prostatic cancer cells in lymph node, soft tissue and bone metastases from untreated and hormonally failing patients [9]. These data demonstrate that normal prostatic glandular cells have an extremely low (i.e., $< 0.20\%$ per day), but balanced, rate of cell proliferation and death producing a turnover time of 500 ± 79 day for these cells. Initial transformation of these cells into high grade intraepithelial neoplasia (PIN), the lesion believed to be the precursor for prostate cancer, results in an increased Kp value with no change in the Kd value. As these early lesions continue to grow into late stage high grade PIN, their Kd increases to a point equaling Kp. This results in cessation of net growth while inducing a six–fold increase in the turnover time (i.e., 56 ± 12 days) of these cells increasing their risk of further genetic changes. The transition of late stage high grade PIN cells into growing localized prostatic cancer cells involves no further increase in Kp but is due to a decrease in Kd resulting in a mean doubling time of 479 ± 56 days. Metastatic prostatic cancer cells within lymph nodes of untreated patients have a 100% increase in their Kp and 40% decrease in their

Kd values as compared to localized prostatic cancer cells producing a mean doubling time of 33 ± 4 days. Metastatic prostatic cancer cells in the bony untreated patients have a 36% increase in Kp and a 50% decrease in Kd, resulting in a mean doubling time of 54 ± 5 days. In hormonally failing patients, there is no further change in Kp. An increase in the Kd for androgen independent prostatic cancer cells is observed within soft tissue or bone metastases with resulting mean doubling times of 126 ± 21 and 94 ± 15 days, respectively, in these metastatic sites. These data demonstrate that the proliferation rate for androgen independent metastatic prostatic cancer cells is very low (i.e., < 3.0%/day) [9], explaining why anti–proliferative chemotherapy has been of such limited value against metastatic prostatic cells. Based upon this realization, what is needed is some type of cytotoxic therapy which induces the death of androgen–independent prostate cancer cells without requiring the cells to proliferate.

Role of Cell Proliferation in PCD Process Induced by Castration

Using standard in vivo ^{3}H–thymidine pulse labeling, the percent of glandular cells entering the S–phase during the period of enhanced prostatic cell death occurring during the first week postcastration was determined. Within 1 day following castration, there is an 80% decrease ($p < 0.05$) in the percent of glandular cells entering S–phase. By 4 days following castration, there is more than a 90% reduction in this value. Comparing the data demonstrates that greater than 98% of prostatic glandular cells die following castration without entering the proliferative cell cycle. These results confirm the previous studies of Stiens and Helpap [45] and Evans and Chandler [46] which likewise demonstrated a decrease in the percent of prostatic glandular cells in S–phase following castration.

During PCD activated by castration double stranded DNA fragmentation of genomic DNA occurs and induces a futile process of DNA repair while cells remain in G_0. This futile process of G_0 DNA repair has been shown to only be associated with, but not causally required for prostatic cell death. This was demonstrated by treating intact male rats with tridaily hydoxyurea for one week which inhibits both prostate specific DNA synthesis and unscheduled G_0 DNA repair by more than 90% for 8 h following an IP injection [7]. Castration of these rats resulted in similar reductions in DNA content and identical glandular morphologic changes as compared to untreated, castrated controls. These results confirm that PCD of prostatic glandular cells induced by androgen ablation does not require progression through S phase or G_0 DNA repair.

To determine whether androgen ablation induced PCD of prostatic glandular cells involves recruitment of nonproliferating cells into early portion of G_1 of a perturbed proliferative cell cycle, rat ventral prostates were assessed temporally following castration for several stereotypical molecular stigmata of entry into the proliferative cell cycle [47]. Northern blot analysis was used to assess levels of transcripts from genes characteristically activated: (1) during the transition from quiescence (G_0) into G_1 of the proliferative cell cycle (cyclin D_1, and cyclin C); (2) during the transition from G_1 to S (cyclin E, cdk2, thymidine kinase, and H4 histone); and 93) during progression through S (cyclin A). While levels of each of these transcripts increased as expected in prostatic glandular cells stimulated to proliferate by administration of exogenous androgen to previously castrated rats, levels of the same transcripts decreased in prostatic glandular cells induced to undergo PCD following androgen withdrawal [47]. Likewise, androgen ablation induced PCD of prostatic glandular cells was not accompanied by retinoblastoma (Rb) protein phosphorylation characteristic of progression from G_1 to S. This is consistent with a decrease in the number of cells entering S cells using ^{3}H–thymidine radioautography. Nuclear run on assays demonstrated that there is no increase in the prostatic rate of transcription of the c–*myc* and c–*fos* genes following castration. Northern and Western blot analysis also demonstrated that there is no increase in the prostatic p53 mRNA or protein content per cell following androgen ablation. Likewise, following castration there is no enhanced prostatic expression of the WAF1/CIP1 gene, a gene whose expression is known to be induced by either increased p53 protein levels or entrance into G_1 [47]. Furthermore, when compared to wild type, p53 deficient, knockout mice showed no difference in the extent of programmed cell death of androgen dependent cells following castration [48]. These results demonstrate that prostatic glandular cells undergo PCD in G_0 without recruitment into G_1 phase of a defective cell cycle and that p53 protein or its function is not required for this death process to occur [7, 47].

Therapeutic Implication of PCD For Prostatic Cancer

Androgen–independent prostatic cancer cells do not undergo programmed cell death following androgen ablation. These cells are unable to initiate the apoptotic cascade when androgens are removed [49]. Even with this defect, however, androgen–independent prostatic cancer cells retain the basic cellular machinery to undergo this programmed cell death pathway. This was demonstrated by using a variety of chemotherapeutic agents which arrest proliferating androgen independent prostatic cancer cells in various phases of the proliferative cell cycle (e.g., G_1, S, or G_2) and which subsequently induce their programmed (i.e., apoptotic) death [49].

Elevations in intracellular calcium may play a role in triggering the PCD cascade upon androgen withdrawal [21]. Androgen–independent prostatic cancer cells may not undergo PCD secondary to androgen ablation because such withdrawal does not induce an elevation of intracellular Ca^{2+} in these cells. To test this possibility, androgen–independent, highly metastatic Dunning R–3327 AT–3 rat prostatic cancer cells were chronically exposed in vitro to varying concentrations of the calcium ionophore ionomycin to sustain various levels of elevation in the their Ca_i [50]. These studies demonstrated that an elevation of Ca_i from a starting value of 35 nM to a value as small as only 3–fold above baseline (i.e., 100 nM) while not inducing immediate toxicity (i.e., death within ≤ hours) can induce the death of the cells if sustained for > 12 hours. Temporal analysis demonstrated that elevation in Ca_i results in these cells arresting in G_0 within 6–12 hours following ionomycin exposure. Over the next 24 hours, these cells begin to fragment their genomic DNA initially into 300–50 Kb size pieces which are further degraded into nucleosome–sized pieces and during the next 24–48 hours these cells undergo cellular fragmentation in apoptotic bodies [50]. Associated with this programmed cell death is an epigenetic reprogramming of the cell in which the expression of a series of genes (to be presented later) is specifically modified. These results demonstrate that even nonproliferating androgen–independent prostatic cancer cells can be induced to undergo programmed cell death if a modest elevation in the intracellular free Ca^{2+} is sustained for a sufficient time. Combining these latter ionomycin data with the chemotherapy data demonstrates that programmed death of androgen independent prostatic cancer cells can be induced in any phase of the cell cycle and does not necessarily require progression through the proliferation cell cycle (i.e., proliferation independent).

A second agent capable of increasing intracellular Ca^{2+} (Ca_i) to induce PCD, and one that could potentially be useful as therapy, is thapsigargin [51, 52], a sesquiterpene γ–lactone isolated from the root of the umbelliferous plant, *Thapsia garganica* [53, 54]. Studies have demonstrated that the Ca^{2+} dependence for TG effects is due to the fact that this highly lipophilic agent enters cells and interacts with the Ca^{2+}–ATpase present in the endoplasmic reticulum (ER) and inhibits its enzymatic activity with an IC_{50} value of 30 nM [55]. This produces an elevation in the Ca_i which is sustainable (i.e., min–hours) if the TG inhibition is maintained [55]. Based upon this background, the ability of thapsigargin to activate programmed cell death in androgen independent prostate cancer cells was tested.

Chronic exposure of each of the four distinct androgen independent prostatic cancer cell lines to 500 nM thapsigargin was found to arrest these cells in the G_0/G_1 phase of the cell cycle. This G_0/G_1 arrest was complete by 24 hours of continuous 500 nM TG exposure. It was further demonstrated that after a 24 hour lag period, the cells begin to fragment their DNA (i.e., to sizes ≤ 300 Kb) and that by 96 hours of treatment ≥ 95% of the cells have fragmented their DNA regardless of cell line tested [52]. The temporal pattern of DNA fragmentation was tightly correlated with the loss of clonogenic ability by the cells for each of the four cell lines (i.e., 72 hours of TG treatment required for 50% of the cells to fragment their DNA and 50% loss of their clonogenic ability) [52]. Time–lapse videomicroscopy studies demonstrated that morphological changes begin occurring within 3–6 hours of initial TG exposure. By 24 hours of TG treatment, cells are smaller in size and rounded in morphology. Between 72–120 hours TG treatment, the cells undergo a period of plasma membrane hyperactivity characterized by the production of plasma membrane blebbing [52]. These surface blebs are highly dynamic, coming and going on the surface and giving appearance of the membrane boiling previously reported for ionomycin induced programmed cell death of AT–3 prostatic cancer cells [21]. These combined results demonstrate that the initiation of DNA fragmentation is occurring in viable nonproliferating (i.e., G_0/G_1) cells from each of the four distinct androgen independent prostatic cancer cell lines tested, 24–48 hours before these cell lyse and that this DNA fragmentation is not the result of a loss of metabolic viability (i.e., loss of mitochondrial or plasma membrane function). In contrast, the data are consistent with the initiation of DNA fragmentation as the irreversible commitment step in the TG induced programmed death of nonproliferating androgen independent rodent or human prostatic cancer cells.

Analysis of a mRNA expression of the series of genes previously demonstrated to be enhanced during the programmed cell death of normal prostatic cells induced by androgen ablation demonstrated that TG treatment of androgen independent prostatic cancer cells likewise leads to an epigenetic reprogramming of the cells. AT–3 rat prostatic cancer cells were treated from 0–36 hours with either 500 nM TG, 10 μM ionomycin, or 100 μM 5–fluordeoxyuri–dine (5–FrdU). Previously, we have demonstrated that prostatic cancer cells must progress through the proliferative cell cycle in order for 5–FrdU to induce their programmed cell death [49]. In contrast, TG and ionomycin induce the proliferation independent programmed death of G_0 cells. These results demonstrate that within 1 hour of either TG or ionomycin treatment, expression of several of these genes is already elevated (e.g., α–prothymosin, calmodulin, ornithine decarboxylase (ODC)) and that by 6 hours additional genes expression is enhanced (e.g., glucose–regulated protein–78 (GRP), c–*myc*). Many of these enhancements are acute with expression decreasing at 24 hours of treatment. There are major differences in gene expression during the proliferation independent programmed death induced by TG or ionomycin and the proliferation dependent death induced by 5–FrdU (e.g., in the latter, c–*myc*, calmodulin, prothymosine are not induced while H–*ras* and TRPM–2 are induced) [51, 52]. These results demonstrate that the programmed death induced by all of these agents involve an active epigenetic reprogramming of the cell and the pathway induced by TG is essentially identical to that induced by ionomycin, but distinct from that induced by 5–FrDU.

Thapsigargin treatment of rapidly proliferating androgen–independent prostatic cancer cells arrests such cells in G_0 and induces their programmed death. This raises the issue of whether this programmed cell death is simply a response to rapid G_0 growth arrest, making the response still dependent upon the rate of cell proliferation. If this were the case, TG would be another type of antiproliferative agent, albeit with a unique mechanism of action. In an attempt to resolve this issue, androgen–independent AT–3 rat prostatic cancer cells were co–treated with TG and the intracellular free Ca^{2+} chelator BAPTA to prevent the rise in Ca_i normally induced by TG [56]. This co–treatment prevented growth arrest of these cells as monitored by DNA flow cytometry and failure to induce mRNA and protein for the *gadd*153 growth arrest gene. Such failure to induce growth arrest did not prevent the co–treatment from inducing programmed cell death of these cells monitored by morphological criteria or clonogenic survival assays [56].

In addition, human prostatic cells were established in serum free defined medium as primary cultures. After an initial proliferation period lasting ~10 days, these cells go out of cell cycle and enter the proliferatively quiescent G_0 state. This was demonstrated by determining the growth fraction of these cultures 10 days postplating using immunoflowcytometeric analysis to determine the percentage of cells expressing the Ki–67 antigen. This Ki–67 antigen is expressed by human cells in all phases of the cell cycle except G_0 [57]. Based upon these analysis, 2.1 ± 0.5% of the primary culture cells express Ki–67 and are thus in cycle at 3 weeks post plating (i.e., of ~ 98% are out of cycle in a proliferatively quiescent state of G_0). That these cells are in G_0 was further demonstrated by videomicroscopic observation that when such stationary primary cultures are scrape wounded to allow an unoccupied culture surface, cells at the edge of the wound migrated slowly to fill the void over the next 5–7 days. During this period however, no mitotic figures, nor enhanced incorporation of ^{14}C–thymidine, nor enhanced expression of Ki–67 in the region of the wound were observed. These results demonstrated that while these cells remain viable, they are nearly exclusively out of cycle in G_0 and that this G_0 state is not due simply to contact inhibition [56].

Such stationary G_0 cultures can be maintained for more than 6 weeks with the spontaneous rate of cell death being 2.1 ± 5% per day. When stationary cultures which have been maintained for 3 weeks were exposed continuously for 1 week to an effective dose (i.e., 0.1 μM) of a cell proliferation dependent chemotherapeutic agent like 5–FrdU, there was no activation of programmed cell death determined either morphologically by videomicroscopic evaluation or quantitatively by DNA fragmentation (i.e., 14 ± 3% of ^{14}C–DNA was fragmented over 1 week observation period in control cultures vs. 15 ± 5% in culture maintained for 1 week in the presence of 5–FrdU). In contrast, exposure of such 21 day old stationary cultures to 500 nM TG results in morphological changes (i.e., cell rounding and membrane boiling) within 24–48 hours with fragmentation of cells into apoptotic bodies starting by day 4 of TG exposure. Quantitatively, 85 ± 15% of ^{14}C–DNA was fragmented within 1 week of TG exposure vs. 14 ± 3% being fragmented during 1 week in control cultures [56].

In conclusion, these studies demonstrate that TG induces the arrest of proliferating androgen independent prostatic cancer out of cycle in G_0 and their subsequent programmed death. TG induced growth arrest is not, however, required for programmed death of these cells. Additionally it has been demonstrated that it is possible for TG to induce the programmed death of proliferatively quiescent G_0 human prostatic cancer cells with no requirement for cell cycle entrance or progression. Thus, TG possess the ability to induce the programmed death of prostatic cancer cells in a totally cell proliferation independent manner. Based upon these findings, TG analogs warrant clinical evaluation for the treatment of androgen ablation failing patients with metastatic prostatic cancer.

References

1 Wingo PA, Tong T, Bolden S. Cancer statistics, 1995. CA Cancer J Clin 1995; 45:8–30.

2 Kerr JFR, Wyllie AG, Currie AR. Apoptosis: a basic biological phenomenon with wide ranging implications in tissue kinetics. Br J Cancer 1972; 26:239–257.

3 Bowen ID, Lockshin RA. Cell Death in Biology and Pathology. London:Chapman and Hall, London, 1981.

4 Wyllie AH. Glucocorticoid induces in thymocytes a nuclease–like activity associated with the chromatin condensation of apoptosis. Nature 1980; 284:555–556.

5 Kyprianou N, Isaacs JT. Activation of programmed cell death in the rat ventral prostate after castration. Endocrinology 1988; 122:552–562.

6 Umansky SR, Korol BA, Nelipovich PA. In vivo DNA degradation in thymocytes of γ–irradiated or hydrocortisone–treated rats. Biochim Biophys Acta 1981; 655:9–17.

7 Berges RS, Furuya Y, Remington L, English HF, Jacks T, Isaacs, JT. Cell proliferation, DNA repair and p53 function are not required for programmed death of prostatic glandular cells induced by androgen ablation. Proc Natl Acad Sci USA 1993; 90:8910–8914.

8 Isaacs JT. Antagonistic effect of androgen on prostatic cell death. The Prostate 1984; 5:545–557.

9 Berges RS, Vukanovic J, Epstein JI, Carmichel M, Cisek L, Johnson DE, Veltri RW, Walsh PC, Isaacs JT. Implication of the cell kinetic changes during the progression of human prostatic cancer. Clin Cancer Res 1995; 1:473–480.

10 Kyprianou N, Isaacs JT. Quantal relationship between prostatic dihydrotestosterone and prostatic cell content: critical threshold concept. The Prostate 1987; 11:41–50.

11 English HF, Kyprianou N, Isaacs JT. Relationship between DNA fragmentation and apoptosis in the programmed cell death in the rat prostate following castration. Prostate 1989; 15:233–251.

12 Prins GS, Birch L, Greene GL. Androgen receptor localization in different cell types of the adult rat prostate. Endocrin 1991; 1229:3187–3199.

13 Yan G, Fukabori Y, Nikolaropoulus S, Wang F, McKeehan WL. Heparin–binding keratinocyte growth factor is a candidate stromal to epithelial cell andromedin. Mol Endocrinol 1992; 6:2123–2128.
14 Kyprinaou N, Isaacs JT. Biological significance of measurable androgen levels in the rat ventral prostate following castration. The Prostate 1987; 10:313–324.
15 Kerr JFR, Searle J. Deletion of cells by apoptosis during castration–induced involution of the rat prostate. Virchows Arch (Cell Pathol) 1973; 13:87–102.
16 English HF, Drago JR, Santen RJ. Cellular response to androgen ablation and repletion in the rat ventral prostate: autoradiography and morphometric analysis. The Prostate 1985; 7:41–51.
17 Coffey DS, Shimazaki J, Williams–Ashman HG. Polymerization of deoxyribonucleotides in relation to androgen–induced prostatic growth. Arch Biochem Biophys 1968; 124:184–198.
18 Lesser B, Bruchovsky N. The effects of testosterone, 5a–dihydrotestosterone and adenosine 3',5'–monophosphate on cell proliferation and differentiation in rat prostate. Biochem Biophys Acta 1973; 308:427–437.
19 Isaacs JT, Lundmo PI, Berges R, Martikainen P, Kyprianou N, English, HF. Androgen regulation of programmed death of normal and malignant prostatic cells. J Andrology 1992; 13:457–464.
20 Kyprianou N, English HF, Isaacs JT. Activation of a Ca^{2+}–Mg^{2+}–dependent endonuclease as an early event in castration–induced prostatic cell death. Prostate 1988; 13:103–117.
21 Martikainen P, Isaacs JT. Role of calcium in the programmed death of rat prostatic glandular cells. The Prostate 1990; 17:175–187.
22 Kyprianou N, Isaacs JT. Identification of a cellular receptor for transforming growth factor–β in rat ventral prostate and its negative regulation by androgens. Endocrinol 1988; 123:2124–2131.
23 Pegg AE, Lockwood DH, Williams–Ashman HG. Concentrations of putrescine and polyamines and their enzymic synthesis during androgen–induced prostatic growth. Biochem J 1970; 117:17–31.
24 Chung LWK, Coffey DS. Biochemical characterization of prostatic nuclei I. androgen–induced changes in nuclear proteins. Biochimica Biophysica Acta 1971; 247:570–583.
25 Synder RD. Polyamine depletion is associated with altered chromatin structure in HeLa cells. Biochem J 1989; 260:697–704.
26 Brüne B, Hartzell P, Nicotera P, Orrenius S. Spermine prevents endonuclease activation and apoptosis in thymocytes. Exp Cell Res 1991; 195:323–329.
27 Lazebnik YA, Kaufman SH, Desnoyers S, Poirer GG, Earnshaw WC. Cleavage of poly(ADP–ribose) polymerase by a proteinase with properties like ICE. Nature 1994; 371:346–347.
28 Nicholson DW, Ali A, Thornberry NA et al. Identification and inhibition of the ICE/CED–3 protease necessary for mammalian apoptosis. Nature 1995; 376:37–43.
29 Fesus L, Thomazy V, Falus A. Induction and activation of tissue transglutaminase during programmed cell death. FEBS Lett 1989; 224:104–108.
30 Savill JS, Dransfield I, Hogg N, Haslett C. Vitronectin receptor–mediated phagocytosis of cells undergoing apoptosis. Nature 1990; 343:170–173.
31 Quarmby VE, Beckman WC Jr, Wilson EM, French FS. Androgen regulation of c–*myc* messenger ribonucleic acid levels in rat ventral prostate. Mol Endocrinol 1987; 1:865–874.
32 Buttyan R, Zakeri Z, Lockshin R, Wolgemuth D. Cascade induction of c–*fos*, c–*myc* and heat shock 70K transcripts during regression of the rat ventral prostate gland. Mol Endocrinol 1988; 2:650–657.
33 Chang C, Saltzman AG, Sorensen NS, Hiipakka RA, Liao S. Identification of glutathione S–transferase Yb_1 mRNA as the androgen repressed mRNA by cDNA cloning and sequence analysis. J Biol Chem 1987; 262:11901–11903.
34 Montpetit ML, Lawless KR, Tenniswood M. Androgen repressed messages in the rat ventral prostate. The Prostate 1986; 8:25–36.
35 Kyprianou N, Isaacs JT. Expression of transforming growth factor–β in the rat ventral prostate during castration induced programmed cell death. Mol Endocrinol 1989; 3:1515–1522.
36 Furuya Y, Isaacs JT. Differential gene regulation during programmed death (apoptosis) versus proliferation of prostatic glandular cells induced by androgen manipulation. Endocrinol 1993; 133:2660–2666.

37 Buttyan R, Olsson CA, Pintar J, Chang C, Bandyk M, Ng PY, Sawczuk IS. Induction of the TRPM–2 gene in cells undergoing programmed death. Mol Cell Biol 1989; 9:3473–3481.
38 Dowd DR, MacDonald PN, Komm BS, Haussler MR, Miesfeld R. Evidence for early induction of calmodulin gene expression in lymphocytes undergoing glucocorticoid–mediated apoptosis. J Biol Chem 1991; 266:18423–18426.
39 Crawford ED et al. A control randomized trial of leuprolide with and without flutamide in prostatic cancer. New Engl J Med 1989; 321:419–424.
40 Isaacs JT. Hormonally responsive *vs* unresponsive progression of prostatic cancer to antiandrogen therapy as studied with the Dunning R–3327–AT and G rat prostatic adenocarcinoma. Cancer Res 1982; 42:5010–5014.
41 Raghavan D. Non–hormone chemotherapy for prostate cancer: principles of treatment and application to the testing of new drugs. Seminars in Oncology 1988; 15:371–389.
42 Shackney SE, McCormack GW, Cuchural GJ. Growth rate patterns of solid tumors and their relationship to responsiveness to therapy. Annals of Internal Medicine 1978; 89:107–115.
43 Isaacs JT, Lundmo PI. Chemotherapeutic induction of programmed cell death in non proliferating prostate cancer cells. Proc Am Assoc Cancer 1992; Res 33:588–589.
44 Tubiana M, Malaise EP. Growth rate and cell kinetics in human tumors: some prognostic and therapeutic implications. In: Symington T, Carter RL, eds. Scientific Foundation of Oncology. Chicago: Year Book Medical Publishers, 1976:126–138.
45 Stiens R, Helpap B. Regressive changes in the prostate after castration. A study using histology, morphometrics and autoradiography with special reference to apoptosis. Pathol Res Pract 1981; 172:73–87.
46 Evans GS, Chandler JA. Cell proliferation studies in the rat prostate: II. The effects of castration and androgen replacement upon basal and secretory cell proliferation. The Prostate 1987; 11:339–352.
47 Furuya Y, Walsh JC, Lin X, Nelson WG, Isaacs JT. Androgen ablation induced programmed death of prostatic glandular cells does not involve recruitment into a defective cell cycle or p53 induction. Endocrinol 1995; 136:1898–1906.
48 Lowe S, Schmitt EM, Smith SW, Osborne BA, Jacks T. p53 is required for radiation induced apoptosis in mouse thymocytes. Nature 1993; 362:847–849.
49 Kyprianou N, Isaacs JT. Thymine–less death in androgen independent prostatic cancer cells. Biochem Biophys Res Commun 1989; 165:73–81.
50 Martikainen P, Kyprianou N, Tucker RW, Isaacs JT. Programmed death of non–proliferating androgen independent prostatic cancer cells. Cancer Res 1991; 51:4693–4700.
51 Furuya Y, Isaacs JT. Proliferation–dependent vs. independent programmed cell death of prostatic cancer cells involves distinct gene regulation. The Prostate 1994; 25:301–309.
52 Furuya Y, Lundmo P, Short AD, Gill DL, Isaacs JT. The role of calcium pH, and cell proliferation in the programmed (apoptotic) death of androgen–independent prostatic cancer cells induced by thapsigargin. Cancer Res 1994; 54:6167–6175.
53 Christensen SB, Norup E, Rasmussen U. Chemistry and structure–activity relationship of the histamine secretagogue Thapsigragin and related compounds. In: Krogsgaard–Larsen P, Christensen SB, Kofod H, eds. Natural Products and Drug Development. 1993:405–418.
54 Rasmussen U, Christensen SB, Sandberg F. Thapsigargin and thapsigargicin, two new histamine liberators from Thapsia garganica. L Acta Pharm Suec 1978; 15:133.
55 Thastrup O, Cullen PJ, Drbak BK, Hanley MR, Dawson AP. Thapsigargin, a tumor promoter, discharges intracellular Ca^{2+} stores by specific inhibition of the endoplasmic reticulum Ca^{2+}–ATPase. Proc Natl Acad Sci USA 1990; 87:2466–2470.
56 Lin XS, Denmeade SR, Cisek L, Isaacs JT. The role of growth arrest in programmed (apoptotic) death of prostatic cancer cells by thapsigargin. (in press).
57 Schwating R, Gerdes J, Jaeschke L, Stein H. Determination of the growth fraction in cells suspensions by flow cytometry using the monoclonal antibody Ki–67. J Immunol Methods 1986; 90:65–70.

Chapter 10

Apoptosis and Cancer, edited by Seamus J. Martin.

Cell Death Regulation in the Kidney and in Renal Neoplasms

Timothy J. McDonnell,[a] Stephen M. Hewitt[b] and Grady F. Saunders[b]

[a] Department of Molecular Pathology, The University of Texas M.D. Anderson Cancer Center, Houston, Texas, U.S.A.

[b] Department of Biochemistry and Molecular Biology, The University of Texas M.D. Anderson Cancer Center, Houston, Texas, U.S.A.

Introduction

The importance of apoptosis in the context of normal embryonic development has been recognized for many years. More recently it has been elegantly demonstrated by the targeted disruption of genes known to function as cell death regulators [1, 2]. The development of the kidney is dependent on complex interactions between cells of the ureteric bud endoderm and the undifferentiated metanephrogenic mesenchyme which will ultimately form the collecting system and nephrons, respectively, of the kidney. These interactions involve a reciprocal induction process in which the metanephrogenic mesenchyme is induced to condense and transform into epithelial structures known as renal vesicles. Conversely, this undifferentiated mesenchyme induces the ureteric bud to branch such that each newly forming nephron is associated with a corresponding renal ampulla and subsequent collecting system. This process commences during week 8–9 of gestation in humans and is ongoing throughout fetal development. A centripetal gradient with respect to differentiation is established within the metanephros during this time such that embryonic induction and formation of nascent glomeruli is occurring subjacent to the renal capsule while progressively more mature nephrons are localized more centrally within the forming kidney.

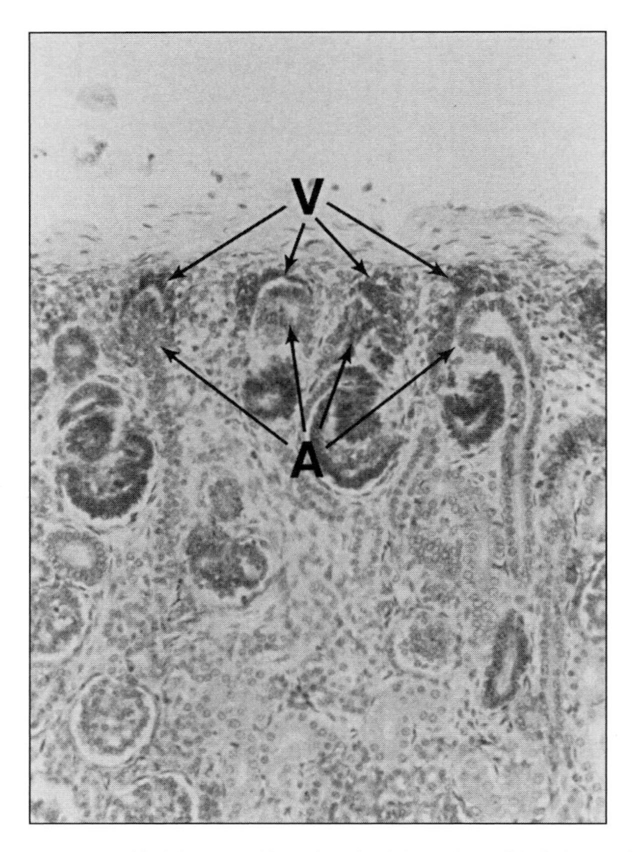

— *Fig. 10.1.* Immunohistochemical detection of *bcl–2* protein in fetal human kidney. Paraffin sections of formal–fixed fetal tissues, 11–20 weeks of gestational age, were immunohistochemically stained for *bcl–2* protein using the alkaline phosphatase anti–alkaline phosphatase (APAAP) technique and a monoclonal anti–*bcl–2* monoclonal antibody (Dako). Renal vesicles (V) are strongly immunoreactive for *bcl–2*. In contrast, renal ampulae (A) are *bcl–2* negative.

The Role of Bcl–2 in the Developing Kidney

The potential contribution of the expression of the *bcl–2* gene, an apoptosis suppressor oncoprotein, to the histogenesis of the kidney is suggested by the immunohistochemical distribution of Bcl–2 protein in fetal tissue [3–5]. Bcl–2 protein is undetectable within the undifferentiated, or uninduced, metanephragenic mesenchyme and the renal ampullae. Bcl–2 is first detected and most abundant in the neogenic, or metanephrogenic, zone of the renal cortex within epithelial–like condensations of metanephrogenic mesenchyme, the nascent renal vesicles

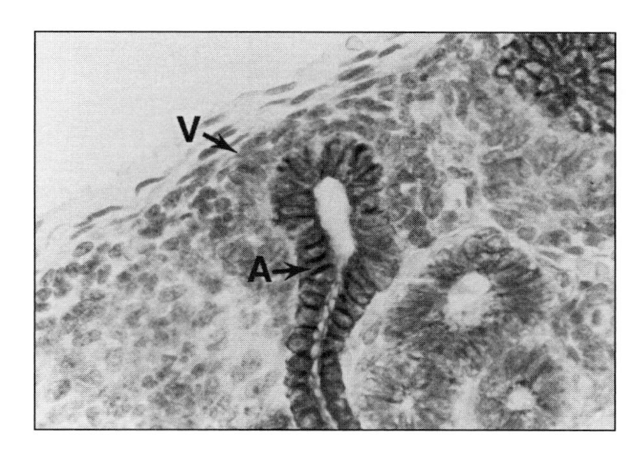

Fig. 10.2. Immunohistochemical detection of cytokeratin in the neogenic zone of the fetal human kidney. The renal ampulae (A) are immunoreactive for cytokeratins, whereas the nascent renal vesicles are cytokeratin negative.

(Fig. 10.1). Commitment to epithelial differentiation is thought to occur within 12 hours of induction and is associated with a marked increase in DNA synthesis and proliferation followed by epithelial differentiation [6]. During metanephrogenic mesenchyme induction the upregulation of Bcl–2 protein is an early event in that it occurs prior to the acquisition of epithelial specific protein markers such as low molecular weight cytokeratins (Fig. 10.2). These findings are consistent with the interpretation that the survival advantage imparted by Bcl–2 is necessary for the conversion of the induced metanephrogenic mesenchyme into the epithelial renal vesicle.

If this viewpoint is correct then the number of glomeruli, or nephrons, would be predicted to be significantly reduced in the absence of *bcl–2* expression. In this regard, it has recently been reported that within the developing kidney of *bcl–2* knock–out mice there is a 5–fold increase in the rate of apoptosis from E–13 to E–16 compared to control littermates [7]. The enhanced rate of apoptosis was associated with renal hypoplasia and an approximately 5–fold reduction in the number of nephrons in *bcl–2* nullizygous embryos by E–17 to E–19. Similar findings were reported in an independently derived *bcl–2* knockout mouse [8]. Additionally, in vitro studies of metanephroi isolated from *bcl–2*–deficient embryos suggest that this developmental defect is cell autonomous [8]. Postnatally, the *bcl–2*–deficient mice develop cystic kidneys soon after birth and succumb to renal failure [9–11].

Initially, the epithelium of the nascent glomerulus is strongly Bcl–2 positive; however, with progressive maturation *bcl–2* expression becomes limited to the parietal layer of Bowman's capsule. Similarly, the nephronic tubules are

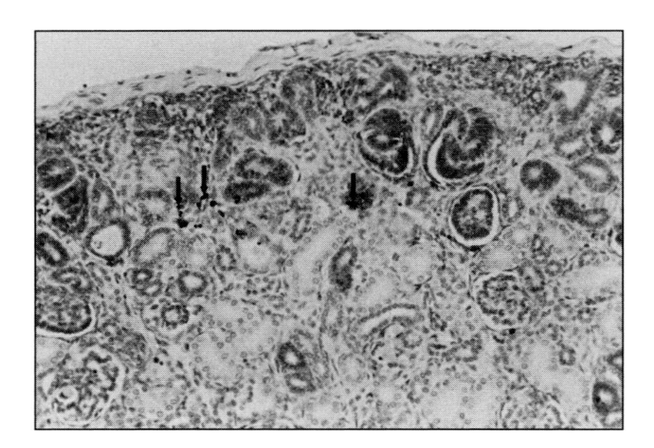

Fig. 10.3. In situ detection of apoptosis in the fetal human kidney. Fragmented DNA within apoptotic cells is end–labeled using the TUNEL technique [12]. Apoptotic cells (arrows) are identified by the presence of a red reaction product. The majority of the TUNEL positive cells are undifferentiated mesenchymal cells.

initially strongly Bcl–2 positive but with maturation become less intensely stained. During development, the collecting ducts exhibit only faint Bcl–2 staining and renal calyces are uniformly Bcl–2 negative.

Cells undergoing apoptotic cell death within the developing kidney can easily be identified using terminal deoxyribonucleotide transferase mediated nick end–labeling techniques, or TUNEL, to detect the presence of fragmented DNA [12]. Apoptotic cells in the developing kidney, in general, localize to an area of the neogenic zone subjacent to the renal ampullae and vesicles where morphogenesis has progressed from the S–phase to newly formed glomeruli (Fig. 10.3). The majority of apoptotic cells within this region of epithelial differentiation consist of Bcl–2 negative undifferentiated mesenchymal cells although occasional glomerular and tubular epithelial cells are also TUNEL positive.

Again, these findings suggest that Bcl–2 may be necessary to maintain the viability of committed mesenchymal cells to enable epithelial differentiation. Apoptosis is not observed immediately subjacent to the renal capsule in the area corresponding to undifferentiated, uninduced metanephrogenic blastemal cells. Thus, the presence of Bcl–2 protein is, in general, inversely correlated with cells undergoing apoptosis within the neogenic zone of the developing kidney. Occasional apoptotic epithelial cells are observed within glomeruli and renal tubules. The majority of apoptotic epithelial cells were observed in the renal vesicles of the neogenic zone which had at least developed to the "S–phase." This observation suggests a potential role for apoptotic cell death during tissue modeling in the developing kidney and appears to be independent of the expression of *bcl–2*.

Immunohistochemical detection of proliferating cell nuclear antigen (PCNA) may be used as an indicator of actively proliferating cells in tissue sections. PCNA positive cells are most abundant in the neogenic zone of the developing kidney. Undifferentiated mesenchymal cells within the neogenic zone as well as the epithelial cells of the renal vesicles and ampullae frequently demonstrated PCNA positivity. Therefore, the distribution of proliferating cells and the presence of Bcl–2 protein in the fetal kidney does not appear to be correlated.

Taken together these observations indicate that more cells are generated than are necessary for the formation of a functional nephron following induction of the metanephrogenic mesenchyme [13]. It has also been demonstrated in the developing rodent kidney that the majority of cells undergoing apoptosis consist of undifferentiated mesenchymal cells [13, 14]. Analogously, in other systems such as the thymus and central nervous system, an excess in the number of cells is generated than ultimately survive during development. Apoptotic cell death is recognized as a common mechanism for the deletion of these excess cells. The contribution of *bcl–2* expression to renal morphogenesis can be summarized in a model in which Bcl–2 is rapidly upregulated in a subset of induced metanephric blastemal cells. These cells subsequently undergo epithelial conversion and form renal vesicles. Those induced blastemal cells which are generated following the burst in proliferation which fail to upregulate Bcl–2 are subsequently deleted by an apoptotic mechanism.

The mechanism(s) by which cell death is regulated during renal morphogenesis is poorly characterized although recent evidence indicates that these deaths are indeed "programmed." In the developing rat kidney the rate of cell death exhibits a predictable time course [13]. Furthermore, inhibition of nascent protein or RNA synthesis with cycloheximide or actinomycin–D has been shown to inhibit cell death in uninduced metanephrogenic mesenchyme [14]. Interestingly, the addition of exogenous epidermal growth factor (EGF) has been shown to markedly diminish the rate of cell death observed in the developing kidney suggesting that the availability of appropriate survival factors may be critically important for normal renal morphogenesis [13, 14]. The potential modulation of *bcl–2* expression, directly or indirectly, by relevant growth (survival) factors during development of the kidney has not been investigated; however, this remains a distinct possibility.

Oncogene and Tumor Suppressor Gene Involvement in Cell Death Regulation in the Kidney

The c–*myc* proto–oncogene also plays a key role in normal metanephric development. In the undifferentiated metanephric blastemal cells, c–*myc* is expressed at high levels and is downregulated as the mesenchyme undergoes

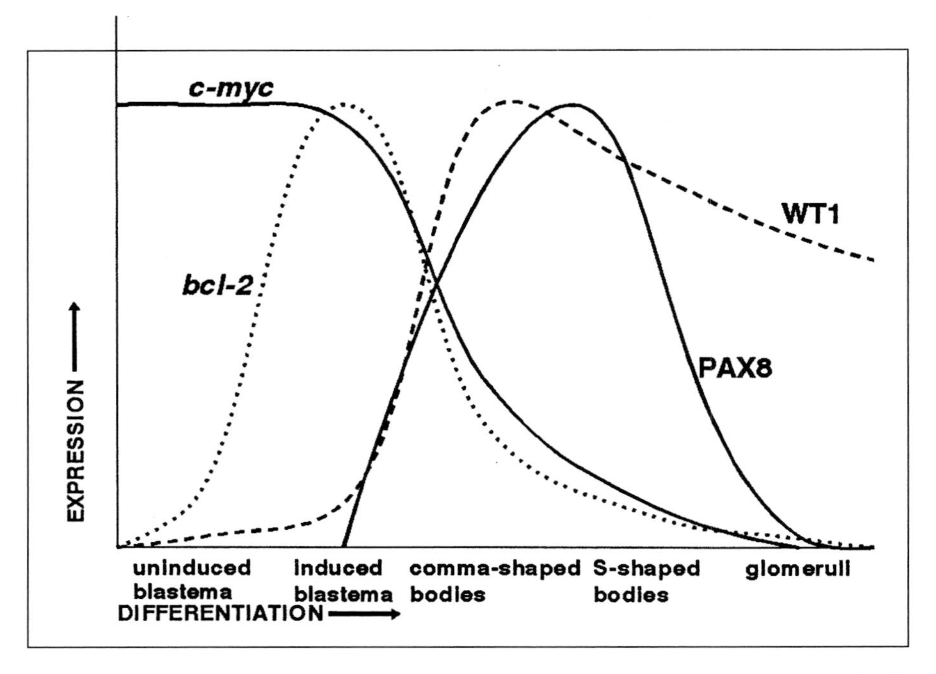

Fig. 10.4. Graphic representation of the relative levels of expression of c–myc, *bcl–2*, WT1 and PAX8 during differentiation in the kidney.

epithelial conversion (Fig. 10.4) [15]. Interestingly, c–*myc* transgenic mice, similar to the *bcl–2* knock–out mice, develop polycystic kidneys and subsequent renal failure [16].

Analysis of the function of the Wilms' tumor suppressor gene WT1 has demonstrated that WT1 regulation of apoptosis is essential in normal renal development and plays a role in tumorigenesis [17]. WT1 encodes a zinc finger transcription factor that regulates the transcription of several genes important in kidney development [18–21]. WT1 binds DNA through the interaction of four cys–cys his–his type zinc fingers in a sequence specific fashion [22, 23]. Mice that are deficient for WT1 expression die in the prenatal period with multiple anomalies [17]. The ureteric bud from the WT1–/– mouse is defective in its ability to induce blastemal cells to undergo mesenchymal–epithelial conversion. Additionally, the blastema appears unable to be induced even when cocultured with normal spinal cord, a potent inducer of metanephric blastemal cells. In the absence of this inductive signal the metanephric blastema undergoes apoptotic cell death.

Interestingly, the immunohistochemical localization of Bcl–2 within the developing kidney corresponds closely to that observed for the transcription factors Pax8, encoded for by a paired box gene [24] and WT1 [25], suggesting that these factors may play a role in the transcriptional regulation of *bcl–2* expression (Fig. 10.4). This possibility is further supported by the presence a GCGGGGGCG WT1/EGR–1 binding element within the human *bcl–2* gene as well as a TGCCC PAX8 consensus binding element in the 5' flanking sequence of *bcl–2* [26, 27]. In addition, the c–*myc* promoter contains three potential WT1 DNA binding motifs suggesting that the coordinate regulation of *bcl–2* and c–*myc* expression could be mediated through WT1. Cotransfection assays utilizing reporter gene constructs that contained fragments of the putative WT1 binding sites in the *bcl–2* or c–*myc* promoters demonstrated that WT1 can repress transcription of both promoters [28]. Furthermore, electrophoretic mobility shift assays confirmed that WT1 is capable of binding oligonucleotides containing the WT1 binding motifs in the *bcl–2* and c–*myc* promoters [28].

PAX8 is a paired box transcription factor expressed in the kidney and several other tissues during development. The paired box family of transcription factors contains nine members that share homology in their DNA binding domain, and several family members also contain a full or partial homeobox. Mutations in PAX genes have been associated with multiple genetic syndromes. Both PAX2 and PAX8 are expressed in the developing kidney [29, 30]. The function of PAX genes in kidney development has not been well–defined but it is considered that PAX2 plays a role in the terminal differentiation of epithelial cells in the mature nephron [31]. PAX8 has also been shown to transcriptionally activate the *bcl–2* promoter and bind to a consensus PAX8 binding element [32].

In the adult kidney, the parietal layer of Bowman's capsule exhibits high levels of Bcl–2 protein whereas the capillary tuft of the glomerulus is Bcl–2 negative (Fig. 10.5). With the exception of widely scattered Bcl–2 positive cells, the proximal convoluted tubules (PCT) are negative. In contrast, the epithelium of the distal convoluted tubules expresses moderate levels of Bcl–2 protein (Fig. 10.5). In addition, the loop of Henle shows moderate levels of Bcl–2 protein with the thin segments, in general, exhibiting more intense staining relative to the thick segments. Widely scattered cells are observed within the collecting ducts are weakly immunoreactivity for Bcl–2 protein. The remainder of the collecting system including the renal calyces and renal pelvis shows no detectable Bcl–2 protein.

Bcl–2 and Renal Neoplasms

Because *bcl–2* has recently been shown to contribute to the pathogenesis of epithelial malignancies [33] and the finding that *bcl–2* is normally expressed in renal epithelial cells [5] it was of interest to examine primary renal cell neoplasms for evidence of *bcl–2* expression. Bcl–2 protein was commonly detected in renal cell carcinomas examined using immunohistochemical techniques [5]. The

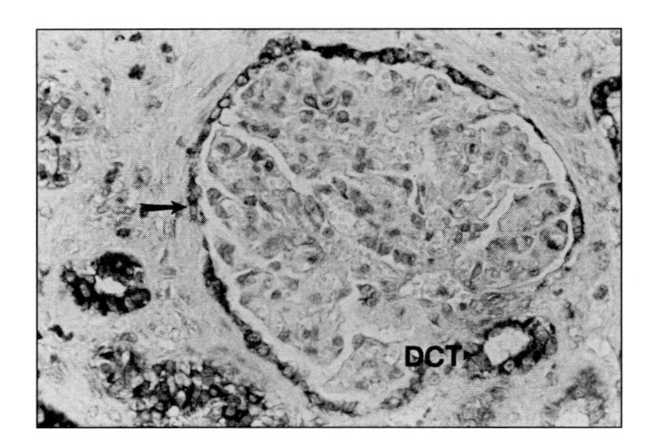

Fig. 10.5. Immunohistochemical detection of *bcl–2* protein in the adult kidney. Paraffin section of formal fixed adult human kidney immunohistochemically stained for *bcl–2* protein using the APAAP technique. Relatively high levels of *bcl–2* protein is present within the parietal layer of Bowman's capsule (arrow) and distal convoluted tubules (DCT).

level of expression was approximately equivalent to that of infiltrating lymphocytes which suggests that the presence of Bcl–2 protein within the tumor cells is of functional significance. Clear cell carcinomas characteristically were intensely immunoreactive for Bcl–2 protein (Fig. 10.6). Other histologic variants of renal cell cancers, including papillary and granular cell forms, also exhibit detectable levels of Bcl–2. Bcl–2 was also found to be variably expressed in Wilms' tumors. Similar results have been reported by other investigators [34]. Steady state levels of *bcl–2* mRNA were also elevated in the renal neoplasms in comparison to the normal adult kidney. Therefore, elevations of Bcl–2 protein within renal cell carcinomas could be accounted for on the basis of increased transcription of *bcl–2* or increased message stability.

It is unclear at present whether *bcl–2* expression in renal neoplasms is a consequence of genetic alterations which result in transcriptional upregulation of the *bcl–2* gene or simply reflect the cell of origin within the kidney. However, it is reasonably certain that *bcl–2* gene deregulation in itself is insufficient for the determination of a malignant phenotype in that carcinogenesis is recognized as a multistep process. Furthermore, the t(14;18) translocation, the cytogenetic hallmark of follicular lymphoma which results in *bcl–2* gene deregulation has been demonstrated to occur in lymphocytes in hyperplastic lymphoid tissue and does not, *a priori,* indicate the presence of lymphoma [35, 36].

Although, the molecular alterations associated with the development of renal cell cancers are incompletely characterized it may be anticipated that the survival advantage imparted by *bcl–2* expression would be expected to enable

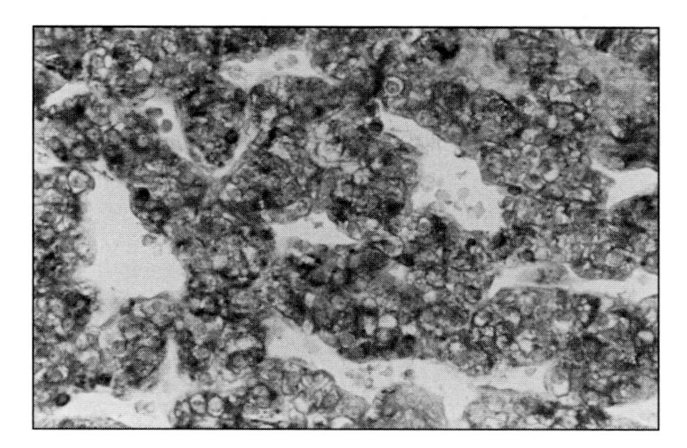

Fig. 10.6. Immunohistochemical detection of *bcl–2* protein in kidney cancer. Frozen section of a clear cell variant of renal cell carcinoma immunohistochemically stained for *bcl–2* protein using the APAAP technique. These tumors characteristically exhibit diffuse and high levels of *bcl–2* protein.

the accumulation of complementing genetic alterations which would result in transformation and tumor progression [37]. The nature of the putative molecular lesions capable of complementing *bcl–2* during multistep renal carcinogenesis are unknown; however, mutations involving *H–ras*, *p53* and *Rb* are uncommon events in these tumors [38–40]. It is of interest that the expression of the *N–myc* proto–oncogene is transcriptionally upregulated during the conversion of metanephrogenic mesenchyme into the epithelial renal vesicles [41–44]. Furthermore, *N–myc* has been shown to genetically complement activated *ras* in primary cultures of porcine kidney tubular epithelial cells [44]. It remains to be determined whether the simultaneous deregulation of *bcl–2* and *N–myc* is relevant to multistep renal carcinogenesis; however, the genetic complementation of c–*myc* and *bcl–2* has been previously demonstrated during multistep lymphomagenesis [45–48]. The basis of this complementation was shown to be the ability of Bcl–2 to abrogate Myc–mediated apoptosis without altering the rates of Myc–mediated proliferation [48]. The clinical or predictive significance of bcl–*2* expression in renal tumors will necessitate further studies with extended follow up of patients.

Acknowledgments

T.J.M. is a Pew Scholar in the Biomedical Sciences.

References

1 Knudson C M, Tung K S, Tourtellotte W G et al. Bax–deficient mice with lymphoid hyperplasia and male germ cell death. Science 1995; 270:96–99.

2 Motoyama N, Wang F, Roth KA et al. Massive cell death of immature hematopoietic cells and neurons in Bcl–x–deficient mice. Science 1995; 267:1506–1510.

3 Lu Q–L, Poulsom R, Wong L et al. *bcl–2* expression in adult and embryonic non–hematopoietic tissues. J Pathol 1993; 169:431–437.

4 LeBrun DP, Warnke RA, Cleary ML. Expression of *bcl–2* in fetal tissues suggests a role in morphogenesis. Am J Pathol 1993; 142:743–753.

5 Chandler D, El–Naggar AK, Brisbay S et al. Apoptosis and expression of the *bcl–2* proto–oncogene in the fetal and adult human kidney: evidence for the contribution of *bcl–2* expression to renal carcinogenesis. Hum Pathol 1994; 25:789–796.

6 Saxen L, Salonen J, Ekblom P et al. DNA synthesis and cell generation cycle during determination and differentiation of the metanephric mesenchyme. Dev Biol 1983; 98:130–138.

7 Nagata M, Nakauchi H, Nakayama K–I et al. Apoptosis during an early stage of nephrogenesis induces renal hypoplasia in *bcl–2*–deficient mice. Am J Pathol 1996; 148:1601–1611.

8 Sorenson CM, Rogers SA, Korsmeyer SJ et al. Fulminant metanephric apoptosis and abnormal kidney development in *bcl–2*–deficient mice. Am J Physiol 1995; 268:F73–F81.

9 Nakayama K, Nakayama K, Negishi I et al. Targeted disruption of *bcl–2* ab in mice: occurrence of gray hair, polycystic kidney disease, lymphocytopenia. Proc Natl Acad Sci USA 1994; 91:3700–3704.

10 Veis DJ, Sorenson CM, Shutter JR et al. Bcl–2–deficient mice demonstrate fulminant lymphoid apoptosis, polycystic kidneys, and hypopigmented hair. Cell 1993; 75:229–240.

11 Kamada S, Shimono A, Shinto Y et al. *bcl–2* deficiency in mice leads to pleiotropic abnormalities: accelerated lymphoid cell death in thymus and spleen, polycystic kidney, hair hypopigmentation, and distorted small intestine. Cancer Res 1995; 55:354–359.

12 Gavrieli Y, Sherman Y, Ben–Sasson SA. Identification of programmed cell death in situ via specific labeling of nuclear DNA fragmentation. J Cell Biol 1992; 119:493–501.

13 Coles HSR, Burne JF, Raff MC. Large–scale normal cell death in the developing rat kidney and its reduction by epidermal growth factor. Development 1993; 118:777–784.

14 Koseki C, Herzlinger D, Al–Awgati Q. Apoptosis in metanephric development. J Cell Biol 1992; 119:1327–1333.

15 Magrauer G, Ekblom P. Contrasting expression patterns of three members of the myc family of protooncogenes in the developing and adult mouse kidney. J Cell Biol 1991; 112:13–25.

16 Trudel M, D'Agati V, Costantini F. c–myc as an inducer of polycystic kidney disease in transgenic mice. Kidney Int 1991; 39:665–671.

17 Kreidberg JA, Sariola H, Loring JM et al. WT1 is required for early kidney development. Cell 1993; 74:679–702.

18 Call KM, Glaser T, Ito CY et al. Isolation and characterization of a zinc finger polypeptide gene at the human chromosome 11 Wilms' tumor locus. Cell 1990; 60:509–520.

19 Drummond IA, Madden SL, Rohwer–Nutter P et al. Repression of the insulin–like growth factor II gene by the Wilms' tumor suppressor WT1. Science 1992; 257:674–678.

20 Wang ZY, Madden SL, Deuel TF et al. The Wilms' tumor gene product, WT1, represses transcription of the platelet–derived growth factor A–chain gene. J Biol Chem 1992; 267:21999–22002.

21 Hewitt SM, Fraizer GC, Wu YJ et al. Differential functions of the Wilms' tumor gene WT1 splice isoforms in transcriptional regulation. J Biol Chem 1996; 271:8588–8592.

22 Madden SL, Cook DM, Morris JF et al. Transcriptional repression mediated by the WT1 Wilms' tumor gene product. Science 1991; 153:1550–1553.

23 Paveletich NP, Pabo CO. Zinc finger–DNA recognition: crystal structure of the Zif268–DNA complex at 2.1 A. Science 1991; 252:809–817.

24 Plachov D, Chowdhury K, Walther C et al. Pax8, a murine paired box gene expressed in the developing excretory system and thyroid gland. Development 1990; 110:643–651.

25 Pritchard–Jones K, Fleming S, Davidson D et al. The candidate Wilms' tumor gene is involved in genitourinary development. Nature 1990; 346:194–197.
26 Zannini M, Francis–Lang H, Plachov D et al. Pax–8, a paired domain–containing protein, binds to a sequence overlapping the recognition site of a homeodomain and activates transcription of two thyroid–specific promoters. Mol Cell Biol 1992; 12:4230–4241.
27 Rauscher FJ, Morris JF, Tournay OE et al. Binding of the Wilm's tumor locus zinc finger protein to the EGR–1 consensus sequence. Science 1990; 250:1259–1262.
28 Hewitt SM, Hamada S, McDonnell TJ et al. Regulation of the proto–oncogenes *bcl–2* and c–myc by the Wilms' tumor suppressor gene WT1. Cancer Res 1995; 55:5386–5389.
29 Poleev A, Fickenscher H, Munlos S et al. PAX8, a human paired box gene: isolation and expression in developing thyroid, kidney and Wilm's tumors. Development 1992; 116:611–623.
30 Eccles MR, Wallis LJ, Fidler AE et al. Expression of the PAX2 gene in human fetal kidney and Wilms' tumor. Cell Growth Differ 1992; 3:279–289.
31 Dressler GR, Wilkinson JE, Rothenpieler UW et al. Deregulation of PAX–2 expression in transgenic mice generates severe kidney abnormalities. Nature 1993; 362:65–67.
32 Hewitt SM, Hamada S, Monarres A et al. Transcriptional activation of the *bcl–2* apoptosis suppressor gene by PAX8 in the kidney. Manuscript submitted.
33 McDonnell TJ, Troncoso P, Brisbay SM et al. Expression of the protooncogene *bcl–2* in the prostate and its association the emergence of androgen–independent prostate cancer. Cancer Res 1992; 52:6940–6944.
34 Paraf F, Gogusev J, Chretien Y et al. Expression of *bcl–2* oncoprotein in renal cell tumors. J. Pathol 1995; 177:247–252.
35 Limpens J, deJong D, vanKrieken JHJM et al. Bcl–2/J_H rearrangements in benign lymphoid tissues with follicular hyperplasia. Oncogene 1991; 6:2271–2276.
36 Aster, JC, Kobayashi Y, Shiota M et al. Detection of the t(14;18) at similar frequencies in hyperplastic lymphoid tissues from American and Japanese patients. Am J Pathol 1992; 141:291–299.
37 McDonnell TJ, Korsmeyer SJ. Progression from lymphoid hyperplasia to high–grade malignant lymphoma in mice transgenic for the t(14;18). Nature 1991; 349:254–256.
38 Nanus DM, Mentle IR, Motzer RJ et al. Infrequent ras oncogene point mutations in renal cell carcinoma. J Urol 1990; 143:175–178.
39 Ishikawa J, Xu H–J, Hu S–X et al. Inactivation of the retinoblastoma gene in human bladder and renal cell carcinomas. Cancer Res 1991; 51:5736–5743.
40 Suzuki Y, Tamura G, Satodate R et al. Infrequent mutation of p53 gene in human renal cell carcinoma detected by polymerase chain reaction single–strand conformation polymorphism analysis. Jpn J Cancer Res 1992; 83:233–235.
41 Hirvonen H, Sandberg M, Kalimo H et al. The N–myc proto–oncogene and IGF–II growth factor mRNAs are expressed by distinct cells in human fetal kidney and brain. J Cell Biol 1989; 108:1093–1104.
42 Mugrauer G, Ekblom P. Contrasting expression patterns of three members of the myc family of protooncogenes in the developing and adult mouse kidney. J Cell Biol 1991; 112:13–25.
43 Mugrauer G, Alt FW, Ekblom P. N–myc proto–oncogene expression during organogenesis in the developing mouse as revealed by in situ hybridization. J Cell Biol 1988, 107:1325–1335.
44 Bendit I, Rich MA, Moldwin R et al. N–myc oncogene expression in porcine renal development and oncogenesis. Pediatr Res 1991; 29:268–271.
45 Strasser A, Harris AW, Bath ML et al. Novel primitive lymphoid tumors induced in transgenic mice by cooperation between myc and *bcl–2*. Nature 1990; 348:331–333.
46 Fanidi A, Harrington EA, Evan GI. Cooperative interaction between c–myc and *bcl–2* proto–oncogenes. Nature 1992; 359:554–556.
47 Bissonnette RP, Echeverri F, Mahboudi A et al. Apoptotic cell death induced by c–myc is inhibited by *bcl–2*. Nature 1992; 359:552–554.
48 Marin MC, Hsu B, Stephens LC et al. The functional basis of c–myc and *bcl–2* complementation during multistep lymphomagenesis in vivo. Exp Cell Res 1995; 217:240–247.

Chapter 11

Apoptosis and Cancer, edited by Seamus J. Martin.

Apoptosis in Leukemia

Sharon L. McKenna and Thomas G. Cotter

Tumour Biology Laboratory, Department of Biochemistry, University College, Lee Maltings, Prospect Row, Cork, Ireland

Introduction

Normal hematopoiesis is a finely tuned equilibrium between stem cell proliferation, differentiation and apoptotic cell death. These processes are directed by a combination of both internal gene expression and external signals from cell–cell contacts, hematopoietic cytokines and growth factors. Disruption of this homeostasis by excessive proliferation, differentiation blocks or delayed cell death, can lead to an accumulation of leukemic cells and complications primarily associated with abrogation of normal hematopoiesis.

The past several years have seen a rapid expansion in interest and information regarding the biological significance of apoptosis and the genetic mechanisms underlying its regulation. Recent attention has been focused on the role of apoptosis in leukemia, particularly as it is apparent that many types of leukemic cells are not highly proliferative compared to their normal counterpart. The deregulation of apoptosis is not only likely to influence the pathogenesis of leukemia, but may also have serious implications for treatment.

This article will review the main morphological, biochemical and molecular aspects of apoptosis which have been observed in eukaryotic cells. The role of apoptosis in hematopoiesis will be discussed, followed by evidence which suggests that deregulation of apoptosis may play a significant role in leukemogenesis and response to cytotoxic therapy.

Morphological and Biochemical Features of Apoptosis

Cell death can occur by either of two morphologically and biochemically distinct modes: necrosis or apoptosis [1]. Necrotic cell death occurs following severe environmental trauma (e.g., hyperthermia, hypoxia, lytic viral infection, exposure to respiratory poisons) and involves cytoplasmic swelling and disruption of internal and external membranes. Irregular clumping of hyperchromatic nuclei is apparent at an early stage, and random degradation of the DNA results in a smear when analyzed by agarose gel electrophoresis. Necrotic lysis results in the liberation of cellular contents into the extracellular fluid and thus is usually accompanied in vivo by the development of inflammation.

Apoptosis is a physiological mode of cell death where the cell actively participates in its demise by providing the necessary endogenous enzymes and energy for the process. Apoptotic cells are characterized by very specific morphological and biochemical changes in the cell. Morphological features include cell shrinkage, condensation of chromatin and cytoplasm, and blebbing of the nuclear and plasma membrane. Membrane integrity is not lost in the initial stages of apoptosis. During the later stages of apoptosis, the cell fragments into apoptotic bodies which consist of nuclear and cytoplasmic fragments enveloped in plasma membrane [2]. In in vitro conditions, apoptotic cells and bodies undergo secondary necrosis [3].

Isolated apoptotic cells are rarely visualized in tissue specimens as they display alterations in surface molecules which are recognized by phagocytic cells leading to rapid ingestion. Apoptotic cell death in vivo is therefore not associated with the induction of an inflammatory response. In mammals it is thought that macrophages are the key effectors in the removal of apoptotic cells, although their removal by other cells such as fibroblasts and endothelial cells has also been observed [4, 5]. Several groups have investigated the nature of the surface changes enabling phagocyte recognition of apoptotic cells. Lectin–like receptors have been suggested to mediate macrophage recognition of changes in surface carbohydrates on apoptotic cells [4, 6]. The vitronectin receptor ($\alpha_v\beta_3$) and CD36 macrophage surface receptor, have also been shown to cooperate in the recognition of apoptotic human neutrophils [7, 8], although their ligand on these cells has yet to be identified. Loss of phospholipid asymmetry of the plasma membrane, resulting in the inappropriate externalization of phosphatidylserine, can influence the recognition of apoptotic lymphocytes by macrophages [9, 10], and a 75 kDa polypeptide (61D3 antigen) on monocyte–derived macrophages has been shown to be important for recognition of apoptotic neutrophils, B lymphocytes and T lymphocytes [11]. It is possible that the mechanism used by macrophages for recognition of apoptotic cells may depend on the subpopulation of macrophages, their cellular environment and the presence or absence of specific cytokines.

Numerous biochemical alterations accompany the degradation phase of apoptosis. These include the disruption of energy metabolism, alterations in ion fluxes, activation of proteases and tissue transglutaminase, and disorganization of the cytoskeleton [12]. The most characteristic biochemical event in apoptotic cells is DNA fragmentation into multiples of nucleosome–sized units (180–200 bp) which are visible as DNA "ladders" on agarose gels. DNA cleavage is one of the most commonly used parameters for assessment of apoptotic cell death. Apoptosis may be quantified by assessing the extent of DNA fragmentation on agarose gels, or by measuring the loss of DNA from apoptotic cells using DNA stains, or by assessing the ability of specific enzymes to tag nicks or double strand breaks in DNA with fluorescent markers [13, 14]. An additional parameter such as apoptotic morphology is always necessary to validate such methods, as cell lines have been reported which do not undergo internucleosomal DNA cleavage [15, 16]. Analysis of higher molecular weight DNA from apoptotic cells has indicated that the formation of large DNA fragments (300, 50 kb) precedes oligonucleosomal DNA fragmentation. The sizes of these fragments have been suggested to arise from the release of loops (50 kb) or rosettes (300 kb) of chromatin as they become detached from the nuclear scaffold [17]. High molecular weight fragments may be present in apoptotic cells in the absence of the low molecular weight oligonucleosomal–sized fragments [15].

The identity of the endonuclease(s) responsible for DNA degradation in apoptosis remains controversial. A Ca^{2+}/Mg^{2+} dependent endonuclease was reported to be involved in DNA degradation in apoptotic thymocytes [18, 19]. A later study suggested that this endonuclease is functionally and antigenically identical to DNase I [20]. Other reports have suggested that DNase II is the predominant enzyme in apoptotic Chinese hamster ovary cell [21] and apoptotic human neutrophils [22]. Several novel endonucleases have also been reported in apoptotic thymocytes [23–25] and cytotoxic T lymphocytes [26]. Thus available evidence suggests that DNA digestion in apoptotic cells is an ordered multistep process which may involve the concerted action of a number of endonucleases and proteases, and that the nature of the endonucleases activated may be cell type specific.

Molecular Mechanisms in Apoptosis

Apoptosis is a genetically encoded cell death program which has been highly conserved throughout multicellular evolution. This has undoubtedly facilitated its genetic analysis as some of the most useful information regarding the regulation of cell death has come from studies of developmental mutants of the nematode worm *Caenorhabditis elegans*. Two genes are required for the 131 cell deaths that occur during normal development of this hermaphrodite: *ced–3* and *ced–4*. Another gene, *ced–9,* acts antagonistically to *ced–3* and *ced–4* to suppress cell death. Human homologs have now been identified for both *ced*–9 (*bcl–2*) and *ced*–3 (the caspases) gene products (see reviews by E. White, 1993 and 1996) [27, 28]. The identification of "death genes" combined with earlier evidence suggesting that apoptosis required novel protein synthesis [29], led to the popular conception that death signals induce the transcription of "killer genes" and consequent activation of a death program. It is now apparent however from other studies that de novo protein synthesis is not always required [30] and that apoptosis gene products are constitutively present in many cells, awaiting either activation, or the removal of a negative inhibitor.

A number of genes have been described which can either positively or negatively influence apoptosis in mammalian systems and may cooperate in its overall regulation. Some of these are already familiar as oncogenes or suppressor genes which have been traditionally associated with proliferation or cell cycle regulation (e.g., *myc*, *ras* and *p53*).

bcl–2 Gene Family

bcl–2 is the prototype of a family of genes which are involved in the regulation of apoptosis. The gene was originally cloned from the t(14:18) translocation breakpoint associated with follicular lymphoma and has subsequently been identified as a negative regulator of apoptosis analogous to its homolog, *ced–3* in *C. elegans*. Overexpression of *bcl–2* has been shown to protect mammalian cell lines from apoptosis induced by a variety of signals including growth factor deprivation, UV irradiation, cytotoxic lymphokines (TNF) and heat shock (reviewed in ref. 31). Bcl–2 is an integral membrane protein associated with the mitochondria, endoplasmic reticulum and nuclear envelope. It has been shown to protect cells from apoptosis in the absence of a nucleus, indicating that its effects are mediated in the cytoplasm [32]. Bcl–2 has been shown to interact with other proteins with which it shares amino acid homology, including Bax, Bcl–x_S, Bak and Bad.

Bax is a Bcl–2 related protein which has been shown to promote cell survival. Bax and Bcl–2 are capable of forming both homodimers and heterodimers [33]. Bax homodimers render a cell more susceptible to apoptosis, whereas binding of Bcl–2 to Bax neutralizes its activity and thus protects cells from apoptosis. It has been proposed that Bcl–2 negatively regulates apoptosis by sequestering the positive effector Bax [34].

Another *bcl–2* gene family member *bcl–x* produces two products by alternative splicing. The long product Bcl–x_L is a negative regulator of apoptosis, whereas the short product Bcl–x_S is a positive regulator of apoptosis. Like Bcl–

2, Bcl–x_L can inhibit cell death induced by growth factor withdrawal from an IL–3 dependent cell line [35]. Bcl–x_L can also heterodimerize with Bax and is thought to act in a similar fashion to, but independently of, Bcl–2 [34, 36].

Bcl–x_S can antagonize cell death inhibition by Bcl–2 and Bcl–x_L, but does not interact with Bax [35, 36]. Bcl–x_S may therefore promote apoptosis by sequestering these negative regulators and preventing their inhibition of Bax. The positive regulator Bad can dimerize with both Bcl–2 and (more strongly) Bcl–x_L. Bad can antagonize apoptosis protection by displacing Bax from Bcl–x_L in vivo [37].

Bak is another Bcl–2 homolog that promotes cell death, and has been shown to interact with both Bcl–2 and Bcl–x_L. It is not known whether Bak induces apoptosis by directly activating the apoptotic machinery, or by antagonizing its anti–apoptotic homologs [38].

Thus, a complex interplay of Bcl–2 family members are involved in the regulation of susceptibility to apoptosis. The biochemical functions of these gene family members remains to be established.

p53

p53 is a tumor suppressor gene that functions at least in part as a transcription factor. It has been referred to as the "guardian of the genome", as it will induce cell cycle arrest in response to damage by various agents such as UV irradiation or cytotoxic drugs, and mitigate the decision to either repair damage or undergo apoptosis [39] (see Fig. 11.1).

Expression of wild type *p53* has been shown to induce apoptosis in a number of cell lines. Ectopic expression of wild type *p53* in a myeloid leukemia cell line, M1, induced apoptosis which could be inhibited by IL–6 [40]. In the myeloid 32D cell line endogenous *p53* expression was shown to significantly accelerate apoptosis upon withdrawal of IL–3 [41]. Thus, p53 can mediate the response of myeloid cell lines to hematopoietic cytokines (Fig. 11.1).

The mechanisms by which p53 induces apoptosis are poorly understood compared to the mechanisms involved in its induction of cell cycle arrest. Reporter gene assays have suggested that p53 can influence the expression of *bcl–2* family members. A p53 negative response element has been mapped in the *bcl–2* gene, and a p53 binding site in the *BAX* gene has been shown to transactivate the *bax* promoter [42, 43]. Thus p53 may promote cell death by increasing the Bax:Bcl–2 ratio. p53 has also been reported to induce *Fas* expression which can also increase cell susceptibility to apoptosis [44].

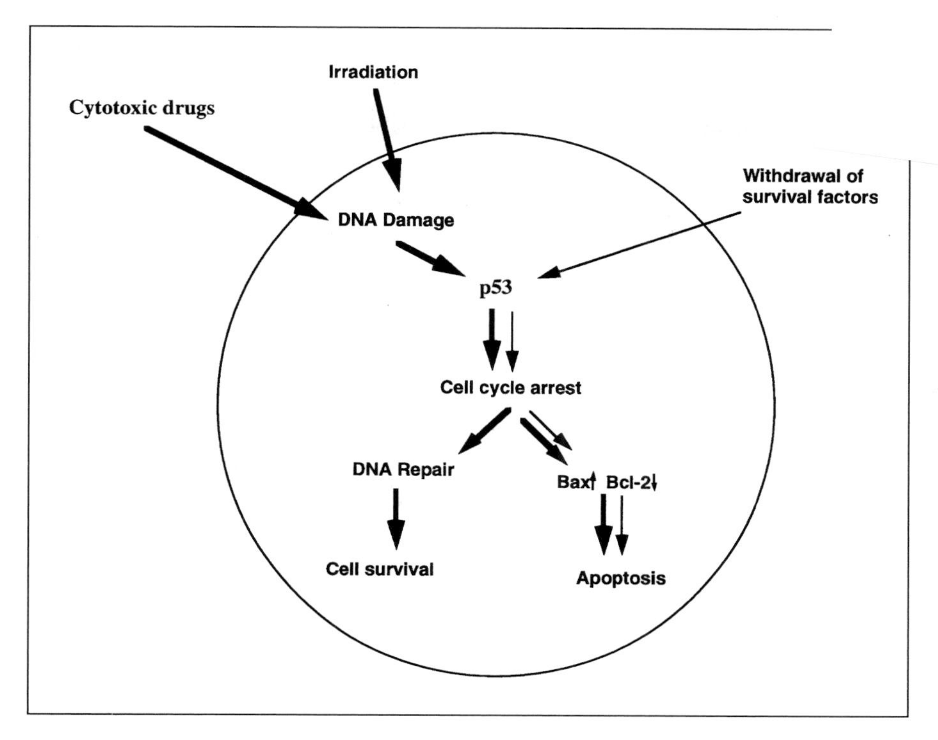

Fig. 11.1. p53 and the choice between life and death. p53 will induce cell cycle arrest in response to DNA damage caused by irradiation or cytotoxic drugs. The DNA may then be repaired, or in the case of extensive damage, p53 will direct the cell to undergo apoptosis. In hematopoietic cells, p53 can also mitigate the decision to undergo apoptosis in the absence of survival factors.

p53 is not required for glucocorticiod–mediated apoptosis of thymocytes [45], and cell lines which do not express p53 are still capable of undergoing apoptosis. Apoptosis may therefore proceed by both p53–dependent and independent pathways.

myc

The protein product of the *c–myc* gene is a transcriptional activator which has been implicated in the control of normal proliferation in many studies. *c–myc* is one of the early response genes induced following mitogenic stimulation. *myc* expression is maintained throughout the cell cycle and is rapidly downregulated following mitogen withdrawal, leading to cell cycle arrest in G_1

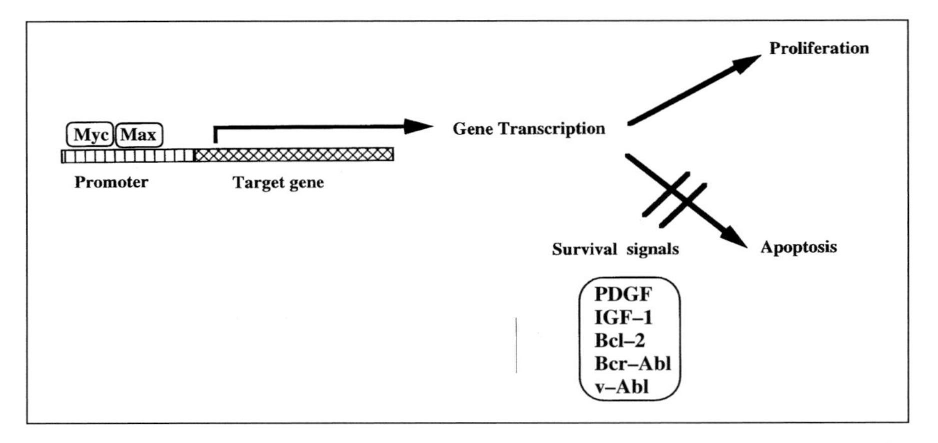

Fig. 11.2. The dual function model for *myc*. Myc:Max heterodimers bind to specific target sequences in a variety of genes, resulting in transcriptional activation. Unregulated expression of c–Myc can result in concomitant proliferation and apoptosis. In the presence of survival factors (PDGF, IGF–1, Bcl–2, Bcr–Abl, v–Abl) apoptosis is blocked and excessive proliferation leads to an expansion of the cell population.

[46]. Constitutive unregulated expression of *c–myc* abolishes cell cycle arrest in fibroblasts following serum deprivation and results in concomitant proliferation and apoptosis [47]. *myc*–dependent apoptosis has also been demonstrated in hematopoietic cells deprived of growth factors [48].

Both induction of apoptosis and cell proliferation require hetero–dimerization of Myc with its partner protein Max [49]. DNA binding of Myc and subsequent transactivation is strictly dependent on dimerization with Max, yet Myc–dependent apoptosis can proceed in the absence of protein synthesis in fibroblasts. This suggested that Myc must be continuously implementing the molecular machinery required for apoptosis, but that the execution of apoptosis is dependent upon other factors. It was subsequently shown that a restricted group of cytokines, principally the insulin–like growth factors and PDGF could inhibit Myc–induced apoptosis in low serum. The suppression of apoptosis by these cytokines was unlinked to any mitogenic activity [50].

Other studies have shown that Bcl–2 can block apoptotic cell death induced by Myc and can cooperate in cellular transformation [51, 52]. Similarly, v–Abl has anti–apoptotic activity and can cooperate with Myc in transformation [53].

Thus "survival signals," such as anti–apoptotic cytokines, or overexpression of a negative regulator of apoptosis are necessary to inhibit a default apoptosis program induced by Myc. The combination of deregulated Myc and survival signals promotes cell proliferation in the absence of apoptosis and may explain the requirement for oncogene cooperation in cellular transformation by *myc* (see Fig. 11.2) [54].

Ras *Genes*

Ras oncogenes have also been implicated in both proliferation and apoptosis. Three mammalian *ras* genes (Ha–*ras*, Ki–*ras* and N–*ras*) encode related G–proteins, which can become oncogenic when activated by mutation [55]. *Ras* can cooperate with other oncogenes, such as mutant *p53*, *myc*, *E1A* and *SV40*, in the transformation of primary cells [56].

Another nuclear protein which can cooperate in *ras* transformation is the transcription factor IRF–1 (interferon regulatory factor 1). Embryonic fibroblasts lacking the *IRF–1* gene can be transformed by activated H–*ras*. These cells also showed resistance to anticancer drugs and ionizing radiation. In the presence of *IRF–1* cDNA, cells with activated *ras* undergo apoptosis. These results suggest that *IRF–1* may act like a tumor suppressor gene as it can mediate the death of an oncogene activated cell [57]. Other groups have reported that activated forms of H–*ras,* or over–expression of *ras* can have a protective effect on apoptosis [58, 60]. The effects of activated H–*ras* may therefore be largely dependent on the presence of other proteins.

R–Ras is a Ras related protein which was isolated in a screen for Bcl–2 interacting proteins [61]. R–Ras has been shown to promote apoptosis in growth factor deprived cells by a mechanism which is suppressible by Bcl–2 [62].

Abl Tyrosine Kinases

Several lines of evidence suggest that the Abl tyrosine kinase is a negative regulator of apoptosis (reviewed in Samali et al, 1996) [63]. Both v–Abl and Bcr–Abl can rescue growth factor–dependent cell lines from apoptosis induced by factor withdrawal [64–66]. Both kinases have also been shown to protect myeloid cells from apoptosis induced by a variety of cytotoxic drugs [67] and Fas–mediated apoptosis [68].

Bcr–Abl has been reported to prevent apoptosis in factor–dependent murine hematopoietic cells by inducing a *bcl–2* expression pathway. Bcr–Abl expressing cells reverted to factor dependence and nontumorigenicity after *bcl–2* expression was suppressed [69].

Fas/Apo–1

The Fas antigen (also known as Apo–1 and CD95) is a 48 kDa cell surface protein which belongs to the TNF superfamily of surface receptors [70]. Stimulation of the Fas by its natural ligand (FasL) or with anti–Fas antibodies induces apoptosis. The interaction of Fas with FasL leads to the production of ceramides by activation of an acidic sphingomyelinase [71] and the activation of IL–1 converting enzyme (ICE)–like protease [72, 73]. Mutations in Fas or its ligand are responsible for the autoimmune disorders in *lpr* and *gld* mice [74]. Fas is widely expressed on normal and malignant hematopoietic cells and on cells outside the immune system [74].

Apoptosis in Hematopoiesis

All hematopoietic cells with the exception of pluripotent stem cells have a finite life span. Apoptosis follows terminal differentiation, negative selection of autoreactive lymphocytes and the removal of activated lymphocytes at the end of an immune response. The life span of hematopoietic cells is regulated by a combination of extracellular survival factors and internal genetic programs. Evidence from various studies, particularly those involving transgenic animals, suggests that many of the genes which have been implicated in the regulation of apoptosis play an important role in the maintenance of normal hematopoiesis.

Apoptosis in Myeloid Cells

Granulocytic leukocytes (neutrophils, eosinophils, basophils) and monocyte/macrophages are important effectors in the innate immune system. They are involved in the destruction of micro–organisms by phagocytosis and in the initiation and control of inflammatory responses. Specialized macrophages (antigen–presenting cells APCs) play a role in processing antigens before presenting them to B and T lymphocytes.

Granulocytes differentiate from progenitor cells in the bone marrow over a period of 6–10 days. Following their release from the bone marrow, granulocytes spend 6–10 hours in the circulation before moving into the tissues to perform their phagocytic function. Mature neutrophils are the most abundant and shortest lived leukocytes. Neutrophils survive for 1–2 days in tissues after which they undergo apoptosis and are engulfed by macrophages. Evidence suggests that cytokines which have been associated with the differentiation and activation of granulocytic cells can also mediate their viability.

Studies with progenitor cell lines have shown that the cytokines IL–3, granulocyte colony–stimulating factor (G–CSF) and granulocyte–macrophage colony stimulating factor (GM–CSF) are essential for the maintenance of cell viability [75, 76]. Cytokines which have been shown to promote viability of differentiated myeloid cells include G–CSF, GM–CSF, IFNγ, TNF, IL–1 and IL–2 [77–80]. Many of these cytokines, for example the colony stimulating factors (CSFs), are also capable of inducing growth and differentiation of myeloid cells. Several studies suggest that growth and viability are separately regulated processes. Low concentrations of macrophage CSF promotes cell viability without proliferation of macrophages, whereas higher concentrations promote both survival and proliferation [81]. IL–1, IL–6, and Steel factor can promote viability in immature myeloid cells without inducing growth [82, 83], and viability of mature neutrophils can be promoted by CSFs in the absence of proliferation [84]. Evidence also suggests that cytokines are more important for survival in some cells than differentiation. Transfection of *bcl–2* into a multipotent hematopoietic cell line alleviated growth factor dependency. The transfected cells were capable of undergoing granulocyte or erythroid differentiation in serum deprived conditions suggesting that growth factors are not obligatory for differentiation but may modulate lineage commitment [85].

Several groups have investigated gene expression in myeloid cell lines and in normal hematopoietic cells. Expression of *c–myc* was shown to be high in proliferating myeloblasts and HL60 cells but dropped upon differentiation [86, 87], suggesting that c–Myc may not be required for apoptosis following terminal differentiation. It is clear however from other studies that *c–myc* expression levels should not be considered in isolation as its effect can be mediated by other genes and survival factors (see previous section). *bcl–2* is also expressed in normal immature myeloid cells but then declines upon cell maturation, becoming almost undetectable in terminally differentiated neutrophils [88, 89]. Similarly, in HL60 cells *bcl–2* expression is high but then declines upon differentiation, suggesting that the downregulation of Bcl–2 may be important for apoptosis of terminally differentiated cells [89]. In addition it has been shown that constitutive overexpression of *bcl–2* prolonged the life of terminally differentiated HL60 cells [90]. *bcl–x_L* is also expressed in immature hematopoietic cells and is prevalent in primitive hematopoietic precursors [91]. In contrast to mice deficient in *bcl–2* which can progress through development, *bcl–x_L* knock–out mice die at embryonic day 13, suggesting that *bcl–x_L* may be an important survival gene in the early stages of hematopoiesis [92]. Mature human neutrophils (but not monocytes or eosinophils) express Fas ligand on their surface and are susceptible to Fas–induced cell death. The coexpression of both Fas and FasL may contribute to their relatively rapid turnover [93].

Thus a number of different cytokines and genes have been implicated in the regulation of apoptosis in myeloid cells. Further research is necessary to establish the possible interactions between various gene products and extracellular survival signals.

Apoptosis in B Lymphocytes

B lymphocytes are the principal mediators of the specific humoral response to infection. B lymphocyte maturation can be divided into two stages: antigen–independent and antigen–dependent. Antigen–independent stages of differentiation take place in the bone marrow and are influenced by cell–cell interactions and an array of growth factors. During this stage of differentiation immunoglobulin heavy chain and light chain genes are rearranged. Successful rearrangement results in expression of IgM on the cell surface. B cell precursors which fail to produce IgM undergo apoptosis in the bone marrow. The IgM^+ B cell then migrates from the bone marrow to the periphery (spleen and lymph nodes) where it undergoes antigen–dependent differentiation. Immature B cells expressing receptors which bind to self antigens become functionally silent (anergy) or undergo apoptosis. Further maturation of the IgM^+ B cell results in the coexpression of IgD. The IgM^+/IgD^+ cell is then regarded as a mature B cell. Activation of mature B cells by antigen leads to proliferation and differentiation into antibody secreting plasma cells, or differentiation into memory B cells depending on the presence of T cell derived cytokines. Thus ligation of B cell receptors can illicit either a negative response (cell death) or a positive response (proliferation), depending on the maturation status of the cell. The mechanisms by which this is achieved is the subject of intensive investigation. Mature B cells may escape apoptosis following stimulation with cytokines [94] or by ligation of other surface molecules such as CD40 ligand on activated T cells [95, 96] or CD2 (see review by Baixeras et al, 1994) [97].

The *bcl–2* gene product has been implicated in mediating B cell survival. Overproduction of Bcl–2 can prevent apoptosis of B cells in vitro [98]. In addition, many of the stimuli which rescue B cells from apoptosis, including CD40 and CD2 also upregulate Bcl–2 expression [97, 99]. Targeted overexpression of *bcl–2* in mice extends normal B cell survival [100], produces B cell lymphomas [101] and causes production of auto–antibodies [102]. *bcl–2* overexpression has been reported not to affect deletion of bone marrow B cells, but to inhibit clonal deletion of self–reactive B cells in the periphery [103]. Mice with loss of function mutations in *bcl–2* (knock–out mice) can progress through development but have severe immune function failure due to loss of mature B and T cells [104, 105].

Ligation of CD40 rapidly induced the appearance of the *bcl–*x_L protein in B cell lymphoma cells and rendered the cells refractory to anti–Ig–induced cell death [106]. Chimeric mice which lacked *bcl–x* expression in the lymphoid system indicated that Bcl–x expression is required for maintaining the life span of immature but not mature lymphocytes [92]. Bax–alpha mRNA and protein expression were shown to be upregulated upon sIg induction of apoptosis in a human B cell line [107]. *Bax* knockout mice show normal development but display lymphoid hyperplasia consistent with a role for Bax in the promotion of apoptosis [108]. A number of *bcl–2* family members are therefore important in the regulation of cell viability in the B lymphoid lineage.

Expression of Fas has been implicated in the elimination of self–reactive B lymphocytes in a state of anergy. When these cells reach the spleen they are killed by T cells in a manner which is dependent upon Fas expression by B cells and FasL expression by T cells [109, 110].

Apoptosis in T Lymphocytes

T lymphocytes are responsible for antigen specific cell–mediated immunity. When antigen is presented to a T cell, the cell becomes activated and performs a variety of functions depending on the subpopulation of T cell.

The T cell receptor (TCR) recognizes small peptides contained within the cleft of antigen–presenting molecules called the major histocompatibility complex (MHC) on the surface of other cells. CD8 expressing T cells (mostly killer T cells) recognize peptides contained within MHC I clefts, and CD4 expressing T cells (mostly helper cells) recognize peptides contained in clefts of MHC class II molecules. The TCR complex is associated with the CD3 complex which couples the binding of antigen/MHC to intracellular signaling. Activated $CD4^+$ T cells secrete cytokines which mediate the B cell immune response, T cell inflammation and differentiation of killer T cells. Killer T cells are negative regulators of immunity (see review by Weissman, 1994) [111].

Immature T cells originate in the bone marrow and migrate to the thymus where they undergo gene rearrangements, proliferation and differentiation. A large number of thymocytes with a broad spectrum of receptor specificity are produced; the majority of which will undergo apoptosis in the thymus. T cells which recognize self MHC I and II molecules are positively selected [112], and T cells which recognize self MHC plus other self antigens are eliminated by negative selection [113]. Extensive apoptosis ensures that the only T cells that reach the periphery are those that have produced receptors which recognize self MHC molecules harboring nonself antigens.

Various studies have demonstrated that autoreactive immature T cells are induced to undergo apoptosis by the same mechanism that activates mature cells, i.e., ligation of the TCR–CD3 complex with specific antibodies [114–116]. The mechanism by which the same signal produces different effects in immature and mature cells is unknown, but evidence suggests that the coupling of TCR and

CD3 is incomplete in immature cells. In both cell types, antibodies to the total TCR–CD3 signaling complex induce a large calcium ion flux, whereas anti–TCR antibodies only induce a calcium flux in mature T cells and not immature thymocytes [117].

Downregulation of Myc with antisense oligonucleotides demonstrated that Myc expression is required for activation–induced apoptosis in a T cell hybridoma but does not affect another outcome of activation: the production of lymphokines [118]. bcl–2 gene family members may also influence thymocyte viability. *bcl–2* is expressed in developmentally early thymocytes, but diminishes at the stage of negative selection [119]. In addition Bcl–x_S peaks the same stage as Bcl–2 is downregulated [35]. In transgenic mice overexpressing *bcl–2* in T cells, self–reactive T cells showed prolonged survival but eventually died by negative selection. T cells also showed resistance to apoptosis induced by radiation, glucocorticoids and anti–CD3 [120]. In *bcl–2* knock–out mice, both B and T lymphocytes are severely depleted through excessive apoptosis [105].

Recent studies have suggested that *Fas* expression may also play a role in T cell viability. Loss of function mutations in *lpr* of *gld* mice result in severe lymphoproliferative disorders. Despite high expression of *Fas* in the thymus, mutations in these genes do not appear to affect elimination of self reactive T cells in the thymus. A deficiency in Fas does, however, affect activation induced cell death of autoreactive T cells which reach the peripheral lymph nodes and spleen. These T cells can proliferate unchecked and have the potential to promote autoantibody production by self–reactive B cells (reviewed in Rathmell and Goodnow, 1995) [110].

Cytotoxic T Lymphocytes

Cytotoxic T lymphocytes may induce apoptosis in target cells by two separate pathways, secretory and ligand induced. In the secretory pathway cytotoxic granules are released from the lymphocyte and enter the target cell through perforin pores. Apoptosis is then initiated by an as yet unknown mechanism. Ligand induced cell death results from cross–linking Fas receptors on target cells with FasL expressed on the surface of cytotoxic lymphocytes (reviewed in Berke, 1995) [121].

It is apparent therefore that regulation of apoptosis is central to the development and function of the hemopoietic system. Deregulation of apoptosis in such a dynamic system can have severe consequences, including the development of autoimmune disease and hematological neoplasia.

Apoptosis and Leukemogenesis

Leukemogenesis is considered to be a consequence of the clonal expansion of hematopoietic cells which have sustained a number of genetic changes, resulting in their inappropriate proliferation, defective differentiation or prolonged lifespan. Considerable interest has recently been generated in the role of prolonged cell survival in leukemogenesis, particularly in the chronic leukemias as the abnormal cells in these disorders are characterized by an excessive life span. What follows is an overview of studies which suggest that genetic changes which deregulate survival factors (apoptosis genes or cytokines) may play a significant role in the pathogenesis of human leukemia.

B Lymphoma

The first indication that deregulation of apoptosis may be important in the development of hematological neoplasia came from studies with the *bcl–2* proto–oncogene. Bcl–2 promotes cell survival by blocking apoptosis (see previous sections). The *bcl–2* gene is overexpressed as a result of the t(14:18) chromosome translocation associated with follicular lymphoma. The contribution of Bcl–2 to the development of neoplasia was assessed in transgenic mice. Expression of high levels of *bcl–2* initially resulted in a polyclonal lymphoid hyperplasia. The development of clonal B lymphomas after a long latent period (~15 months) was associated with secondary genetic changes. Half of the high–grade lymphomas also had a *c–myc* translocation. Direct evidence for the cooperation of *myc* and *bcl–2* was provided by *myc/bcl–2* double transgenic mice. These mice rapidly developed malignant lymphomas which appeared to be derived from a lymphomyeloid progenitor cell (reviewed by Cory et al, 1994) [122]. Mice transgenic for m*yc* alone did not show hyperplasia despite an augmentation in proliferation. The increased rate of proliferation was effectively counterbalanced by an increased rate of cell death [123]. Thus a genetic lesion which promotes cell viability can interfere with apoptosis induced by deregulated Myc and allow only proliferation. In vitro studies have also provided evidence for this cooperation between *myc* and *bcl–2*. Deregulated Myc in fibroblast cells was shown to promote both apoptosis and proliferation in the absence of survival signals. In the presence of Bcl–2, apoptosis was inhibited and continuous proliferation resulted in the accumulation of cells. (see previous section on *myc*).

Chronic Myeloid Leukemia

An anti–apoptotic event may also be the primary (or at least very early) lesion in the development of chronic myeloid leukemia. Chronic myeloid leukemia (CML) is a disease characterized by the progressive accumulation of granulocytic cells in the peripheral blood and bone marrow. The hallmark of CML, found in over 95% of patients is the Philadelphia chromosome, which results from a reciprocal translocation between chromosomes 9 and 22. This translocation produces the *bcr–abl* fusion gene and constitutive expression of the Bcr–Abl tyrosine kinase. Overexpression of Bcr–Abl in transgenic mice induces a CML–like illness [124]. The role of Bcr–Abl in the etiology of CML has not been defined; however, recent evidence suggests that it may cooperate in cellular transformation by negatively influencing the ability of cells to undergo apoptosis [63]. Thus the constitutive anti–apoptotic activity of the Bcr–Abl tyrosine kinase may play a major role in prolonging the life span of myeloid cells in CML.

The indolent/benign stage of CML (chronic phase) is relatively easy to treat and many patients maintain a near–normal quality of life. After a median interval of ~4 years the disease becomes more aggressive, and the myeloid cells lose their ability for terminal differentiation. About one–third of patients undergo lymphoblastic transformation; the majority undergo transformation to acute myeloid leukemia. Secondary acute leukemias derived from CML are notoriously difficult to treat and survival at this stage rarely exceeds 12 months [125]. In about 80% of cases new cytogenetic abnormalities are found at the aggressive stage of the disease which have almost certainly contributed to the evolution of the disease. Abnormalities in *myc* and *ras* are infrequent [126–128], whereas abnormalities in *p53* are found in 15–30% of patients in the accelerated phase of blast crisis [129, 130]. Deletions and altered expression of interferon genes have also been noted in patients in blast crisis [131, 132].

Thus the excessive life span of the initial CML clone allows the accumulation of additional genetic lesions which cooperate with the expression of Bcr–Abl to produce a highly aggressive and drug resistant disease (see Fig. 11.3). This underscores the importance of developing therapy which will eradicate the Philadelphia clone in the chronic phase.

Chronic Lymphocytic Leukemia

Chronic lymphocytic leukemia (CLL) is characterized by the accumulation of large numbers of monoclonal lymphocytes in the peripheral blood and bone marrow. Most cases involve B lymphocytes, although approximately 5% have T cell disease. CLL lymphocytes are morphologically mature but functionally in-

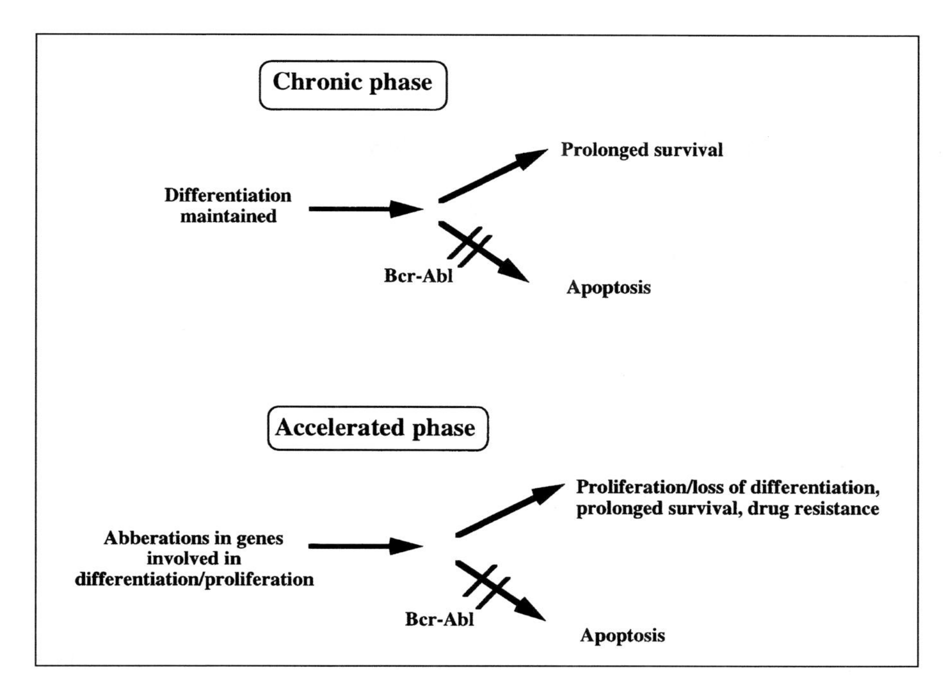

Fig. 11.3. A putative model for the role of Bcr–Abl in the evolution of CML. The constitutive anti–apoptotic activity of the Bcr–Abl tyrosine kinase may be responsible for the prolonged survival, and accumulation of granulocytic cells in CML. As the disease progresses to an accelerated phase and blast crisis, the cells have sustained additional genetic lesions allowing a gain in proliferative capacity and loss of terminal differentiation. The cooperation of these additional genetic lesions, with an early anti–apoptotic lesion, may facilitate the development of a highly aggressive and drug–resistant disease in blast crisis.

active and long–lived. B–CLL is thought to arise by clonal expansion of a rare CD5 positive B cell, usually found in the mantle zone of peripheral lymph nodes [133]. Variable cytogenetic abnormalities are found in 56–65% of CLL patients. Trisomy 12 and a breakpoint involving the long arm of chromosomes 14 (14q+) or 13 (13q+) are frequent abnormalities present in 18%, 13% and 10% of patients, respectively [134]. The molecular mechanisms underlying these chromosome abnormalities are unknown. Involvement of the *ras* gene has been postulated because of its location, but no *ras* gene mutations were found in 93 patients [135, 136]. The 14q+ abnormality involves a breakpoint at 14q32 at the site of the Ig heavy chain gene. Translocations such as t(11;14)(q13;q32) may juxtapose critical genes to the Ig heavy chain gene. Candidate genes located on 11q13 include *bcl–1, cd20, int–2, hst, sea* and *prad–1* [137].

Translocation of the *bcl–2* gene is infrequent in CLL but expression of the gene at the RNA and protein level is increased compared to normal $CD5^+$ B cells [138] and to normal lymphocytes [139, 140]. The upregulation of Bcl–2 may therefore contribute to the excessive lifespan of CLL lymphocytes in vivo.

Hematopoietic cytokines can also promote viability in B–CLL lymphocytes. IL– 4 was shown to enhance the survival of CLL cells in culture, whereas normal peripheral blood cells were unaffected [141]. IL–4, IFNγ and IFNα can also prevent hydrocortisone–induced cell death of CLL lymphocytes [142, 143]. The protective effects of IFNα have been associated with an increase in Bcl–2 expression [143]. Another study has suggested that *IL–7* gene expression may be required for the maintenance of CLL viability in vivo. *IL–7* transcripts were rapidly downregulated preceding apoptosis of CLL lymphocytes in culture. However, *IL–7* gene expression could be retained and apoptosis prevented by culturing the CLL cells on a monolayer of EA.hy926 human umbilical cord endothelial hybrid cells. Cell–cell contact was required as culture supernatants did not prevent apoptosis. Integrins expressed on the cell surface were suggested to play a significant role in CLL cell viability [144].

Several groups have also analyzed Fas (CD95) expression in B–CLL lymphocytes. CD95 is weakly expressed on the surface of most B–CLL cells, but can be upregulated in vitro by stimulation with IL–2 or *Staphylococcus aureus* Cowan (SAC). Upregulation of CD95 allowed Fas–mediated apoptosis and coincided with a downregulation of Bcl–2 mRNA [145]. In contrast upregulation of Fas by IFNα or IFNγ resulted in Fas positive cells which did not undergo apoptosis in the presence of anti–Fas monoclonal antibody. It was suggested that the upregulation of Bcl–2 by these interferons may over ride the cytotoxic signaling from Fas [146].

The progression of CLL has also been associated with an increase in cytogenetic abnormalities [137, 147]. Mutations in *p53* have been found in 10–15% of CLL patients and are associated with a more aggressive form of disease and resistance to chemotherapy [148]. Unlike CML, CLL does not evolve into acute leukemia although in rare cases blastic transformation may occur as a terminal lymphoma (Richters syndrome). Death is usually due to bone marrow failure and immune deficiency [149]. It is conceivable that the progression of CLL, like CML, is facilitated by the prolonged life span of the original CLL clone, allowing the accumulation of additional genetic lesions.

Acute Lymphoblastic Leukemia

The deregulation of apoptosis may also play an important role in the pathogenesis of acute leukemia. The Philadelphia translocation occurs in about 20% and 5% of adult and pediatric ALL patients, respectively. Compared to Ph^-ALL patients who respond very favorably to chemotherapy, Ph+ ALL patients are very difficult to treat and are seldom cured [150]. The anti–apoptotic activity of p190 Bcr–Abl may play an important role in both the development of the disease and the drug resistance associated with Ph+ ALL.

High levels of Bcl–2 expression in acute lymphoblastic leukemia (ALL) cells correlated with prolonged survival when cultured in the absence of growth factors [151]. Elevated Bcl–2 expression was also associated with leukemic growth and survival of a lymphoblastoid cell line in mice [152]. In a SCID mouse model of T–ALL, in vivo treatment with anti–Fas antibodies induced apoptosis in T–ALL cells leading to significantly prolonged survival. A fraction of cells were however resistant to Fas–mediated apoptosis as the leukemia recurred [153]. A later study by the same group found that Fas was constitutively expressed in 21 out of 30 T–ALL patients but that the majority of Fas positive T–ALL were resistant to anti–Fas–mediated apoptosis. Resistance was independent of Bcl–2 expression. Inhibition of protein synthesis induced sensitivity to Fas–mediated apoptosis in most T–ALL, suggesting that resistance to Fas–induced apoptosis in ALL is maintained by an active cellular program [154].

Acute Myeloid Leukemia

Expression of Bcl–2 in AML blast cells has been correlated with poor clinical outcome [155–157]. One study found that Bcl–2 was present in 87.3% of AML cases at presentation and in 100% of relapses. Relapses always had higher percentages of positive cells than those at onset [157]. Autonomous in vitro growth of blast cells from AML patients has been associated with a lower remission rate and a higher relapse rate [158]. A recent study has demonstrated that AML blasts with autonomous growth are relatively resistant to the induction of apoptosis and that this is related to the autocrine production of GM–CSF. Autocrine GM–CSF was found to upregulate expression of Bcl–2 [159]. Another preliminary study has also suggested that Bcl–2 can be upregulated by G–CSF in vivo [160].

Resistance to apoptosis may therefore play an important role in the progression and drug resistance of acute leukemias. However the very nature of acute leukemias (their proliferative capacity and fast clinical progression), suggest that they may have sustained early genetic lesions associated with proliferation and/or differentiation. For example an early genetic lesion in Burkitts lymphoma is activation of the proto–oncogene *myc*. This lymphoma has a high growth rate and clinical progression, and can be fatal within weeks or months without adequate treatment. *myc* translocations have also been associated with B– and T–ALL [161, 162]. It has been previously noted that one of the differences between acute and chronic leukemias is the nature of the genes involved in chromosome translocations, i.e., translocations in acute leukemias more often involve genes which encode nuclear transcription factors (e.g., *myc*, *mml*, *pml*, *plzf*, *eto*, *aml–1*, *cbf*, *rar*, *env–1*), whereas translocations involving cytoplasmic proteins (BCL–2, BCR, ABL) are more often a feature of chronic diseases [161]. Whether this could be extended to *early lesions proliferative/differentiation defective* (acute leukemia), and *early lesions apoptotic* (chronic leukemia), remains to be established.

The term preleukemia (or myelodysplastic syndrome, MDS) refers to a group of hematopoietic disorders which frequently evolve into AML, and rarely into ALL [163]. Recent studies with preleukemia suggest that the peripheral blood cytopenias and hypocellular marrows found in many of these patients may be a direct result of elevated apoptosis [164, 165]. Elevated apoptosis can be a feature of cells with deregulated oncogenes (e.g., *myc* or *ras*) and may indeed be a cellular protection mechanism against oncogene deregulation. The potential involvement of *ras* is suggested by the high incidence of *ras* mutations in MDS—40% [166]. In addition, the frequent loss of 5q31.1 (and thus IRF1, see *ras* section) in MDS and leukemia may cooperate with deregulated *ras* and allow disease progression [57]. Undoubtedly mouse models will provide further insight into the molecular mechanisms involved in multistep evolution of the various types of leukemia.

Apoptosis and Drug Resistance

Resistance to cytotoxic therapy remains a major limiting factor in the treatment of hematological malignancies. In some leukemias it is almost predictable that the disease will be drug resistant (e.g., Ph+ALL, AML with 11q abnormalities, blast crisis CML), suggesting that drug resistance, or "anti–apoptosis," is closely linked to the pathogenesis of the disease. Disruption of the normal signaling pathways leading to cell death may have been an important event in the development of the malignancy and may also play a crucial role in the development of drug resistance. In addition, patients who relapse after previous courses of therapy are characteristically more resistant to their previous therapy and to other types of drug to which they had not been previously exposed (referred to as multidrug resistance, MDR). It is conceivable that cytotoxic therapy selects for an increasingly anti–apoptotic leukemic clone.

It is now apparent that many, if not all, of the cytotoxic drugs used in the treatment of leukemia kill cells by inducing apoptosis. These include cytotoxic drugs with widely disparate modes of action as defined by their primary targets, e.g., dihydrofolate reductase inhibitors [167], topoisomerase I poisons [168, 169]; topoisomerase II poisons [170, 171]; nucleoside analogs (cytosine arabinoside, fludarabine and 2–chloro–2'–deoxyadenosine) [172, 173]; microtubule poisons (vinblastine and vincristine) [174]; and DNA damaging agents, such as cisplatin and alkylating agents [175–177]. Thus unique drug/target interactions appear to be coupled to a conserved pathway for cell death.

Modulation of the expression of genes involved in apoptosis has been shown to affect chemosensitivity. Transfection of the wild–type *p53* gene into a small cell lung cancer cell line increased cellular sensitivity to cisplatin [178]. Transfection of *bcl*–x_s into MCF–7 (breast cancer cells) increased their sensitivity to etoposide 5– to 10–fold [179]. In contrast, transfection of the apoptosis inhibitor *bcl–2* into a murine lymphoid cell line conferred a 2–fold resistance to nitrogen mustard and camptothecin [180], and antisense Bcl–2 oligonucleotides increased sensitivity of AML blasts to ARA–C [181]. Overexpression of Bcl–x_L in tumor cell lines reduced the cytotoxicity of bleomycin, cisplatin, etoposide and vincristine for up to four days in culture [182]. Cells containing a temperature sensitive v–Abl protein tyrosine kinase were sensitive to melphan or hydroxyurea at the permissive temperature, but at the restrictive temperature drug–treated cells died by apoptosis [183], and downregulation of Bcr–Abl with antisense sequences renders K562 cells more susceptible to induction of apoptosis by chemotherapeutic agents [67].

A number of clinical studies have shown an association between the expression of genes involved in apoptosis and response to therapy. *p53* mutations are found in 5–10% of AML patients [184, 185] and have been shown to be a strong prognostic indicator of response to chemotherapy and survival [186]. *p53* mutations are also found in 15–30% of CLL patients [184, 187, 188] and have been associated with poor clinical outcome [147].

B–CLL lymphocytes express high levels of *bcl–2* relative to their normal counterparts [138, 140], and high expression of *bcl–2* in ALL and AML cells has also been associated with poor response to chemotherapy [155, 156].

The nature of the genes that define the capability of cells to undergo apoptosis after drug treatment may, therefore, be an important arbiter of therapeutic response in leukemia.

Future Perspectives

The ability of transfected genes and antisense molecules to modulate drug sensitivity in vitro suggests that such approaches may have therapeutic potential. The use of specific DNA molecules in vivo is however limited at present by technical problems such as gene delivery and target specificity. The uniqueness of the chimeric gene *bcr–abl* makes antisense therapy an attractive option for CML. Antisense Bcr–Abl junction sequences combined with drug therapy achieved efficient elimination of Ph^+ leukemic cells in mice [189]. A Bcr–Abl antisense oligomer has recently been used for autologous marrow purging. Three out of four patients were entirely or predominantly Ph–negative at three months [190]. A number of tyrosine kinase inhibitors have also been shown to inhibit Bcr–Abl and induce differentiation in cell lines [191–193]. The clinical use of such molecules may however be limited by their lack of specificity.

The effect of Bcl–2 is dependent upon its physical interactions with other family members (such as Bax). It may be possible to design small molecules which could specifically interfere with this interaction and drive cells into apoptosis [31]. Again cell specificity would limit such an approach in vivo. The combination of cytokines and cytotoxic therapy may be an avenue worth exploring. For example IL–2 has been shown to inhibit the proliferation of K562 cells and reduce Bcr–Abl transcript levels [194]. Cytokines have also been shown to modulate the expression of Bcl–2 [159, 195].

Survival signals from cytokines may be activating specific tyrosine kinases in lymphoblastic cells [196, 197]. A recent study found that inhibition of Jak–2 by a specific tyrosine kinase blocker, AG–490, inhibited growth of ALL cells in vitro and in vivo by inducing apoptosis, with no deleterious effects on normal hematopoiesis [198].

It is an encouraging thought that although leukemic cells may be highly resistant in vivo, they still retain a death program which is simply inactive. For example leukemic cells isolated from a patient and incubated in culture will eventually undergo apoptotic cell death. It is possible that although the normal signaling pathway to cell death may be blocked in a leukemic cell, other pathways could still be inducible. It is anticipated that a better understanding of the molecular mechanisms involved in the induction of apoptosis will prove useful in the development of new therapeutic options.

Acknowledgments

We are grateful to The Childrens Leukemia Research project and The E.U. Biomedical program for their generous support.

References

1 Kerr JFR, Wyllie AH, Currie AH. Apoptosis, a basic biological phenomenon with wider implications in tissue kinetics. Br J Cancer 1972; 26:239–245.
2 Wyllie AH, Kerr JFR, Currie AR. Cell death: the significance of apoptosis. Int Rev Cytol 1980; 68:251–306.
3 Cotter TG, Lennon SV, Glynn JM, Green DR. An intact microfibular network is necessary for the formation of apoptotic bodies in cells undergoing apoptosis. Cancer Res 1992; 52:997–1005.

4 Hall SE, Savill JS, Henson PM, Haslett C. Apoptotic neutrophils are phagocytosed by fibroblasts with participation of the fibroblast vitronectin receptor and involvement of a mannose/fucose–specific lectin. J Immunol 1994; 153:3218–3227.
5 Dini L, Lentini A, Diez Diez G, Rocha M, Falasca L, Serafino L et al. Phagocytosis of apoptotic bodies by liver endothelial cells. J Cell Sci 1995; 108:967–973.
6 Dini L, Autuori F, Lentini A, Oliverio S, Piacentini M. The clearance of apoptotic cells in the liver is mediated by the asialoglycoprotein receptor. FEBS Lett 1992; 296:174–178.
7 Savill JS, Dransfield I, Hogg N, Haslett C. Vitronectin receptor–mediated phagocytosis of cells undergoing apoptosis. Nature 1990; 342:170–173.
8 Savill JS, Hogg N, Ren Y, Haslett C. Thrombospondin cooperates with CD36 and the vitronectin receptor in macrophage recognition of neutrophils undergoing apoptosis. J Clin Invest 1992; 90:1513–1522.
9 Martin SJ, Reutelingsperger CPM, McGahon AJ, Rader JA, Van Schie RCAA, Laface DM et al. Early redistribution of plasma membrane phosphatidylserine is a general feature of apoptosis regardless of the initiating stimulus: inhibition by overexpression of bcl–2 and abl. J Exp Med 1995; 182:1545–1556.
10 Fadok VA, Voelker DR, Campbell PA, Cohen JJ, Bratton DL, Henson PM. Exposure of phosphatidylserine on the surface of apoptotic lymphocytes triggers specific recognition and removal by macrophages. J Immunol 1992; 148:2207–2216.
11 Flora PK, Gregory GD. Recognition of apoptotic cells by human macrophages: Inhibition by a monocyte/macrophage–specific monoclonal antibody. Eur J Immunol 1994; 24:2625–2632.
12 Kroemer G, Petit P, Zamzami N, Vayssiere JL, Mignotte B. The biochemistry of programmed cell death. FASEB J 1995;9:1277–1287.
13 Carbonari M, Cibati M, Fiorilli M. Measurement of apoptotic cells in peripheral blood. Cytometry 1995; 22:161–167.
14 Darzynkiewicz Z, Bruno S, Del Bino G, Gorczyca W, Hotz MA, Lassota P, Traganos F. Features of apoptotic cells measured by flow cytometry. Cytometry 1992; 13:795–808.
15 Oberhammer F, Wilson JW, Dive C, Morris ID, Hickman JA, Wakeling AE, Walker PR, Sikorska M. Apoptotic death in epithelial cells: cleavage of DNA to 300 and/or 50kb fragments prior to or in the absence of internucleosomal fragmentation. EMBO J 1993; 12(9):3670–3684.
16 Cohen GM, Sun X–M, Snowden RT, Dinsdale D, Skilleter DN. Key morphological features of apoptosis may occur in the absence of internucleosomal DNA fragmentation. Biochem J 1992; 286:331–334.
17 Filipski J, Leblanc J, Youdale T, Sikorska M, Walker PR. Periodicity of DNA folding in higher order chromatin structures. EMBO J 1990; 9:1319–1327.
18 Cohen JJ, Duke RC. Glucocorticoid activation of a calcium–dependent endonuclease in thymocyte nuclei leads to cell death. J Immunol 1984; 132:38–42.
19 Wyllie AH, Arends MJ, Morris RG, Walker SW, Evans G. The apoptosis endonuclease and its regulation. Sem Immunol 1992; 4:389–397.
20 Peitsh MC, Polzar B, Stephan H, Crompton T, MacDonald HR, Mannherz HG, Tschopp J. Characterization of the endogenous deoxyribonuclease involved in nuclear DNA degradation during apoptosis (programmed cell death). EMBO J 1993; 12:371–377.
21 Barry MA, Eastman A. Identification of deoxyribonuclease II as an endonuclease involved in apoptosis. Arch Biochem Biophys 1993; 300:440–450.
22 Gottlieb RA, Giesing HA, Engler RL, Babior BM. The acid deoxyribonuclease of neutrophils: A possible participant in apoptosis–associated genome destruction. Blood 1995; 86:2414–2418.
23 Tanuma S, Shiokawa D. Multiple forms of nuclear deoxyribonuclease in rat thymocytes Biochem Biophys Res Comm 1994; 203:789–797.
24 Montague JW, Gaido ML, Frye C, Cidlowski JA. A calcium–dependent nuclease from apoptotic rat thymocytes is homologous with cyclophilin. J Biol Chem 1994; 269:18877–18880.
25 Nikonova LV, Beletsky IP, Umansky SR. Properties of some nuclear nucleases of rat thymocytes and their changes in radiation–induced apoptosis. Eur J Biochem 1993; 215:893–901.
26 Deng G, Podack ER. Deoxyribonuclease induction in apoptotic cytotoxic T lymphocytes. FASEB J 1995; 9:665–669.

27 White E. Death–defying acts: a meeting reveiw on apoptosis. Genes and development 1993; 7:2277–2284.
28 White E. Life, death, and the pursuit of apoptosis. Genes and Development 1996; 10:1–15.
29 Wyllie AH, Morris RG, Smith AL, Dunlop D. Chromatin cleavage in apoptosis: association with condensed chromatin morphology and dependence on macromolecular synthesis. J Pathol 1984; 142:67–77.
30 Martin SJ. Apoptosis: suicide, execution or murder? Trends Cell Biol 1993; 3:141–144.
31 Reed JC. Bcl–2: Prevention of apoptosis as a mechanism of drug resistance. Hematology/oncology clinics of North America 1995; 9:451–473.
32 Jacobson MD, Burne JF, Raff MC. Programmed cell death and Bcl–2 protection in the absence of a nucleus. EMBO J 1994; 13:1899–1910.
33 Oltvai ZN, Milliman CL, Korsmeyer SJ. Bcl–2 heterodimerizes in vivo with a conserved homolog, Bax, that accelerates programmed cell death. Cell 1993; 74:609–619.
34 Yin XM, Oltvai Z, Korsmeyer S. BH1 and BH2 domains of *Bcl–2* are required for inhibition of apoptosis and heterodimerization with Bax. Nature 1994; 369:321–323.
35 Boise LH, Gonzalez–Garcia M, Postema CE, Ding L, Linsten T, Turka LA, Mao X, Nunez G, Thompson C. Bcl–x, a *bcl–2*–related gene that functions as a dominant regulator of apoptotic cell death. Cell 1993; 74:597–608.
36 Sedlak TW, Oltvai ZN, Yang E, Wang K, Boise LH, Thompson CB, Korsmeyer SJ. Multiple Bcl–2 family members demonstrate selective dimerizations with Bax. Proc Natl Acad Sci USA 1995; 92:7834–7838.
37 Yang E, Zha J, Jockel J, Boise LH, Thompson CB, Korsmeyer SJ. Bad, a Heterodimeric Partner for Bcl–xl and Bcl–2, Displaces Bax and Promotes Cell Death. Cell 1995; 80:285–291.
38 Chittenenden T, Harrington EA, O'Connor R, Flemington C, Lutz RJ, Evan GI, Guild BC. Induction of apoptosis by the Bcl–2 homolog Bak. Nature 1995; 374:733–736.
39 Prokocimer M, Rotter V. Structure and function of *p53* in normal cells and their aberrations in cancer cells: Projection on the hematologic cell lineages. Blood 1994; 84:2391–2411.
40 Yonish–Rouach E, Resnitzky D, Lotem J, Sachs L, Kimchi A, Oren M. Wild–type *p53* induces apoptosis of myeloid leukemic cells that is inhibited by interleukin–6. Nature 1991; 352:345–347.
41 Blandino G, Scardigli R, Rizzo MG, Crescenzi M, Soddu S, Sacchi A. Wild type *p53* modulates apoptosis of normal, IL–3 deprived, hematopoietic cells. Oncogene 1995; 10:731–737.
42 Miyashita T, Harigai M, Hanada M, Reed JC. Identification of a p53–dependent negative response element in the *bcl–2* gene. Cancer Res 1994; 54:3131–3135.
43 Miyashita T, Reed JC. Tumor suppressor *p53* is a direct transcriptional activator of the Human *bax* gene. Cell 1995; 80:293–299.
44 Owen–Schaub LB, Zhang W, Cusack JC, Angelo LS, Santee SM, Fujiwara T, Roth JA, Deisseroth AB, Zhang W–W, Kruzel E et al. Wild–type human *p53* and a temperature–sensitive mutant induce Fas/APO–1 expression. Mol Cell Biol 1995; 15:3032–3040.
45 Clark AR, Purdie CA, Harrison DJ, Morris RG, Bird CC, Hooper ML, Wyllie AH. Thymocyte apoptosis induced by p53–dependent and independent pathways. Nature 1993; 362:849–852.
46 Marcu KB, Bossone SA, Patel,AJ. Myc function and regulation. Ann Rev Biochem 1992; 61:809–860.
47 Evan GI, Wyllie AH, Gilbert CS, Littlewood TD, Land H, Brooks M, Waters CM, Penn LZ, Hancock DC. Induction of apoptosis in fibroblasts by c–Myc protein. Cell 1992; 69:119–128.
48 Askew DS, Ashmun RA, Simmons BC, Cleveland JL. Constitutive c–*myc* expression in an IL–3–dependent myeloid cell line suppresses cell cycle arrest and accelerates apoptosis. Oncogene 1991; 6:1915–1922.
49 Amati B, Littlewood TD, Evan GI, Land H. c–Myc protein induces cell cycle progression and apoptosis through dimerization with Max. EMBO J 1993; 12:5083–5087.
50 Harrington EA, Bennett MR, Fanidi A, Evan GI. c–Myc–induced apoptosis in fibroblasts is inhibited by specific cytokines. EMBO J 1994; 13:3286–3295.
51 Bissonnette RP, Echeverri F, Mahboubi A, Green DR. Apoptotic cell death induced by c–myc is inhibited by bcl–2. Nature 1992; 359:552.
52 Fanidi A, Harrington EA, Evan GI. Cooperative interaction between *c–myc* and *bcl–2* proto–oncogenes. Nature 1992; 359:554–556.

53 Vogt M, Lesley J, Bogenberger JM, Haggblom C, Swift S, Haas M. The induction of growth factor–independence in murine myelocytes by oncogenes results in monoclonal cell lines and is correlated with cell crisis and karyotypic instability. Oncogene Res 1987; 2:49–63.
54 Green DR, Bissonnette RP, Cotter TG. Apoptosis and Cancer. In: De Vita VT, Hellman S, Rosenberg SA, eds. Important Advances in Oncology. Philadelphia: Lippincott Company, 1994.
55 Barbacid M. *Ras* genes. Ann Rev Biochem 1987; 56:779–827.
56 Weinberg RA. The action of oncogenes in the cytoplasm and nucleus. Science 1985; 230:770–776.
57 Tanaka N, Ishihara M, Kitagawa M, Harada H, Kimura T, Matsuyama T, Lamphier MS, Aizawa S, Mak TW, Taniguchi T. Cellular Commitment to Oncogene–Induced Transformation or Apoptosis Is Dependent on the Transcription Factor IRF–1. Cell 1994; 77:829–839.
58 Lin HJL, Eivner V, Prendergast GC, White E. Activated H–ras rescues E1A–induced apoptosis and cooperates with E1A to overcome p53–dependent growth arrest. Mol Cell Biol 1995; 15:4536–4544.
59 Arends MJ, McGregor AH, Toft NJ, Brown EJH, Wyllie AH. Susceptibility to apoptosis is differentially regulated by *c–myc* and mutated *Ha–ras* oncogenes and is associated with endonuclease availability. Br J Cancer 1993; 68:1127–1133.
60 Moore J, Boswell S, Hoffman R, Burgess G, Hromas R. Mutant H–ras overexpression inhibits a random apoptotic nuclease in myeloid leukemia cells. Leukemia Res 1993; 17:703–709.
61 Fernandez–Sarabia MJ, Bischoff JR. Bcl–2 associates with the ras–related protein R–ras p23. Nature 1994; 366:274–275.
62 Wang H–G, Millan JA, Cox AD, Der CJ, Rapp UR, Beck T, Zha H, Reed JC. R–Ras promotes apoptosis caused by growth factor deprivation via a bcl–2 suppressible mechanism. J Cell Biol 1995; 129:1103–1114.
63 Samali A, Gorman AM, Cotter TG. Role of bcr–abl kinase in resistance to apoptosis. Adv Pharmacol 1996; (in press).
64 Rovera G, Valtieri M, Maullo F, Reedy EP. Effect of Abelson murine leukemia virus on granulocytic differentiation and interleukin–3 dependence of a murine progenitor cell line. Oncogene 1987; 1:29–35.
65 Daley GQ, Baltimore D. Transformation of an interleukin 3–dependent hematopoietic cell line by the chronic myelogenous leukemia–specific $p^{210bcr-abl}$ protein. Proc Natl Acad Sci USA 1988; 85:9312–9319.
66 Hariharan IK, Adams JM, Cory S. bcr–abl oncogenes renders myeloid cell line factor independent: Potential autocrine mechanisms in chronic myeloid leukemia. Oncogene Res. 1988; 3:387–399.
67 McGahon A, Bissonnette R, Schmitt M, Cotter KM, Green DR, Cotter TG. Bcr–Abl maintains resistance of chronic myelogenous leukemia cells to apoptotic cell death. Blood 1994; 83:1179–1187.
68 McGahon AJ, Nishioka WK, Martin SJ, Mahboubi A, Cotter TG, Green DR. Regulation of the Fas apoptotic cell death pathway by Abl. J Biol Chem 1995; 270:22625–22631.
69 Sanchez–Garcia I, Grutz G. Tumorigenic activity of the *bcr–abl* oncogenes is mediated by BCL–2. Proc Natl Acad Sci USA 1995; 92:5287–91.
70 Gruss H–J, Dower SK. Tumor necrosis factor ligand superfamily: involvement in the pathology of malignant lymphomas. Blood 1995; 85:3378–3404.
71 Cifone GM, Maria RD, Roncaioli P, Rippo MR, Azuma M, Lanir LL, Santoni A, Testi R. Apoptotic signaling through CD95 (Fas/Apo–1) activates an acidic sphingomyelinase. J Exp Med 1994; 177:1547–1552.
72 Enari M, Hug H, Nagata S. Involvement of an ICE–like protease in Fas–mediated apoptosis. Nature 1995; 375:78–81.
73 Los M, Van de Craen M, Penning LC, Schenk H, Westendorp M, Baeuerle PA, Droge W, Krammer PH, Fiers W, Schulze–Osthoff K. Requirement of an ICE/CED–3 protease for Fas/APO–1–mediated apoptosis. Nature 1995; 375:81–83.
74 Nagata S, Suda T. Fas and Fas ligand: *lpr* and *gld* mutations. Immunol Today 1995; 16:39–42.
75 Williams GT, Smith CA, Spooncer E, Dexter TM, Taylor DR. Hemopoietic colony stimulating factors promote cell survival by suppressing apoptosis. Nature 1990; 343:76–79.

76 Rodriguez–Tarduchy G, Collins M, Lopez–Rivas A. Regulation of apoptosis in interleukin–3–dependent hemopoietic cells by interleukin–3 and calcium ionophores. EMBO J 1990; 9:2997–3002.
77 Lopez AF, Williamson J, Gamble JR, Begley CG, Harlan JM, Klebanoff SJ, Waltersdorph A, Wong G, Clark SC, Vadas MA. Recombinant human granulocyte–macrophage colony–stimulating factor stimulates in vitro mature human neutrophil and eosinophil function, surface receptor expression and survival. J Clin Invest 1986; 78:1230.
78 Mangan DF, Wahl SM. Differential regulation of human monocyte programmed cell death (apoptosis) by chemotactic factors and pro–inflammatory cytokines. J Immunol 1991; 147:3408–3412.
79 Hogquist KA, Nett MA, Unanue ER, Chaplin DD. Interleukin 1 is processed and released during apoptosis. Proc Natl Acad Sci USA 1991; 88:8485–8489.
80 Lotem J, Sachs L. Interferon–g inhibits apoptosis induced by wild–type p53, cytotoxic anticancer agents and viability factor deprivation in myeloid cells. Leukemia 1995; 9:685–692.
81 Tushinski R, Oliver IT, Guilbert LJ, Tynan PW, Warner JR, Stanley ER. Survival of mononuclear phagocytes depends on a lineage specific growth factor that the differentiated cells selectively destroy. Cell 1982; 28:71–81.
82 Lotem J, Sachs L. Selective regulation of the activity of different hematopoietic regulatory proteins by transforming growth factor 1 in normal and leukemic myeloid cells. Blood 1990; 76:1315–1322.
83 Caceres–Cortes J, Rajotte D, Dumouchel J, Haddad P, Hoang T. Product of the Steel Locus Suppresses Apoptosis in Hemopoietic Cells. J Biol Chem 1994; 269:12084–12091.
84 Begley CG, Lopez AF, Nicola NA, Warren DJ, Vadas MA, Sanderson CJ, Medcalf D. Purified CSFs enhance survival of human neutrophils and eosinophils in vitro: a rapid and sensitive microassay for CSFs. Blood 1986; 68:162–166.
85 Fairbairn LJ, Cowling GJ, Reipert BM, Dexter TM. Suppression of Apoptosis Allows Differentiation and Development of a Multipotent Hemopoietic Cell Line in the Absence of Added Growth Factors. Cell 1993; 74:823–832.
86 Liebermann D, Hoffman–Liebermann B. Proto–oncogene expression and dissection of the myeloid growth to differentiation developmental cascade. Oncogene 1989; 4:583–592.
87 Collins SJ. The HL–60 promyelocytic leukemia cell line: Proliferation, differentiation, and cellular oncogene expression. Blood 1987; 70:1233.
88 Hockenbery DM, Zutter M, Hickey W, Nahm M, Korsmeyer SJ. Bcl–2 protein is topographically restricted in tissues characterized by apoptotic cell death. Proc Natl Acad Sci USA 1991; 88:6961–6965.
89 Delia D, Aiello A, Soligo D, Fontella E, Melani C, Pezzella F, Pierotti M, Della Porta G. Bcl–2 proto–oncogene expression in normal and neoplastic human myeloid cells. Blood 1992; 79:1291–1298.
90 Naumovski L, Cleary ML. Bcl2 inhibits apoptosis associated with terminal differentiation of HL–60 myeloid leukemia cells. Blood 1994; 83:2261–2267.
91 Park JR, Bernstein ID, Hockenbery DM. Primitive human hematopoietic precursors express Bcl–x but Not Bcl–2. Blood 1995; 86:868–876.
92 Motoyama N, Wang F, Roth K, Sawa H, Nakayama K, Nakayama K, Negishi I, Senju S, Zhang Q, Satoshi F, Loh D. Massive cell death of immature hematopoietic cells and neurons in Bcl–x–deficient mice. Science 1995; 267:1506–1510.
93 Liles WC, Klebanoff SJ. Regulation of apoptosis in neutrophils–Fas track to death? J Immunol 1995; 155:3289–3291.
94 Holder MJ, Knox K, Gorden J. Factors modifying survival pathways of germinal center B cells. Glucocorticoids and transforming growth factorb, but not cyclosporin A or anti–CD19, block surface immunoglobulin mediated rescue from apoptosis. Eur J Immunol 1992; 22:2725–2728.
95 Tsubata T, Wu J, Honjo T. B cell apoptosis induced by antigen receptor cross linking is blocked by a T cell signal through CD40. Nature 1993; 364:645–648.
96 Clark EA, Ledbetter JA. How B and T cells talk to each other. Nature 1994; 367:425–428.

97 Baixeras E, Bosca L, Stauber C, Gonzalez A, Carrera AC, Gonzalo JA, Martinez–A C. From apoptosis to autoimmunity: Insights from the signaling pathways leading to proliferation or to programmed cell death. Immunol Rev 1994; 142:53–91.
98 Nunez G, London L, Hokenbery D, Alexander M, McKearn JP, Korsmeyer SJ. Deregulated *Bcl–2* gene expression selectively prolongs survival of growth factor–deprived hemopoietic cell lines. J Immunol 1990; 144:3602–3610.
99 Liu YJ, Mason DY, Johnson GD, Abot S, Gregory CD, Hardie DL, Gordon J, MacLennan IC. Germinal center cells express bcl–2 protein after activation by signals which prevent their entry into apoptosis. Eur J Immunol 1991; 21:1905–1910.
100 Nunez G, Hockenbery D, McDonnell TJ, Sorensen CM, Korsmeyer SJ. Bcl–2 maintains B cell memory. Nature 1991; 353:71–73.
101 McDonnell TJ, Korsmeyer SJ. Progression from lymphoid hyperplasia to high–grade malignant lymphoma in mice transgenic for the t(14;18). Nature 1991; 349:254–256.
102 Strasser A, Whittingham S, Vaux DL, Bath ML, Adams MJ, Cory S, Harris AW. Enforced *bcl–2* expression in B lymphoid cells prolonges antibody responses and elicits autoimmune disease. Proc Natl Acad Sci USA 1991; 88:8661–8665.
103 Nisitani S, Tsubata T, Murakami M, Okamoto M, Honjo T. The *bcl–2* gene product inhibits clonal deletion of self–reactive B lymphocytes in the periphery but not in the bone marrow. J Exp Med 1993; 178:1247–1254.
104 Kamada S, Shinto AA, Tsujimura Y, Takahashi T, Noda T, Kitamura Y, Kondoh H, Tsujimoto Y. Bcl–2 deficiency in mice leads to pleiotropic abnormalities: accelerated lymphoid cell death in the thymus and spleen, polycystic kidney, hair hypopigmentation, and distorted small intestine. Cancer Res 1995; 55:354–359.
105 Nakayama KI, Nakayama K, Negishi I, Kuida K, Shinkai Y, Louie MC, Fields LE, Lucas PJ, Stewart V, Alt FW, Loh DY. Disappearance of the lymphoid system in *bcl–2* homozygous mutant chimeric mice. Science 1993; 261:1584–1588.
106 Choi MS, Boise LH, Gottschalk AR, Quintans J, Thompson CB, Klaus GG. The role of bcl–XL in CD40–mediated rescue from anti–mu–induced apoptosis in WEHI–231 B lymphoma cells. Eur J Immunol 1995; 25:1352–1357.
107 Bargou RC, Bommert K, Weinmann P, Daniel PT, Wagener C, Mapara MY, Dorken B. Induction of Bax–alpha precedes apoptosis in a human B lymphoma cell line: potential role of the *bcl–2* gene family in surface IgM–mediated apoptosis. Eur J Immunol 1995; 25:770–775.
108 Knudson CM, Tung K, Brown G, Korsmeyer SJ. Bax deficient mice demonstrate lymphoid hyperplasia but male germ cell death. Science 1995; 270:96–99.
109 Rathmell JC, Cooke MP, Ho WY, Grein J, Townsend SE, Davis MM, Goodnow CC. CD95 (Fas)–dependent elimination of self–reactive B cells upon interaction with CD4+ T cells. Nature 1995; 376:181–184.
110 Rathmell JC, Goodnow CC. The Fas track: the discovery of mutations in the gene encoding Fas in patients with an autoimmune lymphoproliferative syndrome, and studies of Fas–deficient mice provide the first molecular view of the pathogenesis of autoimmunity. Curr Biol 1995; 5:1218–1221.
111 Weissman IL. Development switches in the immune system. Cell 1994; 76:207–218.
112 Von Boehmer H. Positive selection of lymphocytes. Cell 1994; 76:219–228.
113 Nossal GJV. Negative selection of lymphocytes. Cell 1994; 76:229–239.
114 Smith CA, Williams GT, Kingston R, Jenkinson EJ, Owen JJT. Antibodies to CD3/T cell receptor complex induce death by apoptosis in immature T cells in thymic cultures. Nature 1989; 337:181–184.
115 McDonald HR, Lees RK. Programmed cell death of autoreactive thymocytes. Nature 1990; 343:642–645.
116 Shi YF, Bissonnette RP, Parfrey N, Szalay M, Kubo RT, Green DR. In vivo administration of monoclonal antibodies to the CD3 T cell receptor complex induces cell death (apoptosis) in immature thymocytes. J Immunol 1991; 146:3340–3346.
117 Finkel TH, Marrack P, Kappler JW, Kubo RT, Cambier JC. ab T cell receptor and CD3 transduce different signals in immature T cells. Implications for selection and tolerance. J Immunol 1989; 142:3006–3012.

118 Shi Y, Glynn JM, Guilbert LJ, Cotter TG, Bissonnette RP, Green DR. Role for *c–myc* in activation–induced apoptotic cell death in T cell hybdidomas. Science 1992; 257:212–214.

119 Veis DJ, Sentman CL, Bach EA, Korsmeyer SJ. Expression of the Bcl–2 protein in murine and human thymocytes and in peripheral T lymphocytes. Cell 1993; 151:2546–2554.

120 Strasser A, Harris AW, Cory S. *Bcl–2* transgene inhibits T cell death and perturbs thymic self–censorship. Cell 1991; 67:889–899.

121 Berke G. The CTL's Kiss of Death. Cell 1995; 81:9–12.

122 Cory S, Harris AW, Strasser A. Insights from transgenic mice regarding the role of bcl–2 in normal and neoplastic lymphoid cells. Philos–Trans–R–Soc–Lond–B–Biol–Sci. 1994; 345:289–295.

123 Langdon WY, Harris AW, Cory S, Adams JM. The *c–myc* oncogene perturbs B lymphocyte development in *Eμ–myc* transgenic mice. Cell 1986; 47:11–18.

124 Daley GQ, Van Etten RA, Baltimore D. Blast crisis in a murine model of chronic myelogenous leukemia. Proc Natl Acad Sci USA 1991; 88:11335–11338.

125 Spiers ASD. The chronic myeloid leukemias. In: Whittaker JA, ed. Leukemia. Oxford: Blackwell Scientific Publications, 1992:434–467.

126 McCarthy DM, Goldman JM, Rasool FV, Graham SV, Birnie GD. Genomic alterations in the *c–myc* proto–oncogene during the evolution of a case of chronic granulocytic leukemia. Lancet 1984; 2:1362–1365.

127 Liu E, Hjelle B, Bishop JM. Transforming genes in chronic myelogenous leukemia. Proc Natl Acad Sci 1988; 85:1952–1956.

128 Ahuja H, Bar–Eli M, Arlin Z, Advani S, Allen SL, Goldman J, Snyder D, Foti A, Cline M. The spectrum of molecular alterations in the evolution of chronic myelocytic leukemia. J Clin Inves 1991; 87:2042–2047.

129 Nakai H, Misawa S, Toguchida J, Yandell DW, Ishizaki K. Frequent *p53* gene mutations in blast crisis of chronic myelogenous leukemia, especially in myeloid crisis harboring loss of chromosome 17p. Cancer Res 1992; 52:6588–6593.

130 Neubauer A, He M, Schmidt CA, Huhn D, Liu ET. Genetic alterations in the *p53* gene in the blast crisis of chronic myelogenous leukemia. Leukemia 1993; 7:593–600.

131 Neubauer A, Neubauer B, Liu E. Polymerase chain reaction based assay to detect allelic loss in human DNA: loss of β–interferon gene in chronic myelogenous leukemia. Nuc Acids Res 1990; 18:993–998.

132 Wetzler M, Kurzrock R, Lowe DG, Kantarjian H, Gutterman JU, Talpaz M. Alteration in bone marrow adherent layer growth factor expression: A novel mechanism in chronic myelogenous leukemia progression. Blood 1991; 78:2400–2406.

133 Caligaris–Cappio F, Gottardi D, Alfarano A, Stacchini A, Gregoretti MG, Ghia P, Bertero MT, Novarino A, Bergui L. The nature of the B lymphocyte in B–chronic lymphocytic leukemia. Blood Cells 1993; 19:601–613.

134 Juliusson G, Oscier DG, Fitchett M, Ross FM, Stockdill G, Mackie MJ, Parker AC, Castoldi GL, Cuneo A, Knuutila S, Elonen E, Gahrton G. Prognostic subgroups in B cell–chronic lymphocytic leukemia defined by specific chromosome abnormalities. N Engl J Med 1990; 323:720–724.

135 Neri A, Knowles DM, Greco A, McCormick F, Dalla Favera R. Analysis of *RAS* oncogene mutations in human lymphoid malignancies. Proc Natl Acad Sci USA 1988; 85:9268–9272.

136 Browett PJ, Yaxley JC, Norton JD. Activation of Harvey *ras* oncogene by mutation at codon 12 is very rare in hemopoietic malignancies. Leukemia 1989; 3:86–88.

137 O'Brien S, Del Giglio A, Keating M. Advances in the biology and treatment of B cell chronic lymphocytic leukemia. Blood 1995; 85:307–318.

138 Schena M, Larsson LG, Gottardi D, Gaidano G, Carlsson M, Nilsson K, Caligaris–Cappio F. Growth and differention association of *bcl–2* in B–chronic lymphocytic leukemia cells. Blood 1992; 79:2981–2989.

139 Mariano M, Moretti l, Donelli A, Grantini M, Montagnani G, Di Prisco AU, Torelli G, Torelli U, Narni F. *Bcl–2* gene expression in hematopoietic cell differentiation. Blood Rev 1993; 80:768–775.

140 Hanada M, Delia D, Aiello A, Stadtmauer E, Reed JC. *Bcl–2* gene hypomethylation and high level expression in B cell chronic lymphocytic leukemia. Blood 1993; 82:1820–1828.

141 Mainou–Fowler T, Copplestone JA, Prentice AG. Effect of interleukins on the proliferation and survival of B cell chronic lymphocytic leukemia cells. J Clin Pathol 1995; 48:482–487.
142 Fournier S, Yang LP, Delespesse G, Rubio M, Biron G, Sarfati M. The two CD23 isoforms display differential regulation in chronic lymphocytic leukemia. Br J Hematol 1995; 89:373–379.
143 Jewell AP, Worman CP, Lydyard PM, Yong KL, Giles FJ, Goldstone AH. Interferon–alpha upregulates bcl–2 expression and protects B–CLL cells from apoptosis in vitro and in vivo. Br J Hematol 1994; 88:268–274.
144 Long BW, Witte PL, Abraham GN, Gregory SA, Plate JM. Apoptosis and interleukin 7 gene expression in chronic B lymphocytic leukemia cells. Proc Natl Acad Sci USA 1995; 92:1416–20.
145 Mapara MY, Bargou R, Zugck C, Dohner H, Ustaoglu F, Jonker RR, Krammer PH, Dorken B. APO–1 mediated apoptosis or proliferation in human chronic B lymphocytic leukemia: correlation with bcl–2 oncogene expression. Eur J Immunol 1993; 23:702–708.
146 Panayiotidis P, Ganeshaguru K, Foroni L, Hoffbrand AV. Expression and function of the FAS antigen in B chronic lymphocytic leukemia and hairy cell leukemia. Leukemia 1995; 9:1227–1232.
147 Gale RP, Caligaris–Cappio F, Dighiero G, Keating M, Montserrat E, Rai K. Recent progress in chronic lymphocytic leukemia. Leukemia 1994; 8:1610–1614.
148 El Rouby S, Thomas A, Costin D, Rosenberg CR, Potmesil M, Silber R, Newcomb EW. *p53* gene mutation in B cell chronic lymphocytic leukemia is associated with drug resistance and is independent of *MDR1/MDR3* gene expression. Blood 1993; 82:3452–3459.
149 Sawitsky A, Rai KR. The chronic lymphoid leukemias. In: Whittaker JA, ed. Leukemia. Oxford: Blackwell Scientific Publications, 1992:468–494.
150 Clarkson B, Chabner BA, Weinberg RA. Targets for specific therapies in leukemia. Meeting report. Cancer Res 1995; 55:4488–4498.
151 Campana D, Coustan–Smith E, Manabe A, Buschle M, Raimondi S, Behm F, Ashmun R, Arico M, Biondi A, Pui Ch. Prolonged survival of B–lineage acute lymphoblastic leukemia cells is accompanied by overexpression of bcl–2 protein. Blood 1993; 81:1025–1031.
152 Pocock CFE, Malone M, Booth M, Evans M, Morgan G, Greil J, Cotter FE. BCL–2 expression by leukemic blasts in a SCID mouse model of biphenotypic leukemia associated with the t(4;11)(q21;q23) translocation. Br J Hematol 1995; 90:855–867.
153 Lucking–Famira KM, Daniel PT, Moller P, Krammer PH, Debatin KM. APO–1 (CD95) mediated apoptosis in human T–ALL engrafted in SCID mice. Leukemia 1994; 8:1825–1833.
154 Debatin KM, Krammer PH. Resistance to APO–1 (CD95) induced apoptosis in T–ALL is determined by BCL–2 independent anti–apoptotic program. Leukemia 1995; 9:815–820.
155 Campos L, Rouault J–P, Sabido O, Oriol P, Roubi N, Vasselon C, Archimbaud E, Magaud J–P, Guyotat D. High expression of bcl–2 protein in acute myeloid leukemia cells is associated with poor response to chemotherapy. Blood 1993; 81:3091–3096.
156 Maung ZT, MacLean FR, Reid MM, Pearson ADJ, Proctor SJ, Hamilton PJ, Hall AG. The relationship between *bcl–2* expression and response to therapy in acute leukemia. Br J Hematol 1994; 88:105–109.
157 Bensi L, Longo R, Vecchi A, Garagnani L, Bernardi S, Tamassia MG, Sacchi S. Bcl–2 oncoprotein expression in acute myeloid leukemia. Hematologica 1995; 80:98–102.
158 Hunter AE, Rogers SY, Reilly IAG, Barrett AJ, Russell NH. Autonomous growth of blast cells is associated with reduced survival in acute myeloblastic leukemia. Blood 1993; 82:899–903.
159 Russell NH, Hunter AE, Bradbury D, Zhu YM, Keith–F. Biological features of leukemic cells associated with autonomous growth and reduced survival in acute myeloblastic leukemia. Leuk–Lymphoma 1995; 16:223–229.
160 Andreeff M, Jiang S, Consoli U, Brandes J, Sanchez–Williams G, Deisseroth A, Estey E. In vivo regulation of Bcl–2 expression in AML progenitors by granulocyte–colony stimulating factor (G–CSF) and direct evidence for selection of Bcl–2^{++} cells by induction therapy of AML. BLOOD 1995; 86(10):2033(abstr.).
161 Rabbitts TH. Translocations, master genes and differences between the origins of acute and chronic leukemias. Cell 1991; 67:641–644.
162 Rabbitts TH. Chromosomal translocations in human cancer. Nature 1994; 372:143–149.

163 Geary CG, Macheta AT. Preleukemia and myelodysplasia: Morphology, clinical presentation and treatment. In: JA Whittaker, ed. Leukemia. Oxford: Blackwell Scientific Publications, 1992:509–540.

164 Clark DM, Lampert IA. Apoptosis is a common histopathological finding in myelodysplasia: the correlate of ineffective hematopoiesis. Leukemia and Lymphoma 1990; 2:415–418.

165 Raza A, Gezer S, Mundle S, Gao XZ, Alvi S, Borok R, Rifkin S, Iftikhar A, Shetty V, Parcharidou A. Apoptosis in bone marrow biopsy samples involving stromal and hematopoietic cells in 50 patients with myelodysplastic syndromes. Blood 1995; 86:268–276.

166 Jacobs A, Culligan D. Myelodysplasia and preleukemia: Pathogenesis and functional aspects. In: JA Whittaker, ed. Leukemia. Oxford: Blackwell Scientific Publications, 1992:227–250.

167 Lorico A, Toffoli G, Biocchi M. Accumulation of DNA strand breaks in cells exposed to methotrexate or N10–propargyl–5,8–dideazafolic acid. Cancer Res 1988; 48:2036–2041.

168 Solary E, Bertrand R, Kohn KW, Pommier Y. Differential induction of apoptosis in undifferentiated and differentiated HL60 cells by DNA topoisomerase I and II inhibitors. Blood 1993; 81:1359–1368.

169 Gorczyca W, Melamed MR, Darzynkiewicz Z. Apoptosis of S–phase HL60 cells induced by DNA topoisomerase II inhibitors: Detection of DNA strand breaks by flow cytometry using the in situ nick translation assay. Toxicol Lett 1993; 67:249–258.

170 Walker PR, Smith C, Youdale T, Leblanc J, Whitfield JF, Sikorska M. Topoisomerase II–reactive chemotherapeutic drugs induce apoptosis in thymocytes. Cancer Res. 1991; 51:1078–1085.

171 Ling Y–H, Priebe W, Perez–Soler R. Apoptosis induced by anthracycline antibiotics in P388 parent and multidrug–resistant cells. Cancer Res 1993; 53:1845–1852.

172 Tosi P, Visani G, Ottaviani E, Manfroi S, Luigi Zinzani P, Tura S. Fludarabine + ARA–C + G–CSF: Cytotoxic effect and induction of apoptosis on fresh acute myeloid leukemia cells. Leukemia 1994; 8:2076–2082.

173 Robertson LE, Chubb S, Meyn RE, Story M, Ford R, Hittelman WN, Plunkett W. Induction of apoptotic cell death in chronic lymphocytic leukemia by 2–chloro–2'–deoxyadenosine and 9–beta–D–arabinosyl–2–fluoroadenine. Blood 1993; 81:143–150.

174 Martin SJ, Cotter TG. Disruption of microtubules induces an endogenous suicide pathway in human leukemia HL–60 cells. Cell Tissue Kinet 1990; 23:545–559.

175 Eastman A. Activation of programmed cell death by anticancer agents: cisplatin as a model system. Cancer Cells 1990; 2:275–280.

176 Frankfurt OS, Byrnes JJ, Seckinger D, Sugarbaker EV. Apoptosis (programmed cell death) and the evaluation of chemosensitivity in chronic lymphocytic leukemia and lymphoma. Oncology Res 1993; 5:37–42.

177 Begleiter A, Lee K, Israels LG, Mowat MRA, Johnstone JB. Chlorambucil induced apoptosis in chronic lymphocytic leukemia (CLL) and its relationship to clinical efficacy. Leukemia 1994; 8 Supp 1:S103–S106.

178 Fujiwara T, Grimm EA, Mukhopadhyay T, Zhang W–W, Owen–Schaub LB, Roth JA. Induction of chemosensitivity in human lung cancer cells in vivo by adenovirus–mediated transfer of the wild type *p53* gene. Cancer Res 1994; 54:2287–2291.

179 Sumantran VN, Ealovega MW, Nunez G, Clarke MF, Wicha MS. Overexpression of bclx–s, a dominant negative inhibitor of bcl–2 sensitizes MCF–7 cells to chemotherapy induced apoptosis. Proc Am Assoc Cancer Res 1995; 36:111.

180 Walton MI, Whysong D, O'Connor PM, Hockenbery D, Korsmeyer SJ, Kohn KW. Constitutive expression of human *Bcl–2* modulates nitrogen mustard and camptothecin induced apoptosis. Cancer Res 1993; 53:1853–1861.

181 Keith FJ, Bradbury DA, Zhu Y–M, Russell NH. Inhibition of bcl–2 with antisense oligonucleotides induces apoptosis and increases the sensitivity of AML blasts to Ara–C. Leukemia 1995; 9:131–138.

182 Minn AJ, Rudin CM, Boise LH, Thompson CB. Expression of Bcl–x_L can confer a multidrug resistance phenotype. Blood 1995; 86:1903–1910.

183 Chapman RS, Whetton AD, Dive C. The suppression of drug–induced apoptosis by activation of v–ABL protein tyrosine kinase. Cancer Res 1994; 54:5131–5137.

184 Imamura J, Miyoshi I, Koeffler KP. p53 in hematologic malignancies. Blood 1994; 84:2412–2421.

185 Fenaux P, Jonveaux P, Quiquandon I, Lai JL, Pignon JM, Loucheux–Lefebvre MH, Bauters F, Berger R, Kerckaert JP. *p53* gene mutations in acute myeloid leukemia with 17p monosomy. Blood 1991; 78:1652–1657.
186 Wattel E, Preudhomme C, Hecquet B, Vanrumbeke M, Quesnel B, Dervite I, Morel P, Fenaux P. *p53* mutations are associated with resistance to chemotherapy and short survival in hematologic malignancies. Blood 1994; 84:3148–3157.
187 Gaidano G, Ballerini P, Gong J, Inghirami G, Neri A, Newcomb E, Magrath I, Knowles D, Dalla Favera R. *p53* mutations in human lymphoid malignancies: association with Burkitt lymphoma and chronic lymphocytic leukemia. Proc Natl Acad Sci USA 1991; 88:5413–5417.
188 Fenaux P, Preudhomme C, Lai JL, Quiquandon I, Jonveaux PH, Vanrumbeke M, Sartiaux C, Morel P, Loucheux–Lefebvre MH, Bauters F, Berger R, Kerckaert JP. Mutations of the *p53* gene in B cell chronic lymphocytic leukemia. A report on 39 cases with cytogenic analysis. Leukemia 1992; 6:246–250.
189 Skorski T, Nieborowska–Skorska M, Barletta C, Malaguarnera L, Szczylik C, Chen T–S, Lange B, Calabretta B. Highly efficient elimination of Philadelphia leukemic cells by exposure to bcr/abl antisense oligodeoxynucleotides combined with mafosfamide. J Clin Invest 1993; 92:194–202.
190 De Fabritiis P, Amadori S, Petti MC al et. In vitro purging with Bcr–Abl antisense oligodeoxynucleotides does not prevent hematologic reconstitution after autologous bone marrow transplantation. Leukemia 1995; 9:662–664.
191 Honma Y, Obake–kado J, Hozumi M, Uehara Y, Mizuno S. Induction of erythroid differentiation of K562 human leukemic cells by herbimycin A, an inhibitor of tyrosine kinase activity. Cancer Res 1989; 49:331–334.
192 Constantinou A, Kiguchi K, Huberman E. Induction of differentiation and DNA strand breakage in human HL60 and K562 leukemia cells by genistein. Cancer Res 1990; 50:2618–2624.
193 Anafi M, Gazit A, Zehavi A, Ben–Neriah Y, Levitzki A. Tyrphostin–induced inhibition of p210bcr–abl tyrosine kinase activity induces K562 to differentiate. Blood 1993; 82:3524–3529.
194 Dilloo D, Hanenberg H, Lion T, Burdach S. IL–2 inhibits proliferation of K562 cells and reduces accumulation of bcr/abl mRNA and oncoprotein. Leukemia 1995; 9:419–424.
195 Lotem J, Sachs L. Control of sensitivity to induction of apoptosis in myeloid leukemic cells by differentiation and *bcl–2* dependent and independent pathways. Cell Growth Differ 1994; 5:321–327.
196 Grimaldi JC, Meeker TC. The t(5;14) chromosomal translocation in a case of acute lymphocytic leukemia joins the *interleukin–3* gene to the immunoglobulin heavy chain. Blood 1989; 73:2081–2085.
197 Dadi H, Ke S, Roifman CM. Interleukin 7 receptor mediates the activation of phosphatidylinositol–3 kinase in human B cell precursors. Biochem Biophys Res Commun 1993; 192:459–464.
198 Meydan N, Grunberger T, Dadi H, Shahar M, Arpaia E, Lapidot Z, Leeder JS, Freedman M, Cohen A, Gazit A, Levitzki A, Roifman CM. Inhibition of acute lymphoblastic leukemia by a Jak–2 inhibitor. Nature 1996; 379:645–648.

Chapter 12

Apoptosis and Cancer, edited by Seamus J. Martin.

Cell Death In Neuroblastoma Tumors

Gerry Melino,[a] Margherita Annicchiarico–Petruzzelli,[a] Penny Lovat,[b] Maria Grazia Farrace,[c] Lucia Piredda[c] and Mauro Piacentini[c]

[a] University of L'Aquila and IDI-IRCCS Biochemistry Lab, University of Rome "Tor Vergata", Rome, Italy
[b] Department of Child Health, University of Newcastle Upon Tyne, United Kingdom
[c] Department of Biology, University of Rome "Tor Vergata", Roma, Italy

Neuroblastoma: Characteristic and Prognostic Features

Neuroblastoma, first described by Virchow in 1864, is the most common extracranial solid tumor of childhood. As a solid tumor of the sympathetic nervous system it accounts for 9% of all childhood cancers occurring in children less than 15 years of age with an incidence of 1:7000 children before the age of five [1]. Neuroblastoma cells arise during early embryogenesis when neural crest cells are, in some way, blocked from entry into a differentiation phase [2].

Neuroblastoma may originate anywhere along the sympathetic nervous system. The most common sites of primary tumors are within the abdomen, 40% in the adrenal gland or 25% in the paraspinal ganglion. Approximately 50% of infants (children less than one year of age) and 70% of older children present evidence of tumor spread beyond the primary site. The most common sites of metastases are lymph nodes, bone marrow, liver and subcutaneous tissue. The prognostic classification of neuroblastoma is based on the refinement of clinical and biological features; indeed, numerous proposals for the prognostic classification of neuroblastoma have been described including the Evan's staging system [3] and classifications proposed by the Pediatric Oncology Group (POG, 4), the Childrens Cancer Study Group (CCSG) and the International Staging System (INSS) (reviewed by Castleberry, 1992). Current treatment recommendations are largely based on the INSS system which combines selected criteria of both the CCSG and POG.

Neuroblastoma is one of the most biologically interesting tumors of childhood in that not only does spontaneous regression occur, but a maturation to benign ganglioneuroma can be induced pharmacologically, in which both apoptosis and/or differentiation contribute to a benign phenotype. The precise mechanisms of how this regression occurs, however, are essentially unknown. Although spontaneous regression occurs in as much as 7% of cases [2] and high–dose chemotherapy and total body irradiation followed by bone marrow rescue have extended the long–term survival of patients with metastatic disease [5–6], the overall prognosis still remains very poor.

The prognosis for a child with neuroblastoma depends on both age and stage [7], and unfortunately the vast majority of patients are stage IV and older than two years of age. Children with early stage disease can be managed with limited therapy with excellent survival, circa 80–90% survival at five years. In contrast, 80% of children with advanced stage disease at diagnosis have a much worse prognosis, with a five years survival of approximately 20–25%. The overall prognosis in children less than one year of age is significantly better regardless of the extent of tumor at diagnosis [7]. Other important prognostic features include cytogenetic abnormalities as well as the expression of biological markers. Cytogenetic studies of neuroblastoma cells have demonstrated abnormalities in approximately 80% of cases [8–10]. The most consistent abnormality is a deletion or rearrangement of the short arm of chromosome 1. The cellular proto–oncogene N–*myc* is also often amplified in neuroblastoma. N–*myc* is normally found as a single copy on the short arm of chromosome 2 (ref. 11); while in approximately 50% of tumors from patients with disseminated neuroblastoma, N–*myc* is amplified, translocated and multiple copies (up to 300) can be detected. Hence, deletion of chromosome 1 and N–*myc* amplification correlate with a poor prognosis [1, 12]. The high expression of N–*myc* and its relation to prognosis might be a relevant event in controlling the growth and death of the tumor cells, in view of the role played by Myc both in the cell cycle and in programmed cell death.

Phenotypic Regulation of Neuroblastoma In Vitro

Phenotypic relationships within and between human neuroblastoma cell lines are complex, indicating that, clinically, neuroblastoma may not be a single cell disease. On the basis of laboratory studies, the phenotype of neuroblastoma cells may be classified into neuroblastic (N), substrate–adherent (S) or intermediate–types (I) [13–14] (Figs. 12.1*a,c*). I–type cells may be either the precursors to, or an intermediate transition phase of, N– and S–type cells which may, nevertheless, retain the ability to transdifferentiate among the other phenotypes [15–17]. S–type cells represent characteristics similar to schwannian, glial and melanocytic cells, reminiscent of different embryonic stages. These phenotypes appear to be present in all neuroblastoma tumors, and the more aggressive tumors in patients with metastatic disease seem comprised of predominantly N–type cells.

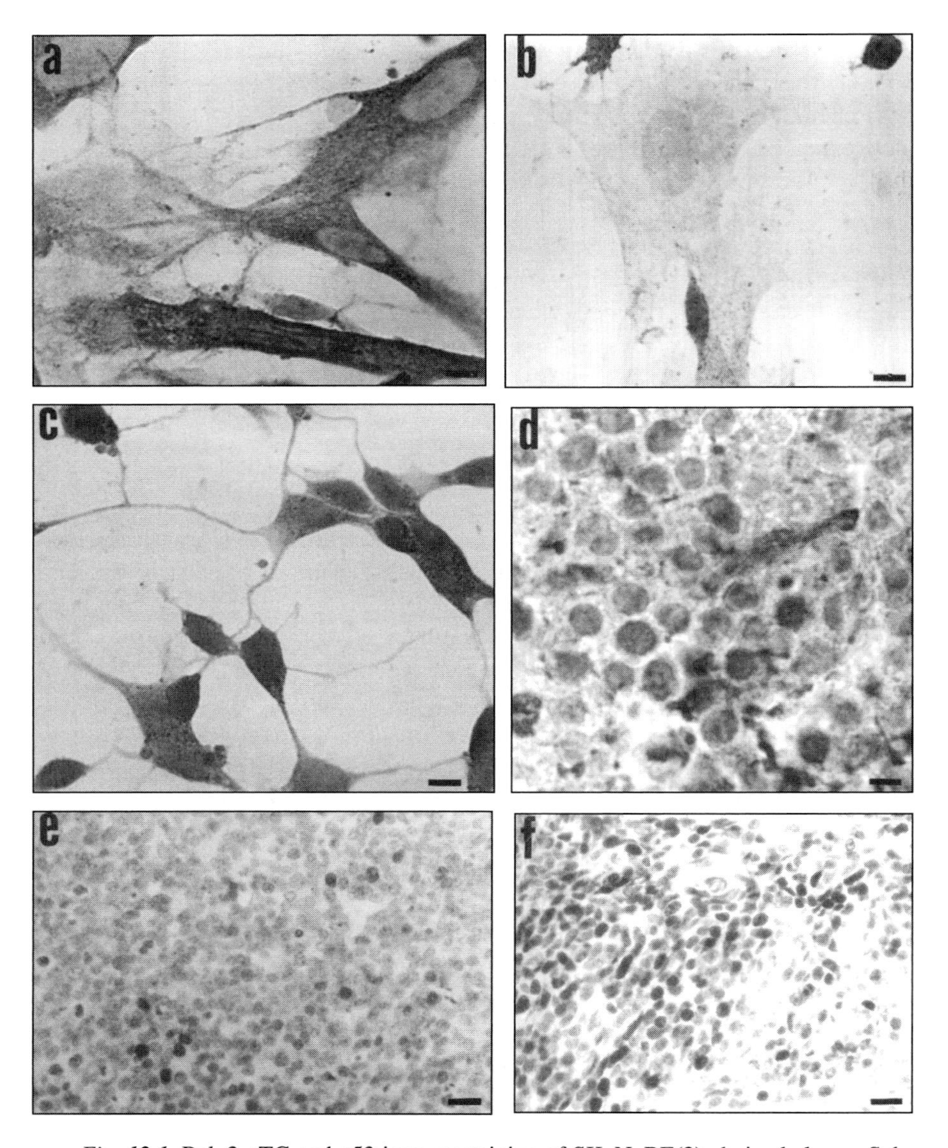

Fig. 12.1. Bcl–2, tTG and p53 immunostaining of SK–N–BE(2)–derived clones. Substrate–adherent BE(2)–SA (*a–b*, *d–e*) and neuroblastic BE(2)–NA cells (*c*, *f*) were grown in vitro (*a–c*) or in SCID mice (*d–f*). Cells were grown on slides, fixed in 2.5% paraformaldehyde and after immunostaining counterstained with hematoxylin. Cells were grown in the presence (*b–c*) or in the absence (*a*) of RA for 3 days. Note the pale Bcl–2 staining in cells showing the substrate–adherent phenotype (*b*), by contrast an intense positivity is present in the same cells for tTG (*a*). Intense staining detected with the Bcl–2 antibody was detected in neuronally differentiated BE(2)–NA cells after treatment with RA (*c*).While tTG was not

N– and S–type neuroblastoma cells are differentially regulated. For example retinoic acid acts on both N– and S–type cells but the resulting phenotype depends on the cell line and the particular sub–clone: S–type cells may differentiate to Schwannian [15] or melanocytic types [16], or undergo apoptosis [18], whereas N–type cells such as the SH–SY–5Y cell line differentiate in response to retinoic acid [19], the extent of differentiation being dependent on the retinoic acid isomer in use [20]. Figure 12.2A schematizes the interplay between all three phenotypes and their respective response to retinoic acid.

Role of Insulin–Like Growth Factors in the Survival of Neuroblastoma

Over the recent past few years, it has become evident that cells require specific signals in order to survive [21]. Insulin–like growth factors (IGF–I and II), previously known as somatomedins, modulate peripheral growth both in embryonal (IGF–I) and adult (IGF–II) life [22–23]. They act as autocrine and/or paracrine factors in a tissue specific manner, driven by the hypothalamic–hypophysis axis mediated by both growth releasing hormone (GRH) and growth hormone (GH).

IGF–II consists of a 67–residue polypeptide chain showing a high degree of homology to IGF–I and proinsulin [24]. In neuroblastoma, IGF–II is the major autocrine/paracrine growth factor, and its action is modulated in a complex manner by exogenous (e.g., eight binding proteins, IGFBs) or endogenous (e.g., two types of receptors and several degrading proteases) regulators.

Several studies have evaluated the role of IGFs [25] and IGFBPs [26] in neuroblastomas, where the effects of IGF–II range from stimulating cell proliferation [27] to rescuing cells from *myc*–induced apoptosis [28]. These pleotrophic effects are generated via a complex modulation of events, acting at different levels. During IGF–II synthesis, alternative splicing and both tissue specific and developmental specific transcription regulators triggered by hormones, such as GH or follicular stimulating hormone (FSH) [29], control both the level and size of the mRNA available for translation [30–31]. The processing of IGF–II then varies, depending on the particular 5' untranslated region present on the mRNA

detected in BE(2)–NA cells, a large number of tTG positive cells were observed in BE(2)–SA–derived tumors (structures with condensed cytoplasm and pycknotic chromatin) (*d*). It is interesting to note that tTG positive apoptotic cell remnants were lining the necrotic areas which may be derived from apoptotic cells undergoing secondary necrosis (*d*). Large numbers of p53 (*f*) positive cells were shown in the substrate–adherent BE(2)–SA cells. By contrast p53 was only weakly positive in the tumors derived from the neuronal BE(2)–NA cells (*e*), Bar unit = 6 mm (A–D); 20 mm (*e–f*).

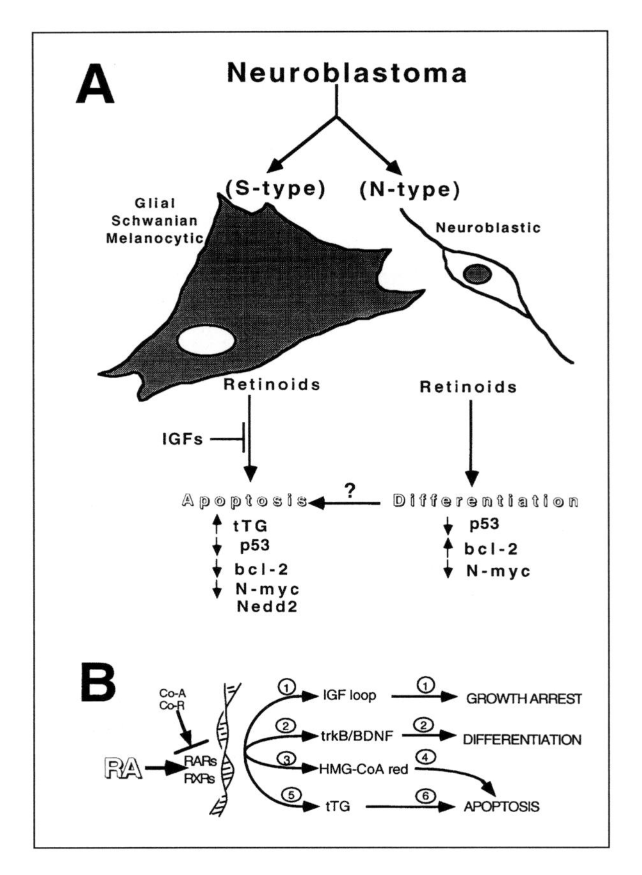

Fig. 12.2. Regulation of growth in neuroblastoma cells. (*a*) Transdifferentiation of neuroblastoma cells between S–, N– and I–phentoypes. Retinoic acid induces apoptosis in S–type cells, while it induces a neuronal differentiation in N–type cells. This is accompanied by a modulation in the expression of genes related to apoptosis. (*b*) Retinoic acid reduces growth by different mechanisms acting on the autocrine IGF loop (arrow 1), on the trkB and BDNF differentiation loop (arrow 2), on the regulation of HMG–CoA reductase (arrows 3–4) and on tTG (arrows 5–6). The transactivation is complex since it depends not only on the expression of an RARs/RXRs panel, but also on the presence of co–activators (Co–A) and corepressors (Co–R) which have seen to be different in different human neuroblastoma cell lines (G. Melino, data not published).

[32]. Once secreted, IGF–I and II may either interact with specific binding proteins, whose exact functions and mechanisms of action are, as yet, not fully understood [33–34] or they may bind with different affinities to IGF–receptors, depending on the particular cellular system [35–37]. Binding to the receptor is followed by the activation of different signal transduction pathways (IRH, IRS

1 and 2) which affect gene regulation. Finally, the action of IGF–II is modulated by degrading endopeptidases. The exact mechanism for the degradation of IGF–II, as yet, has not been elucidated although several enzymes have been proposed to play a role in this process including insulin–degrading enzyme (IDE) [38]. IDE is a neutral thiol metalloendoprotease of 116 kDa (EC 3.4.24.56) with a wide tissue distribution [39–40]. Although IGF–I and II bind preferentially to distinct plasma membrane receptors, all three molecules can interact with the same specific thiol protease of 116 kDa, IDE. However, IDE cleaves insulin, IGF–I and IGF–II at different rates in vitro [38]. Thus, the proliferating activity of IGF–II in neuroblastomas may well be modulated by degrading endoproteases, such as IDE or the recently identified thermolysin–like metallo–endopeptidase (TME).

IGF–II stimulates the growth of several different neuroblastoma cell lines by activating specific receptors coupled to tyrosine kinase activities. Indeed, El–Brady et al [25, 27] have shown that neuroblastoma is dependent on IGF–II either in a paracrine or autocrine manner. Recently, Harrington et al [28] suggested that, whilst c–*myc* activates two distinct pathways leading to either cell progression or programmed cell death, IGF–I and IGF–II are able to block the latter cascade of events, resulting in the maintenance of cell proliferation. The effect of IGFs in the prevention of cell death may therefore be distinct from its mitogenic effect [41]. This concept of IGFs acting as anti–death factors has been supported by several studies. IGF–I or high concentrations of insulin can delay apoptosis induced by antibodies directed against the EGF receptor in a colorectal carcinoma cell line [42]; similarly, IGF–I suppresses the spontaneous onset of apoptosis in cultured pre–ovulatory follicles [43]. This effect is both IGF dose–dependent and type I receptor dependent [41, 44–45]. These data demonstrate a relevant role for IGFs and type I IGF receptor as inhibitors of apoptosis, independent of their mitogenic effect. IGFs may therefore become targets for therapeutic interventions. However, at the molecular level, all of these alleged interactions and mechanisms remain to be clarified. For example, the interaction between the IGFs within the apoptotic machinery is unknown. Moreover, several factors participate in the regulation of action of IGF, its receptor type and availability, the presence of endogenous [26] or serum [43] IGFBPs and of specific proteases.

Finally, new data indicate the existence of crosstalks between membrane–membrane and membrane–nuclear receptors [46]. The cellular display of receptors and the signal transduction components at the time of IGF stimulation determine the type of response of IGF–I. IGF signaling may also interact with nuclear receptors including insulin and IGF–I and II which control growth and differentiation through the estrogen receptor. As supported by transient transfection studies, IGF modulates the estrogen receptor–activated promoters [47]. This mechanism may be relevant during the development of the brain where the

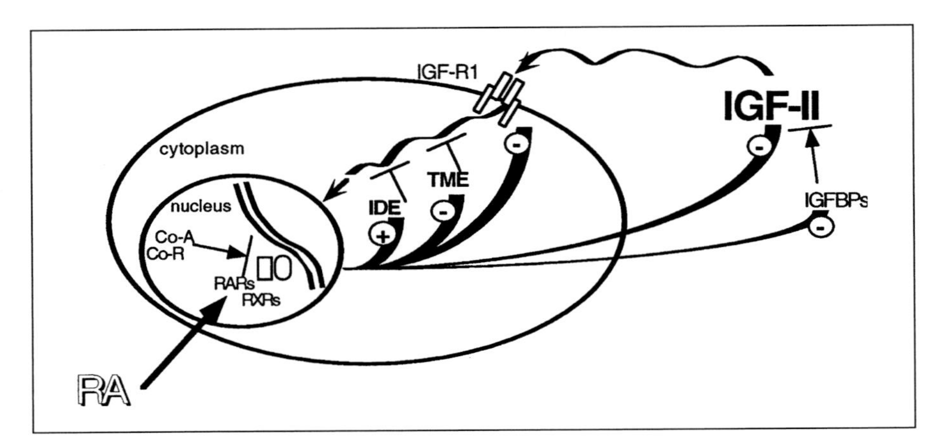

Fig. 12.3. The autocrine/paracrine IGF–II growth loop drives the proliferation of neuroblastoma cells in conjunction with other minor growth loops (β–endorphins, IL–6, vaso–intestinal peptide, FGF, etc.). The effect of the autocrine IGF–II loop is exerted through its eight binding proteins, its two receptors and its degrading proteases (insulin–degrading enzyme and termolysin–like metallo–endoprotease). Retinoids reduce the efficacy of the IGF–II loops at various levels. Resistance to retinoids is achieved also at various levels: loss/modulation of RARs/RXRs receptors, co–activators (Co–A), corepressors (Co–R), IGF–R1, IDE, TME, IGFBPs and IGF–II itself.

estrogen receptor is expressed in various embryonic cells. Similarly, IGF may interact with other receptors of the steroid superfamily, crucial in embryogenesis, including the retinoic acid and retinoid–X receptors (RARs and RXRs).

Figure 12.3 shows a simplistic mechanism of action of IGF–II as growth factor in neuroblastoma cells and how retinoids modulate the IGF loop. Resistance to the growth–inhibitory action of retinoids might result from (i) change in the expression pattern of retinoids receptors, (ii) increased expression of IGF–II or its type I receptor and/or (iii) modulation of IGF–II biological half–life through IGFBPs and the degrading proteases. All these mechanisms might have therapeutic consequences.

Apoptosis in Neuroblastomas

Neuroblastoma cells undergo apoptosis in vivo and this phenomenon may also be reproduced in vitro in response to various stimuli [48] (Table 12.1). In response to these agents neuroblastoma cells undergo typical cytoplasmic (condensation of the cytoplasm followed by blebbing and fragmentation) as well as nuclear (chromatin condensation and fragmentation in large masses) changes

Table 12.1. Agents inducing apoptosis in neuroblastoma cell lines

Agents	Cell Line	Reference
Adriamycin	SH–SY5Y	Lasorella et al, 1995 [92]
Cis–platin	”	Cece et al, 1995 [93]
Colchicine	”	Nuydens et al, 1995 [94]
Cis–platin	SH–N–BE(2)	Piacentini et al, 1993 [95]
Cyclopentenyl cytosine	”	Slingerland et al, 1995 [96]
Retinoids	”	Piacentini et al, 1991 [18]
Somatostatin	”	Candi et al, 1995 [97]
Tamoxifen	”	Candi et al, 1995 [97]
Etoposide	SK–N–MC	Dole et al, 1995 [98]
TNFa	”	Talley et al, 1995 [99]
Sodium butyrate	TR14	Nuydens et al, 1995 [94]
Neocarzinostatin	SK–N–SH	Hartsell et al, 1995 [100]
Retinoids	LA–N–5	Di Vinci et al, 1994 [101]

Table 12.2. Effect of retinoic acid on apoptosis in neuroblastoma cell lines

	Percent apoptotic bodies		
	SK–N–BE(2) (I–type)	BE(2)–NA (N–type)	BE(2)–SA (S–type)
control	9±2	6±2	14±3
RA	21±3	4±2	58±7

Several human neuroblastoma cell lines SK–N–BE(2), BE(2)–M17, BE(2)–NA and BE(2)–SA were incubated with either 0.07% ethanol or 1 μM retinoic acid for 72 hours. Apoptotic cells were evaluated by flow cytometric evaluation. Each value was based on three different sets of experiments [18, 65].

characteristic of cells undergoing programmed cell death [49], therefore suggesting that access to the death program remains intact in these transformed cells. However, it is important to note that the regulation of growth and/or death in cells expressing the N– and S–phenotype is different, occurring in a phenotypic–specific manner (Fig. 12.2*a*). This phenotype–specific regulation is evident in vitro: S–type cells undergo apoptosis when exposed to retinoic acid; in contrast, N–type cells exposed to retinoic acid show a drastic induction of neuronal differentiation (Table 12.2). Both phenomena result in a reduction in growth kinetics.

Increasing evidence over the past few years has shown that cells not only require specific signals to proliferate and differentiate, but also to survive [21]. Under physiological and pathological conditions, in the absence of such survival factors, cells enter a gene–regulated program of self–elimination [49] dependent upon the expression of a specific set of genes [50–56]. These genes can be grouped according to their distinct roles within the *C. elegans* genetic program of cell death [57]. Indeed, genes with functional homology to some *C. elegans* have been described in mammals, such as the Bcl–2 protein [58–59]. *bcl–2*, cooperatively with other homologous genes, *bcl–X, Bax, BAG–1*, and *Bak* [60], suppresses the expression of the cell death program. In addition, important regulatory roles for *p53*, c–*myc* and Fas/Fas ligand [54–55, 61] in the initiation of the cell death program have also been described. The disposal group of genes includes those genes coding for the DNA–degrading enzymes [62–63] as well as those required for the engulfment of the dying cells [51–52]. This list also includes "tissue" transglutaminase (tTG; EC 2.3.2.13; also known as type II, TGase II or C). We demonstrated that this gene is specifically induced and activated in apoptotic cells, catalyzing an acyl–transfer reaction among polypeptide chains—forming ε(γ–glutamyl)lysine and N–N–bis(γ–glutaml)polyamine isodipeptide linkages [48, 50].

Although the list of apoptosis associated genes as well as our understanding of the regulation of the cell death program is ever increasing, there is still relatively little known concerning the specific role of these genes in neuroblastoma.

The identification of both structural and functional similarities between the product of the *C. elegans* cell–death gene *ced–3* and the mammalian interleukin–1β–converting enzyme (ICE) [55], as well as other members of the growing family of proteases including the mouse gene Nedd2 [64], has provided important information on the role of proteases in the program of cell death [55]. In fact, overexpression of Nedd2 in neuroblastoma cells also results in apoptosis, which is suppressed by Bcl–2 [64]. We have recently shown the presence of ICE, ICE relII, ICE relIII, ICH–2, Mch–II, Mch–III and CPP32 in human neuroblastoma cells. Their expression is not related to the sensitivity of the cell lines to undergo apoptosis, with the exception of CPP32 itself, and the consequent degradation of PARP (G. Melino, data not published).

Tissue transglutaminase (tTG) is activated by retinoic acid in SK–N–BE(2) cells, resulting in a 12–fold increase in the enzyme activity [65]. The genetic regulation of tTG mRNA occurs within 24 hours of exposure to retinoic acid and correlates with the downregulation of N–*myc* [66]. However, the level of tTG protein varies in different neuroblastoma cell lines with different phenotypes. Immunohistochemical analysis of the parental SK–N–BE(2) cell line clearly shows that tTG protein is not detectable in all cells (Fig. 12.1), being absent in N–phenotype cells and specifically localized in the large, flat S–type cells (Fig. 12.1*a*). Immunohistochemical analysis also demonstrated that tTG protein was localized in shrunken globular structures containing condensed chromatin which displayed morphologically typical features of mature apoptotic bodies (Fig. 12.1*b,d*). These results further suggest that the ability of neuroblastoma cells to undergo apoptosis is associated with the S–phenotype. In contrast, the expression of the N–phenotype may be linked to the inhibition of the apoptotic program. Furthermore, the expression of tTG and Bcl–2 protein in all neuroblastoma cells shows a negative correlation. In fact, S–cells do not express Bcl–2 protein, instead this is largely expressed in cells displaying the N–phenotype (Fig. 12.1*a* vs. *b*). Whether the expression of these two putative apoptotic genes is mutually exclusive and due to an inverse regulation is a challenging matter for future studies. However, it implies that the serial expression of apoptotic genes reflects the progression of the dying cell through the apoptotic pathway.

The expression of the neuroblastic N–phenotype is associated with the inhibition of the apoptotic program operated at the molecular level by specific survival genes such as Bcl–2 protein [59]. Therefore the ability of neuroblastoma cells to undergo apoptosis is mainly associated with the expression of an S–phenotype. This is consistent with the presence of specific stages in the differentiation of neural crest–derived cell lineages during which these cells are prevented from entry into the cell death program. Thus the expression of the substrate–adherent phenotype may be important in the commitment of those neuroblastoma cells to cell death by apoptosis. Bcl–2 is expressed in N–type cells represented in the more aggressive tumors [67].

Retinoic Acid Induces Apoptosis in Human Neuroblastoma Cell Lines

As a regulatory molecule, retinoic acid has a wide–ranging importance with respect to development, morphogenesis, carcinogenesis and the maintenance of epithelial phenotypes [68]. Retinoic acid regulates gene expression by binding to specific nuclear receptors (RARs) and retinoid X receptors (RXRs) [68–69]. Low molecular–mass cytosolic proteins, cellular retinol–binding protein (CRBP) and cellular retinoic acid–binding protein (CRABP), are involved in regulating the availability of all–trans retinoic acid to the nuclear receptors [70]. RXRs function as auxiliary proteins which enable RARs to bind to specific retinoic acid response elements (RAREs) present in the 5' regulatory sequences of retinoic acid–regulated genes [68]. RARs thus appear to function predominantly as heterodimers with RXRs.

Retinoids Regulate Apoptosis in Neuroblastoma Cells

In most neuroblastomas, retinoic acid affects cell growth by inducing different degrees of either apoptosis or differentiation, Figures 12.2A and 12.2B. It is interesting to note also that decreased expression of N–*myc* precedes both retinoic acid induced differentiation and apoptosis [66]. As displayed in

Table 12.2, in the SK–N–BE(2) cell line, an increase in apoptotic bodies is observed after a three day treatment with 1 μM RA. Since neuroblastoma cells express the different N–, I–, and S–phenotypes in vitro, we investigated two different cell lines expressing the pure N– and S–phenotype. The cell lines were selected from the parental SK–N–BE(2) cell line on the basis of cell–to–matrix adhesiveness [65]. The BE(2)–SA cells showed a high spontaneous rate of apoptosis and were extremely sensitive to RA–induced death. The BE(2)–NA cell line behaved differently to retinoic acid exposure by differentiating into neuronal cells [72]. These observations, in support of previous reports [18], led us to investigate further the possible molecular mechanisms triggering either differentiation or apoptosis in this cell system.

Expression of RARs/ RXRs Receptors in Neuroblastoma Cells and Their Regulation of tTG in Apoptotic Cells

We investigated the presence of the different retinoic acid receptors present in the SK–N–BE(2), BE(2)–NA and BE(2)–SA cell lines. In order to measure RAR/RXR isoform mRNA expression levels, RT–PCR and northern blots were performed on total mRNA extracts from the three neuroblastoma cell lines tested; a summary of the RT–PCR is displayed in Figure 12.4*a*. The distribution of RAR/RXR–receptor mRNA varied significantly amongst the three cell lines studied. Northern blot analysis demonstrated the presence of RARα, RARγ (Fig. 12.4*b*,*d*) and RXRα (Fig. 12.4*c*) receptors in all three cell lines; RAR–α and γ being upregulated by retinoic acid in the S–KN–BE(2) and BE(2)–NA cell lines. Data analysis indicated that both the levels of RARα and RARγ mRNA expression was increased over a 72–hour period in the BE(2)–NA cell line. RAR–β_2 mRNA expression did not appear to be induced by retinoic acid in the N–type cells of this cell line although other reports have shown, using a complete RAR–β probe, that this receptor subtype is, in fact, upregulated in several other N–type neuroblastoma cell lines [47, 73]. In the BE(2)–SA cell line, we observed

Fig. 12.4. Expression of retinoids receptors (RARs and RXRs) in neuroblastoma cells. A large degree of variability in receptor expression has been observed in various cell lines or tumor biopsies. Here we show the expression of receptors in the SK–N–BE(2) cell line, showing an I–phenotype, and in its sub–clones BE(2)–NA, with a pure N–phenotype, and BE(2)–SA, with a pure S–phenotype. (*a*) RT–PCR expression of receptors in the three cell lines before and after 3 days exposure to 1 μM retinoic acid. There is no clear–cut correlation with the phenotype to justify a simplistic differential effect of RA, as shown in Fig. 12.2. All trans RA is able to induce all receptors in nearly every cell line. (*b*) Total mRNA Northern blot hybridized for RARα and RARγ. The effect of 1 μM retinoic acid exposure is shown

A

	S-KN-BE(2)		BE(2)NA		BE(2)-SA	
	Control	RA	Control	RA	Control	RA
RARα	+	+	+	+	+	+
RARβ	-	+	-	-	-	+
RARγ	+	+	+	+	+	+
RXRα	+	+	+	+	-	+
RXRβ	+	+	-	+	-	-

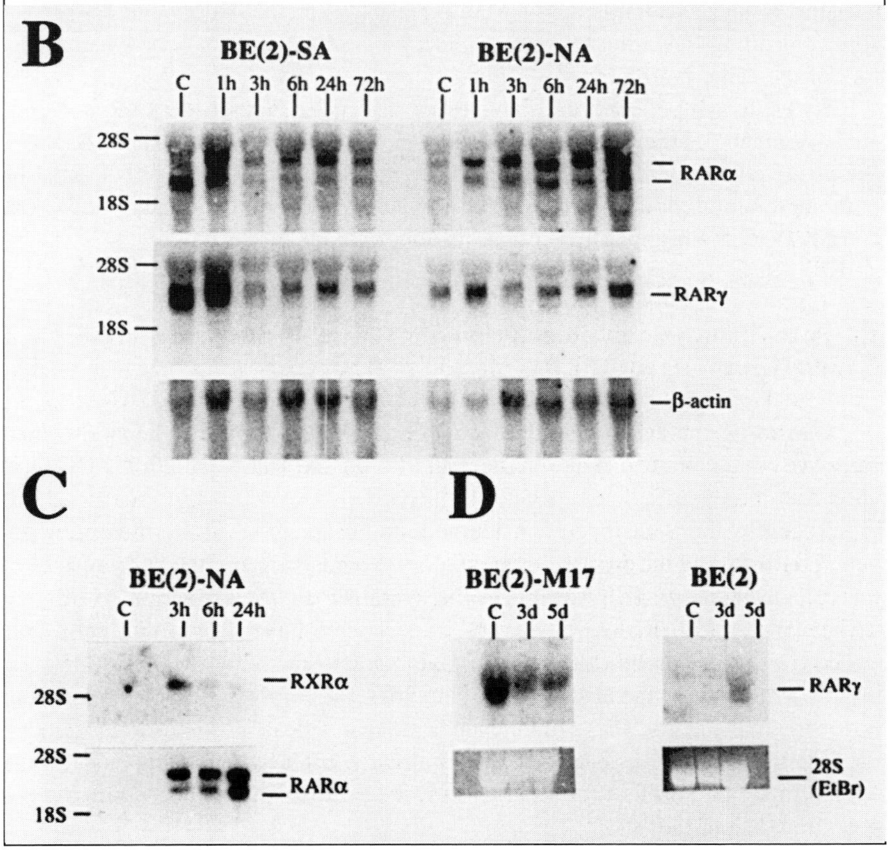

B

BE(2)-SA
C 1h 3h 6h 24h 72h
BE(2)-NA
C 1h 3h 6h 24h 72h
28S
18S
RARα
28S
18S
RARγ
β-actin

C

BE(2)-NA
C 3h 6h 24h
RXRα
28S
28S
RARα
18S

D

BE(2)-M17
C 3d 5d
BE(2)
C 3d 5d
RARγ
28S (EtBr)

at the times indicated (hours). (*c*) Poly A+ mRNA Northern blot for RXR provide evidence for a rapid transient induction by 1 μM retinoic acid. (*d*) Poly A+ mRNA Northern blot for RARγ in the RA–sensitive SK–N–BE(2) cell line versus the RA–resistant cell line BE(2)–M17. Retinoid resistance is accompanied by a distinct RARγ modulation, implying a role for the co–activators and corepressors indicated in Figs. 12.2 and 12.3.

a change in the size of the isoform expressed suggesting probably a change in splicing triggered by RA. To evaluate the modulation of receptors in greater detail, we performed Northern blots on poly–A^+ mRNAs from BE(2)–NA cells (Fig. 12.4*c*,*d*). In these experiments the early induction of RXRα at 3 hours and the early–intermediate induction of RARα mRNA expression levels was particularly striking. The retinoid–resistant cell line BE(2)–M17 failed to modulate RARγ (Fig. 12.4*d*). Using RT–PCR, we showed that, in some of the cell lines studied, $RAR\beta_2$, RXRα and RXRβ mRNA expression were induced by RA. The presence and the ratio of these receptors most probably influences the possible RAR/RXR dimer combination and thus their function as transcriptional factors. The scenario is further complicated by the presence of co–activators and co–repressors controlling RA–transactivation, and by a complex crosstalk between receptors of the same family.

By using synthetic retinoids able to selectively activate RAR/RXR isoforms, we have identified the RAR/RXR receptors involved in the induction of apoptosis. We observed that while the RARα and γ alone were able to induce tTG, only the combined stimulation of both RARα and RARγ induced apoptosis (G. Melino and M. Piacentini, unpublished observations).

Post–Translational Regulation of "Tissue" Transglutaminase: A Hypothesis for the Regulation of Apoptosis

Can work on tTG in neuroblastoma be expanded into more general terms? Here we would like to make some remarks on the fine regulation of tTG in apoptosis in general.

Recently, there has been considerable emphasis given to the hierarchic genetic control provided by various regulatory elements during the early phases of the cell death program, resulting in the synthesis and/or activation of effector "killer" genes [57]. However, several experiments have shown that cells may undergo a process morphologically similar to apoptosis in the absence of de novo protein synthesis, in the absence of the nucleus, or simply by direct activation of downstream elements. These findings indicate that, at least in some circumstances, the machinery able to execute the cell is already in place before the onset of the apoptotic signal and that the final events of apoptosis may be controlled at the post–translational level.

Several laboratories have shown in different experimental settings, both in vivo and in vitro that there is a tight association between the expression of tTG and the onset of programmed cell death. The tTG–dependent formation of stable ε(γ–glutamyl)lysine and N,N–bis(γ–glutamyl)polyamine crosslinks leads to protein polymerization conferring resistance to breakage and chemical attack on those polypeptides involved in the linkage, such as insolubility in SDS and chaotropic agents [50]. The tTG–catalyzed crosslinks found in the apoptotic bodies are irreversible. Although endoproteases capable of hydrolyzing these crosslinks have yet to be identified in vertebrates, lysosomes do not express any catalytic activities and hence are unable to split the ε(γ–glutamyl)lysine bonds [74]. Consequently, the isodipeptide is detectable (after degradation of apoptotic bodies into phagolysosomes) in both the culture medium and plasma [48, 74]. However, only recently has the association of tTG expression with programmed cell death been directly established. Thus, tTG is not only associated with apoptosis, but also plays a crucial role in the killing process both by promoting the condensation of the cytoplasm, its subsequent fragmentation as well as stabilizing the apoptotic cell during the late apoptotic stages preceding its clearance [65, 75]. The overexpression of a tTG cDNA (3.3 kbp) in human neuroblastoma cells induces the characteristic features of cells undergoing apoptosis. tTG transfectants show a large reduction in their growth capacity, not only in vitro, but also when xenografted into SCID mice (Figs. 12.1*d–f* and 12.5). By contrast, transfection of the same neuroblastoma cells with an expression vector containing a 1.1 kbp segment of the human tTG cDNA in antisense orientation resulted in a decrease of both spontaneous as well as retinoic acid–induced apoptosis [65]. These findings indicate that tTG could act as an effector element of programmed cell death.

The tTG protein is undetectable in the majority of cells: its mRNA being transcribed as a consequence of the induction of apoptosis [48]. However, the tTG gene is constitutively expressed in some cell types, including endothelial cells, smooth muscle cells and mesangial cells [48]. The survival of cells constitutively expressing high tTG protein levels is so far unexplained, raising the possibility of its post–translational regulation during programmed cell death. The tTG sequence contains distinct regulatory domains: the tTG active site (residues 275–304 in rat, including cysteine–277) is clearly distinct from the Ca^{2+}–binding region (residues 426–454 in rat). Another TG, TGase–1 or TG–K involved in skin cell death, has also been shown to be activated into a truncated form in vivo (P.M. Steinert, personal communication), suggesting a possible post–translational regulation. Moreover, tTG binds guanine nucleotides and hydrolyzes GTP [76].

A paper by Nakaoka and coworkers [77] provided striking evidence offering a new perspective on the potential regulation of tTG in pre–apoptotic cells. These authors demonstrate that the 74 kDa α subunit ($G_{\alpha h}$) associated with the 50 kDa β subunit ($G_{\beta h}$) of the GTP–binding protein G_h is tTG. The G_h dimer acts in association with the rat liver α_1–adrenergic receptor in a ternary complex also containing an a_1–agonist. Thus, the $G_{\alpha h}$ (i.e., tTG; hereafter referred to as tTG/$G_{\alpha h}$ in order to stress the multiple functions of the enzyme) is a multifunctional protein, which by binding GTP in a $G_{\alpha hGTP}$ complex, can modulate both recep-

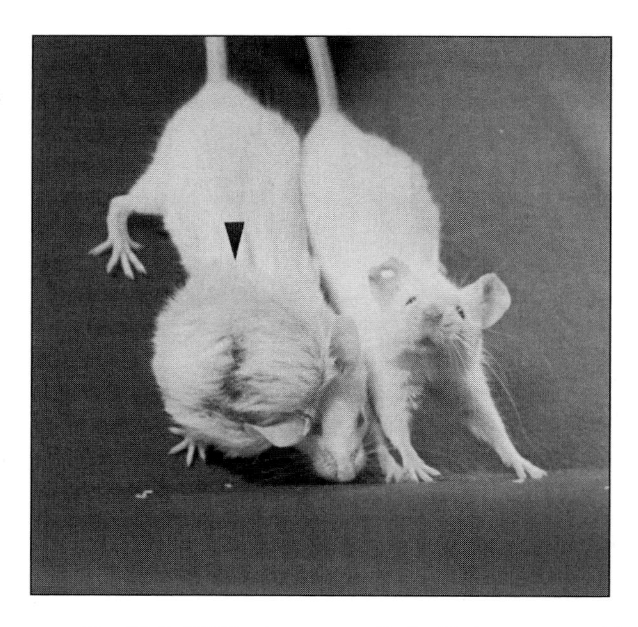

Fig. 12.5. SCID mice xenografted with human neuroblastoma SK–N–BE(2)–derived cell clones. The tumor developing for 60 days in the SCID mice xenografted with the SK–N–BE(2) human neuroblastoma cells transfected with the neomycin resistant gene only is indicated by the arrow. By contrast no tumors were detected in mice xenografted with cells overexpressing tTG (TGA clone; Melino et al, 1994, ref. 65).

tor–stimulated phospholipase–C (PLC) activation and transglutamination (see scheme in Fig. 12.6*a*). $G_{\alpha h}$ represents a novel class of GTP–binding proteins that participate in the receptor–mediated signaling pathway. The novelty of the action of the $G_{\alpha h}$ component is that, through this complex regulatory mechanism, receptor–stimulated GTP binding may prevent the activation of harmful components of the genetic death program. Indeed, the Ca^{2+}–dependent crosslinking activity of tTG is finely tuned by GTP levels which in turn regulate secondary messengers such as the production of inositol–1,4,5–trisphosphate (IP_3) and *sn*–1,2–diacylglycerol (DAG) from phosphatidyl–inositol–4,5–bisphosphate (PIP_2) (Fig. 12.6*a*). The GTP binding activity of tTG/$G\alpha_h$ actively prevents activation of the lethal crosslinking activity of tTG and, moreover, provides a parallel integrated pathway to regulate the free intracellular Ca^{2+} concentration which is required for the activation of tTG. Thus, the fine modulation of the tTG/$G_{\alpha h}$ protein by GTP and Ca^{2+} (and possibly additional molecules such as free putrescine and other polyamines) may explain how cells are able to survive in the presence of high tTG/$G_{\alpha hGTP}$ protein levels in their cytoplasm. Cell death prevention could also be achieved by the DAG–dependent activation of protein kinase C (PKC, 78, see Fig. 12.6*a*). For example, for a cell in a nonapoptotic status, the simple downregulation of the α_1–adrenergic receptor or reduction of its ligand level could be fatal. In keeping with this hypothesis, the cells which constitutively express tTG are localized in particular areas of mammalian tissues which are exposed to environmental and functional stress; hence, in order to avoid harmful consequences, these cells may have the apoptotic machinery ready to act whenever their integrity is compromised. It is therefore tempting to speculate that the tTG/$G_{\alpha hGTP}$ complex may influence a shift in the cell status from viable toward apoptotic (Fig. 12.6*b*). The suggested role of free intracellular Ca^{2+} as a trigger to activate the irreversible cascade of events leading to the establishment of the apoptotic phenotype is in keeping with this hypothesis. Indeed, endonucleases and some proteases, as well as tTG, require for their activation a sustained elevation of the intracellular level of Ca^{2+}. This extreme plasticity of the effector elements confers on the cell the possibility of inhibiting the apoptotic program in its late stages when the "killer machinery" is ready to act, just needing a "final touch."

Is the post–translational regulation of effector element function a general phenomenon? The biochemistry of the growing family of proteases involved in apoptosis is expected to indicate the regulation at the post–translational level of these enzymes. Squier and Cohen [79] reported the activation of Ca^{2+}–dependent neutral protease calpain. On the other hand, dimerization of interleukin (IL–2) by the post–translational action of a Ca^{2+}–dependent transglutaminase converts IL–2 into a factor cytotoxic to oligodendrocytes in regenerating optic nerves from fish [80]. Other findings [81] suggest that the putative apoptotic endonuclease Nuc 18, also involved in the DNA degradation occurring during the late stages of apoptosis, is identical to cyclophillin. Cyclophillin is a well characterized cellular target for cyclosporin–A which also displays an active Ca^{2+} binding activity. Thus, although rather preliminary, these findings would seem to highlight the multifunctional nature of the "killer" elements of the apoptotic program, which may be regulated at a post–translational level.

Apoptosis can be induced or inhibited by a wide spectrum of natural as well as toxic stimuli [82]. We propose that intercellular signals and their receptors which are known to be able to control the process of cell death, might act by regulating tTG and might do so in a very direct way. We propose the hypothesis that growth factors might have acquired in the course of evolution the ability to prevent activation of the effector elements of the physiological death program by post–translationally converting their function into one favoring cell survival. If this hypothesis is confirmed, we will be facing an additional important event in the phylogenetic selection of survival factors, as well as in the cellular control of life or death decisions. We do not propose this model based on tTG as exclusive

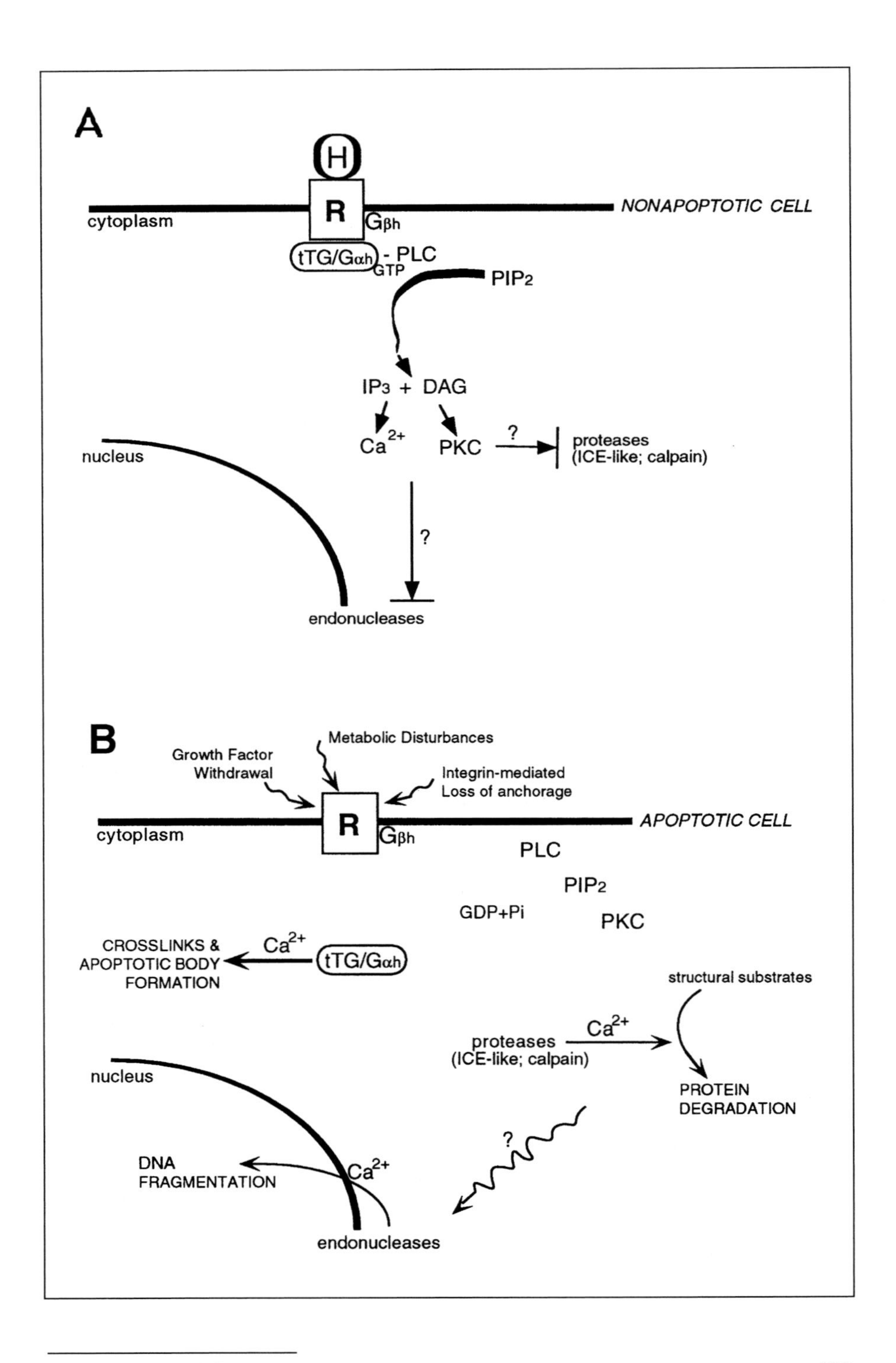
A
H
R
cytoplasm
Gβh
NONAPOPTOTIC CELL
tTG/Gαh - PLC
GTP
PIP2
IP3 + DAG
Ca2+
PKC
?
proteases
(ICE-like; calpain)
nucleus
?
endonucleases
B
Metabolic Disturbances
Growth Factor
Withdrawal
Integrin-mediated
Loss of anchorage
R
cytoplasm
Gβh
APOPTOTIC CELL
PLC
PIP2
GDP+Pi
PKC
CROSSLINKS &
APOPTOTIC BODY
FORMATION
Ca2+
tTG/Gαh
structural substrates
Ca2+
proteases
(ICE-like; calpain)
PROTEIN
DEGRADATION
nucleus
?
DNA
FRAGMENTATION
Ca2+
endonucleases

mechanism, but rather as an additional control to allow more plasticity and flexibility to the cell. The emphasis therefore is now to ascertain whether and how the effector elements may be controlled post–translationally in the programmed cell death pathway and with which consequences for their function.

Clinical Perspectives

In recent years there has been considerable focus directed towards the potential clinical use of drugs able to induce differentiation of neuroblastoma in vitro. However, to date, these agents have only demonstrated limited success. Since apoptosis is clearly an important mechanism of tumor regression of neuroblastoma in vivo, being demonstratable in stage IVs disease [83], an important future direction towards effective therapy therefore should include agents able to also specifically activate the pathway of apoptosis. Some cytotoxic drugs used in the treatment of children with neuroblastoma induce apoptosis, e.g., cisplatin [95], somatostatin [97] and tamoxifen [97], as indicated in Table 12.1. More interestingly, the mechanism of action of other cytotoxic agents, at therapeutic low doses, involve the upregulation of CD95 (Fas/APO–1) and its ligand (G. Melino and K.M. Debatin, unpublished data), raising the likely possibility that abnormalities in this mechanism result in drug resistance.

The precise mechanisms which regulate both differentiation and apoptosis, as yet, have not been fully elucidated. However, on the basis of our present knowledge an obvious approach to take would be to block IGFs' autocrine loop or other survival genes such as the *bcl–2* family. The expression of Bcl–2 in neuroblas-

Fig. 12.6. A model for the post–translational regulation of the effector elements of programmed cell death. (*a*) Nonapoptotic cell. The hormonal growth factor survival signal (H) activates through the receptor (R) the G_h transducer protein, of which the $G_{\alpha h}$ subunit has been shown to be (tTG) (tTG/Gαh; Nakaoka et al, 1994). tTG/Gαh binds GTP and forms a complex with phospholipase C (PLC) in order to activate phosphatidyl–inositol–4,5–bisphosphate (PIP_2) degradation allowing the formation of second messengers such as inositol–1,4,5–trisphosphate (IP_3) and *sn*–1,2–diacyl glycerol (DAG). In the proposed model the activity of effector elements is tightly maintained in an inactive state. tTG is inactive in its crosslinking activity when integrated into the GTP•PLC receptor complex [77]; moreover, protein kinase C (PKC) has been shown to inhibit apoptosis [78]. (*b*) Apoptotic cell. Exogenous elements such as growth factor withdrawal and possibly metabolic disturbances or integrin–mediated loss of anchorage [89–90] may alter the receptor ternary complex, allowing the separation of the tTG/$G_{\alpha h}$ subunit, and its calcium–dependent protein–protein crosslinking activity leading to the formation of the apoptotic body [65]. The lack of PIP_2 signaling leads to the activation of proteases such as ICE–like [55–56, 91] and calpain [79] resulting in protein degradation.

toma tumors has been well described [84–85] in tumors from all stages of disease. However, it appears there is no correlation between *bcl–2* and *bcl–*X_L expression and clinical stage [86] even following chemotherapy [87], except for one contradictory report where *bcl–2* expression was reported to correlate with low–stage disease [88]. Bcl–2 therefore seems to be important in the survival of neuroblastoma tumors.

Retinoids can induce both apoptosis and differentiation in neuroblastoma, resulting in a reduction of growth. Apoptosis is induced in S–type cells but prevented in N–type cells, probably due to the induction of survival genes such as *bcl–2* (Figs. 12.1 and 12.2). Although, neuronal differentiation is induced by retinoic acid in N–type cells this is not an irreversible status, and therefore the subsequent apoptosis of these cells may be an important event.

Future therapeutic strategies for neuroblastoma should therefore be directed towards both the induction of both differentiation and apoptosis.

Acknowledgments

We would like to thank Dr. Muriel Draoui for her scientific support. This work was supported in part by a grant from AIRC, Ministero Sanita' (AIDS project to M. Piacentini and A. Finazzi–Agrò; Progetti Finalizzati IRCCS to G. Melino) and CNR (ACRO and bilateral projects to G. Melino).

References

1 Castleberry RP. Clinical and biological features in the prognosis and treatment of neuroblastoma. Curr Opin Oncol 1992; 4:116–123.
2 Hall AK. Influence of cyclic AMP and serum factors upon expression of a retinoid responsive gene in neuroblastoma cells. J Mol Neurosci 1992; 3:155–163.
3 Evans AE, D'Angio GJ, Randolph J. A proposed staging system for neuroblastomas Cancer 1971; 27:374–378.
4 Hayes FA, Green A, Hustu HO et al. Surgicopathological staging of neuroblastoma:prognostic significance of regional lymph node metastases. J Ped 1983; 102:59–62.
5 Hata Y, Haruhiko N, Famiaki S et al. Fifteen years experience of neuroblastoma: a prognostic evaluation according to the Evans and UICC staging systems. J Ped Surg 1990; 25:325–329.
6 Pinkerton CR. Where next with therapy in advanced neuroblastoma? Brit J Cancer 1990; 61:351–353.
7 Woods WG, Tuchman M. Neuroblastoma: The case for screening infants in North America Ped 1987; 79:869–873.
8 Brodeur GM, Sekhon GS, Goldstein MN. Chromosomal aberations in human neuroblastomas. Cancer 1977; 40:2256–2263.
9 Brodeur GM, Green AA, Hayes FA. Cytogenetic studies of primary human neuroblastomas. In: AE Evans, ed. Advances In Neuroblastoma I. New York: Raven Press, 1980:173–180.
10 Brodeur GM, Fang CT. Molecular biology and genetics of human neuroblastoma. Cancer Gen Cytogen 1989; 41:153–174.

11 Schwab M, Varmus H, Bishop J. Human *N–myc* gene contributes to neoplastic transformation of cells in culture. Nature 1985; 316:106–112.

12 Rosen N, Reynolds CP, Thiele C et al. Increased *N–myc* expression following progressive growth of human neuroblastoma. Cancer Res 1986; 46:4139–4142.

13 Biedler JL, Casals D, Chang TD et al. Multidrug resistant human neuroblastoma cells are more differentiated than controls and retinoic acid further induces lineage–specific differentiation. Prog Clin Biol Res 1991; 366:181–191.

14 Ross RA, Bossart E, Spengler BA et al. Multipotent capacity of morphologically intermediate (I–type) human neuroblastoma cells after treatment with differentiation inducing agents. Prog Clin Biol Res 1991; 366:193–201.

15 Tsokos M, Scarapa S, Ross RA et al. Differentiation of human neuroblastoma recapulates neural crest development. Study of morphology, neurotransmitter enzymes, and extracellular matrix proteins. Am J Pathol 1987; 128:484–496.

16 Slack R, Lach B, Gregor A et al. Retinoic acid and staurosporin induced bidirectional differentiation of human neuroblastoma cell lines. Exp Cell Res 1992; 202:17–27.

17 Matsushima H, Bogenmann E. Bi–modal differentiation pattern in a new human neuroblastoma cell line in vitro. Int J Cancer 1992; 51:250–258.

18 Piacentini M, Annicchiarico–Petruzzelli M, Oliverio S et al. Phenotypic specific tissue transglutaminase regulation in human neuroblastoma cells in response to retinoic acid: correlation with cell death by apoptosis. Int J Cancer 1992; 52:271–278.

19 Pahlman S, Ruusala AI, Abrahamsson L et al. Retinoic acid induced differentiation of cultured human neuroblastoma cells: a comparison with phorbol–ester–induced differentiation. Cell Diff 1984; 14:135–144.

20 Lovat PE, Lowis SP, Pearson ADJ et al. Concentration dependent effects of 9–cis retinoic acid on neuroblastoma differentiation and proliferation in vitro. Neurosci Lett 1994; 182:29–32.

21 Raff MC. Social controls on cell survival and cell death. Nature 1992; 356:397–399.

22 Han VKM, D'Ercole AJ, Lund PK. Localization of somatomedin messenger RNA in the human fetus. Science 1987; 236:193–196.

23 Brice AL, Cheetham JE, Bolton VN et al. Temporal changes in the expression of insulin–like growth factor II gene associated with tissue maturation in the fetus. Development 1989; 106:543–549.

24 Humble RE. Insulin–like growth factor, somatomedines and multiplication–stimulating activity: chemistry. Eur J Biochem 1990; 190:445–462.

25 El–Brady O, Romanus JA, Helman LJ et al. Autonomous growth of a human neuroblastoma cell line is mediated by Insulin–like Growth factor II. J Clin Invest 1989; 84:829–839.

26 Bernardini S, Cianfarani S, Spagnoli A et al. Expression and down regulation by retinoic acid of IGF binding protein–2 and –5 in medium from human neuroblastoma cells. Neuroendocrinol 1994; 6(4):409–413.

27 El–Brady O, Helman LJ, Chatten J et al. Insulin–like growth factor II–mediated proliferation of human neuroblastoma. J Clin Invest 1991; 87:648–657.

28 Harrington EA, Bennett MR, Fanidi A et al. *c–myc*–induced apoptosis in fibroblasts is inhibited by specific cytokines. EMBO J 1994; 13:3286–3295.

29 Voutilaine R, Miller WL. Coordinate trophic hormone regulation of mRNAs for insulin growth factor II and the cholesterol side–chain–cleavage enzyme, p450ssc, in human steroidgenic tissues. Proc Natl Acad Sci USA 1987; 84:1590–1594.

30 Depagter–Holthuizen P, Jansen M, Van Schaik FMA et al. The human insulin–like growth factor–II gene contains two development–specific promoters. FEBS Lett 1987; 214:259–264.

31 Gray A, Tam AW, Dull TJ et al. Tissue–specific and developmentally regulated transcription of the insulin–like growth factor 2 gene. DNA 1987; 6:283–295.

32 de Moor CH, Jansen M, Sussenbach JS et al. Differential polysomal localization of human insulin–like growth factor II mRNAs in cell lines and foetal liver. Eur J Biochem 1994; 222:1017–1024.

33 Shimasaki S, Ling N. Identification and molecular characterization of insulin–like growth factor binding protein (IGFBP1–6). Prog Growth Factor Res 1991; 3:243–266.

34 Andress DL, Birnbaum RS. Human osteoblast–derived IGFBP–5 stimulates osteoblast mitogenesis and potentiates IGF action. J Biol Chem 1992; 267:22467–22472.

35 Steele–Perkins G, Turner J, Edman JC et al. Expression and characterization of a functional human insulin–like growth factor I receptor. J Biol Chem 1988; 263:11486–11492.
36 McDonald RG, Pfeffer SR, Coussens L et al. A single receptor binds both insulin–like growth factor II and mannose 6–phosphate. Science 1988; 239:1134–1137.
37 Rechler MM, Zapf J, Nissley SP et al. Interactions of insulin–like growth factors I and II and multiplication–stimulating activity with receptors and serum carrier proteins. Endocrinology 1980; 107:1451–1459.
38 Roth RA, Mesirow ML, Yokono K et al. Degradation of insulin growth factors I and II by a human insulin degrading enzyme. Endocr Res 1984; 10:101–112.
39 Duckworth WC. Insulin metabolism and degradation. Endocr Reviews 1988; 9(3):319–345.
40 Authier F, Rachubinski RA, Posner BI et al. Endosomal proteolysis of insulin by an acid thiol metalloprotease unrelated to insulin degrading enzyme. J Biol Chem 1994; 269(4):3010–3016.
41 Sell C, Baserga R, Rubin R. Insulin–like growth factor I (IGF–I) and the IGF–I receptor present etoposide–induced apoptosis. Cancer Res 1995; 55(2):303–312.
42 Wu X, Fan Z, Masui H et al. Apoptosis induced by an anti–epidermal growth factor receptor monoclonal antibody in a human colorectal carcinoma cell line and its delay by insulin. J Clin Invest 1995; 95(4):1897–1905.
43 Chun SY, Billig H, Tilly JL et al. Gonadotrophin suppression of apoptosis in cultured preovulatory follicles: mediatory role of endogenous insulin–like growth factor I. Endocrinology 1994; 135(5):1845–1853.
44 Resnicoff M, Abraham D, Yutanawiboochai W et al. The insulin–like growth factor I receptor protects tumor cells from apoptosis in vivo. Cancer Res 1995; 55(1):2463–2469.
45 Baserga R. The insulin–like growth factor I receptor: a key to tumor growth? Cancer Res 1995; 55(2):249–252.
46 Jones JI, Clemmons DR. Insulin–like growth factors and their binding proteins: biological actions. Endocrine Reviews 1995; 16(1):3–34.
47 Redfern CPF, Lovat PE, Malcolm AJ et al. Gene expression and neuroblastoma cell differentiation in response to retinoic acid: differential effects of 9–*cis* and all–*trans* retinoic acid. Eur J Cancer 1995; 31A:486–494.
48 Piacentini M, Davies PJA, Fesus L. In: Tomei LD, Cope FO, eds. Apoptosis: The Molecular Basis of Apoptosis in Disease. Cold Spring Harbor Laboratory Press, New York 1994:143–163.
49 Wyllie AH, Kerr JFR, Currie AR. Cell death the significance of apoptosis. Int Rev Cytol 1980; 68:251–306.
50 Fesus L, Thomazy V, Autuori F et al. Hepatocytes undergoing programmed death become insoluble in detergent and chaotropic agents. FEBS Lett 1989; 245:150–154.
51 Savill JS, Dransfield I, Hogg C et al. Vitronectin receptor–mediated phagocytosis of cells undergoing apoptosis. Nature 1990; 343:170–173.
52 Dini L, Autuori F, Lentini A et al. The clearance of apoptotic cells in the liver is mediated by the asialoglycoprotein receptor. FEBS Lett 1992; 296:174–178.
53 Yonish–Rouach E, Resnitzky D, Lotem J et al. Wild–type p53 induces apoptosis of myeloid leukemic cells that is inhibited by interleukin–6. Nature 1991; 352:345–348.
54 Evan GI, Wyllie AH, Gilbert CS et al. Induction of apoptosis in Fibroblasts by c–*myc* protein. Cell 1992; 69:119–128.
55 Yuan J, Shaham S, Ledoux S et al. The *C. elegans* cell death gene *ced–3* encodes a protein similar to mammalian interleukin–1b–converting enzyme. Cell 1993; 75:641–652.
56 Nicholson DW, Ali A, Thornberry NA et al. Identification and inhibition of the ICE/CED–3 protease necessary for mammalian apoptosis. Nature 1995; 376:37–43.
57 Ellis RE, Yuan J, Horvitz HR. Mechanism and functions of cell death. Annu Rev Cell Biol 1991; 7:663–698.
58 Hockenbery D M, Zutter W, Hickey M et al. Bcl–2 protein is an inner mitochondrial membrane protein that blocks topographically programmed cell death. Nature 1990; 348:334–336.
59 Hockenbery D, Nunez G, Milliman C et al. Bcl–2 protein is topographically restricted in tissues characterized by apoptotic cell death. Proc Natl Acad Sci USA 1991; 88:6891–6891.
60 Vaux DL. A boom for necrobiology. Current Biol 1993; 3:877–888.
61 Nagata S, Golstein P. The Fas death factor. Science 1995; 267:1445–1449.

62 Eastman A. Deoxyribonuclease II in apoptosis and the significance of intracellular acidification. Cell Death Diff 1994; 1:7–9.
63 Hughes FM, Cidlowski JA. Apoptotic DNA degradation: evidence for novel enzymes Cell Death Diff 1994; 1:11–17.
64 Kumar S, Kinoshita M, Noda M et al. The induction of apoptosis by mouse Nedd2 gene, which encodes a protein similar to the product of the *Caenorhabditis elegans* cell death gene *ced–3* and the mammalian IL–1 bete converting enzyme. Genes Dev 1994; 8(14):1613–1626.
65 Melino G, Annicchiarico–Petruzzelli M, Pireda L et al. "Tissue" transglutaminase and apoptosis: sense and antisense transfection studies in human neuroblastoma cells. Mol Cell Biol 1994; 14(10): 6584–6596.
66 Thiele CJ, Reynolds CP, Israel MA. Decreased expression of N–*myc* precedes retinoic acid–induced morphological differentiation of human neuroblastoma. Nature 1985; 313:404–406.
67 Hanada M, Krajewski S, Tanaka S et al. Regulation of Bcl–2 oncoprotein levels with differentiation of human neuroblastoma cells. Cancer Res 1993; 53:4978–4986.
68 Guddas LJ, Sporn MB, Roberts AB. Cellular biology and biochemistry of the retinoids. In: Sporn MB, Roberts AB, Goodman DS, eds. The Retinoids–Biology, Chemistry and Medicine. New York: Raven Press, 1994:443–520.
69 Redfern CPF. Retinoic acid receptors. Pathobiology 1992; 60:254–263.
70 Manglesdorff DJ, Evans RM. Vitamin A receptors, new insights on retinoid control of transcription. In:G.Morriss–Kay, ed. Retinoids in normal development and teratogenesis. Oxford University Press. 1992; 27–50.
71 Napoli JL. Biosynthesis ad synthesis of retinoic acid: current roles for cellular retinol binding protein (CRBP) and cellular retinoic acid binding protein (CRABP) in retinoic acid homeostasis. J Nutrition 1993; 362:30–36.
72 Melino G, Knight RA, Thiele CJ. New insight in the biology of neuroectodermal tumors. Cancer Res 1993; 53:926–929.
73 Lovat PE, Pearson ADJ, Malcolm A et al. Retinoic acid receptor expression during the in vitro differentiation of human neuroblastoma. Neurosci Lett 1993; 162:109–113.
74 Fesus L, Tarsca E, Kedei N et al. Degradation of cells dying by apoptosis leads to accumulation of ε(γ–glutamyl) lysine isodipeptide in culture fluid and blood. FEBS Lett 1991; 284:109–112.
75 Gentile V, Thomazy V, Piancentini M et al. Expression of tissue transglutaminase in BALB–C 3T3 fibroblasts: effects on cellular morphology and adhesion. J Cell Biol 1992; 119:463–474.
76 Achyuthan KE, Greenberg CS. Identification of a guanosine triphosphate–binding site on guinea pig liver transglutaminase: role of GTP and calcium ions in modulating activity. J Biol Chem 1987; 262:1901–1906.
77 Nakaoka H, Perez DM, Baek KJ et al. Gαh: A GTP–binding protein with transglutaminase activity and receptor signaling function. Science 1994; 264:1593–1596.
78 McConkey DJ, Orrenius S, Jondal M. Cellular signaling in programmed cell death (apoptosis). Immunol Today 1990; 11:120–121.
79 Squier MK, Cohen JJ. Calpain and cell death. Cell Death Differ 1996; 3: in press.
80 Eithan S, Soloman A, Lavie V et al. Recovery of visual response of injured adult rat optic nerves treated with transglutaminase. Science 1994; 264:1764–1768.
81 Montague JW, Gaido ML, Frye, C et al. A calcium–dependent nuclease from apoptotic rat thymoctyes is homologous with cyclophillin. J Biol Chem 1994; 269:18877–18880.
82 Arends MJ, Wyllie AH. Apoptosis. Mechanism and role in pathology. Int Rev Exp Pathol 1991; 32:223–254.
83 Iwata M, Koshinaga T, Okabe I et al. Biological characteristics of neuroblastoma with spontaneous tumor reduction: a case report. J Pediatr Surg 1995; 30:722–723.
84 Koizumi H, Wakisaka M, Nakada K et al. Demonstration of apoptosis in neuroblastoma and its relationship to tumor regression. Virchows Arch 1995; 427:167–173.
85 Riboni L, Prinetti A, Bassi R et al. A mediator role of ceramide in the regulation of neuroblastoma Neuro2a cell differentiation. J Biol Chem 1995; 270:26868–26875.
86 Ikeda H, Hirato J, Akami M et al. Bcl–2 oncoprotein expression and apoptosis in neuroblatoma. J Pediatr Surg 1995; 30:805–808.

87 Ramani P, Lu–QL. Expression of *bcl–2* gene product in neuroblastoma. J Pathol 1994; 172:273–238.
88 Hoehner JC, Hedborg F, Wiklund HJ et al. Int J Cancer 1995; 62:19–24.
89 Frish SM, Franceis H. Disruption of epithelial cell–matrix interactions induces apoptosis. J Cell Biol 1994; 214:619–626.
90 Meredith JE, Fazeli B, Schwartz MA. The extracellular matrix as a cell survival factor. Mol Biol Cell 1993; 13(4):953–961.
91 Miura M, Zhu H, Rotello R et al. Induction of apoptosis in fibroblasts by IL–1β–converting enzyme, a mammalian homolog of the *C. elegans* cell death gene *ced–3*. Cell 1993; 75:653–660.
92 Lasorella A, Iavarone A, Israel MA. Differentiation of neuroblastoma enhances Bcl–2 expression and induces alterations of apoptosis and drug resistance. Cancer Res 1995; 55:4711–4716.
93 Cece R, Barajon I, Tredici G. Cisplatin induces apoptosis in SH SY 5Y human neuroblastoma cell line. Anticancer Res 1995; 15:777–782.
94 Nuydens R, De Jong M, Nuydens R et al. Neuronal kinase stimulation leads to aberrant tau phosphorylation and neurotoxicity. Neurobiol Aging 1995; 16:564–475.
95 Piacentini M, Fesus L, Melino G. Multiple cell cycle to the apoptotic death program in human neuroblastoma cells. FEBS Letters 1993; 320:150–154.
96 Slingerland RJ, Van Gennip AH, Bodlaender JM et al. The effect of cyclopentenyl cytosine on human SK–N–BE(2)–C neuroblastoma cells. Biochem Pharmacol 1995; 50:277–279.
97 Candi E, Melino G, De–Laurenzi V et al. Tamoxifen and somatostatin affect tumors by inducing apoptosis. Cancer Lett 1995; 96:141–145.
98 Dole MG, Jasty R, Cooper MJ et al. Bcl–xL is expressed in neuroblastoma cells and modulates chemotherapy–induced apoptosis. Cancer Res 1995; 55:2576–2582.
99 Tally AK, Dewhurst S, Perry SW et al. Tumor necrosis factor alpha–induced apoptosis in human neuronal cells: protection by the antioxidant N–acetylcysteine and the genes *bcl–2* and *crmA*. Mol Cell Biol 1995; 15:2359–2366.
100 Hartsell TL, Yalowich JC, Ritke MK et al. Induction of apoptosis in murine and human neuroblastoma cell lines by the enediyne natural product neocarzinostatin. J Pharmacol Exp Ther 1995; 275:479–485.
101 Di–Vinci A, Geido E, Infusini E et al. Neuroblastoma cell apoptosis induced by the synthetic retinoid N–(4–hydroxyphenyl) retinamide. Int J Cancer 1994; 59:422–426.

Chapter 13

Apoptosis and Cancer, edited by Seamus J. Martin.

TNF–Based Strategies for Manipulating Apoptosis: Adjuncts to Cancer Therapy

Grace H. W. Wong,[a] Gordon Vehar[a] and Roger L. Kaspar

[a] Department of Molecular Oncology, Genentech Inc., South San Francisco, California, U.S.A.

[b] Department of Chemistry and Biochemistry, Brigham Young University, Provo, Utah, U.S.A.

Introduction

Tumor necrosis factor (TNF) and lymphotoxin (LT, also called TNF–β [1] or LT–α [2]) are closely related multipotent cytokines that are produced predominantly by activated macrophages and lymphocytes, respectively [3]. Recently, a TNF alpha–converting enzyme (TACE), which converts physiologically–inactive proTNF into the biologically active form, was isolated and cloned [4, 5]. Although LT and TNF share many overlapping activities, there are distinct and important differences. Unlike TNF, LT may not be induced during septic shock since TNF monoclonal antibodies can effectively block LPS–induced sepsis [6]. Thus, either LT is not induced during sepsis or LT is not involved in toxic shock [6]. Similarly, TNF but not LT, is involved in rheumatoid arthritis.[7]

Both TNF and LT bind the p55 and p75 TNF receptors eliciting similar, but not identical, biological effects [3, 8–12]. The differences in activity might be explained by differences in trimer stability and/or distinct interactions with the p55 and p75 receptors [13]. Human TNF appears to be more toxic than human LT in mice [11]. Since human TNF and human LT bind only the p55 receptor in these animals, human TNF must activate the p55 receptor differently from human LT. Alternatively, the decreased toxicity of LT may be due to LTß receptor binding via a $LTaß_2$ heterotrimer [14]. Unlike TNF, LT does not exist as a transmembrane homotrimer form that can bind to TNF receptors [15]. The membrane–bound form of TNF is more effective than soluble TNF (and possibly LT) in activating the p75 receptor and can trigger stronger and/or different responses.

Human Colo205 tumor cells are resistant to soluble TNF but are sensitive to killing by membrane–bound TNF, possibly through increased activation of p75 [16]. Experiments are underway to examine whether mice lacking membrane–bound TNF are less sensitive to TNF killing than their wild–type counterparts (G. Kollias and G. Wong, unpublished).

Most cells express TNF receptors [1, 17]. The extracellular domains of the p55 and p75 receptors share some homology whereas the intracellular domains do not, supporting the idea that these receptors have nonredundant functions [18]. Agonist antibodies against the p55 receptor have been shown to enhance antiviral activity [19, 20], induce fibroblast proliferation [21], increase the production of oxygen free radicals [22] and upregulate cellular genes, such as the antioxidant enzyme manganese superoxide dismutase (MnSOD) [23, 24] and protease inhibitors [23]. On the other hand, agonist antibodies to p75 stimulate proliferation of lymphocyte [25] as well as inducing NF–κB [26]. Another function of the p75 receptor, proposed by Tartaglia and Goeddel, is to act as a carrier by binding low levels of TNF which are then "passed" to the p55 receptor [27]. TNF receptor is shed from the cell surface under some circumstances. The physiologic function of soluble TNF receptor is unknown, but it has been proposed to either neutralize TNF toxicity or act as a carrier that enhances TNF activity [28]. One way to clarify its function would be to construct transgenic mice that lack the ability to shed TNF receptor.

Knockout and transgenic mouse experiments have shown the physiological importance of LT/TNF and their receptors. LT knockout mice show abnormal development in their peripheral lymphoid organs, having no detectable lymph nodes, Peyer's patches nor germinal centers [29, 30]. Similarly, TNF knockout mice have no germinal centers; in addition, they are resistant to LPS–induced toxicity and more susceptible to *Listeria monocytogenes* infection [31]. In p55, but not p75 receptor knockout mice, peripheral lymphoid germinal centers did not develop [32]. Furthermore, the absence of p55 receptor makes mice less sensitive to endotoxic shock but more susceptible to *Listeria* infection [33, 34]. Intriguingly, p75 receptor knockout mice show decreased sensitivity to TNF killing [35]. Along these lines, it is interesting to note that mouse TNF is more toxic in mice than is human TNF; mouse TNF binds both mouse p55 and p75 whereas human TNF only binds the murine p55 receptor [17]. Taken together, these results suggest that the p75 receptor is involved in toxicity. Transgenic mice overexpressing TNF but not LT, show systemic toxicity [36]. However, mice overexpressing membrane-bound TNF develop arthritis [37]. It will be interesting to see if crossing these mice with p75 knockout mice will diminish this arthritis.

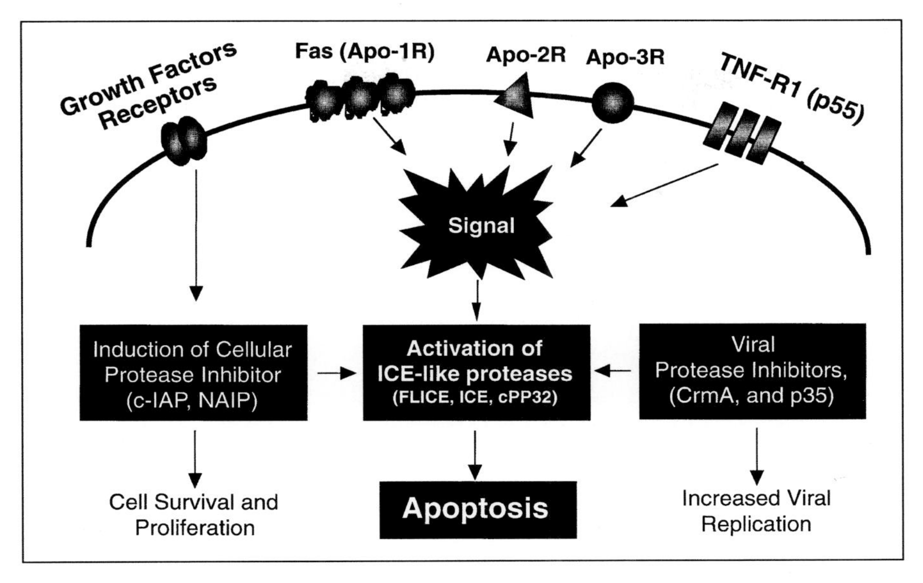

Fig. 13.1. Apoptosis is triggered by activation of TNF–R1, Fas or other unknown receptors. The pathway leading to cell death involves activation of apoptosis–associated proteases. These proteases and apoptosis can be blocked by overexpression of the viral protease inhibitors, *CrmA* and p35, and possibly by the cellular protease inhibitors, c–IAP or NAIP. We propose that cellular protease inhibitors can be induced by growth factors, such as IGF–1, GH and cytokines TNF/LT or IL–1. The growth–promoting actions of these factors are mediated in part through downregulation of apoptosis–associated protease activity and subsequent cell death.

TNF Signaling of Apoptosis

Apoptosis or programmed cell death is an essential component of physiological development and homeostasis. It is characterized by cellular shrinkage, chromatin condensation, blebbing of the membrane and inter–nucleosomal cleavage of nuclear DNA [38]. Apoptosis can be triggered by many stimuli, including the activation of certain cell surface receptors such as the TNF receptor (p55/TNF–R1) by binding of the ligand or agonist antibodies [39, 40].

The p55 receptor contains an intracellular motif (death domain) that is related to a similar cytoplasmic domain of Fas, a receptor whose sole function appears to be to mediate apoptosis. These death domains appear to be involved in protein-protein interactions and are required to transmit the death signal [41–43]. The signal transduction pathways initiated by TNF or Fas appear to activate proteases that are directly involved in cell killing (Fig. 13.1).

A number of protein factors that associate with the p55 receptor and Fas intracellular death domains have been identified using immunoprecipitation and the yeast two-hybrid system [44–48]. TRADD [44] and FADD [48] associate with the death domains of the p55 TNF receptor and Fas, respectively. Interestingly, the FADD-TRADD association is stronger than that of FADD–Fas; this association appears to be where the TNF and Fas cell death pathways intersect [47, 49]. Another protein, TRAF–2, also associates with TRADD but does not appear to be involved in apoptosis; rather, TRAF–2 transduces a signal to activate the transcription factor NF–κB [47], which has been proposed to mediate protection against oxidative stress [50–52].

Paradoxical Action of TNF/LT: Protecting Normal Tissue and Sensitizing Tumor Cells to Chemotherapy or Radiation Treatment

TNF/LT may protect normal cells from damage due to irradiation or chemotherapy by inducing the synthesis of protective proteins. Both TNF/LT increase expression of MnSOD in normal cells but not tumor cells [53, 54]. In addition, overexpression of MnSOD protects cultured cells from killing by irradiation or cytotoxic drug treatment [55, 56]. Therefore, the TNF/LT protective effect may be due to their ability to induce MnSOD and/or other protective proteins. TNF/LT pretreatment can protect mice against otherwise lethal doses of irradiation [55, 57] (Wong et al, submitted). LT pretreatment prior to irradiation enhances the recovery of platelets, erythrocytes and leukocytes, and also reduces damage to bone marrow precursor and intestinal crypt cells. Furthermore, LT protects rats against doxorubicin–mediated toxicity and hair loss induced by chemotherapy (Wong et al, submitted).

Paradoxically, TNF/LT treatment not only protects normal tissue but seems to sensitize some tumor cells to killing by irradiation or chemotherapeutics. In vitro, TNF/LT enhanced doxorubicin killing in 20 of 30 tumor cell lines but had no effect on five primary cell cultures including fibroblasts and PBMCs. In vivo, massive necrosis, induction of apoptosis and reduced tumor size occurred within the tumors of tumor–bearing nude mice pretreated with TNF/LT followed by irradiation or chemotherapy (Wong et al, submitted). Only TNF and LT appear to have the dual properties of radio– or chemoprotecting normal tissue while simultaneously radio– and chemosensitizing tumor cells (Table 13.1). Although interleukin–1 (IL–1) [58] and stem cell factor [59] can also radioprotect mice,

Table 13.1. In vivo radioprotective and radiosensitization effects of various cytokines

Pretreatment	MnSOD Induction	Radioprotector in Mice	Sensitize Tumor to Radiation
IL–1	+	+	–
TNF	+	+	+++
LT	+	+	+++
Stem cell factor	–	+	–
GM–CSF	–	–	–
IFN–γ	–	–	–
IL–2/IL–3/IL–6	–	–	–

Comparison of cytokine induction of MnSOD with its ability to radioprotect normal tissue while simultaneously radiosensitizing tumor cells. BALB/c (8 week old) female mice were injected with murine Meth–A fibrosarcoma cells (2 x 10^5 cells/mouse), and the tumors were then allowed to grow for 14 days and mice bearing similar sized tumors were randomly assigned (10 mice/group) to different treatment groups. These tumor bearing mice were pretreated with either IL–1, LT, TNF, stem cell factor, GM–CSF, IFN–γ, IL–2, IL–3, IL–6 or PBS on two consecutive days prior to 750 cGy whole–body irradiation using 137 Cs γ–rays. Survival of the mice was measured daily. Tumor size was measured twice per week. The minimum lethal dose for BALB/c mice (100% die within 30 days of irradiation) was approximately 750 cGy. — is designated as no induction of MnSOD (< 1.5–fold), no radioprotection (< 20% of mice survived, compared to 0% of control mice surviving) or no tumor sensitization (no decrease in tumor size). + is designated as a detectable positive response in MnSOD expression (with 2– to 100–fold induction), a dramatic increase in radioprotection and a significant decrease in tumor size.

they have no tumor–sensitizing ability. Other cytokines such as IL–2, IL–3, IL–6, and IFNγ have neither protective nor tumor sensitizing effects. Understanding the mechanism(s) of how TNF/LT protects normal cells while sensitizing tumor cells will enable the discovery of better therapeutic agents.

Possible Mechanisms of LT–Protection Against Oxidative Stress: Induction of Protective Proteins

Since overexpression of MnSOD increases protection against oxidative stress [55, 56], and as its expression can be induced by TNF/LT [53], we believe that MnSOD appears to be one of the proposed protective proteins [53]. In addition to MnSOD, other potential protective proteins may include Bcl–2 [60], A20 [61], heat shock proteins [62, 63], p53, p21, ferritin [64], DNA repair enzymes including PARP [65], metallothionein [66, 67] and protease inhibitors [68, 69].

We also speculate that growth factors and cytokines can induce higher levels of protease inhibitors which may be essential for normal inhibition of apoptosis (Fig. 13.1). This may explain why deprivation of growth factors leads to cell death. Indeed, growth hormone and TNF have been shown to induce protease inhibitors [70, 71]. We have recently observed that pretreatment of human A549 lung carcinoma cells with growth hormone (GH), TNF, or IL–1 can protect these cells from H_2O_2–mediated apoptosis (G.H.W. Wong et al, unpublished data). Other growth factors such as IL–2, IL–3, EGF, IGF–1 and NGF, may also be capable of inducing the expression of these protease inhibitors in their specific target cells and thereby inhibiting the activation of proteases as well as apoptosis.

Interestingly, the two known viral protease inhibitors, the CrmA gene product of the cowpox virus [72] and the p35 baculovirus protein [73, 74] can block TNF–, Fas–, APO–2L– [75] and possibly DR3–[76] mediated apoptosis (Fig. 13.1), presumably through inhibition of apoptosis–associated proteases. In spite of intensive effort, the receptor for APO–2L (also called TRAIL) [77] has not been cloned nor has the ligand for DR3 [76]. The identification of which would shed light on whether these molecules signal through the same or distinct pathways as TNF and Fas ligand.

Potential Mechanisms for Tumor Sensitization: Activation of Apoptosis–Associated Proteases

Proteases such as interleukin 1β converting enzyme (ICE) and other ICE–like proteases (caspases) [78], including CPP32 [79] and MORT [45]/FLICE [46], have been, shown to be involved in apoptosis. These ICE–like proteases (or caspases) are specific ASPases that are produced as inactive proenzymes. Several lines of evidence implicate ICE–like proteases in apoptosis. First, they are homologous to the well–defined *ced–3* cell death gene, which is required for proper apoptotic cell death during *C. elegans* development [80]. Second, antisense RNA to ICE [81], as well as specific peptide substrate inhibitors of ICE and CPP32 [82–84] can block apoptosis initiated by Fas and TNF. Third, overexpression of each of the ICE–like proteases as well as the Fas/TNF receptor–associated proteins TRADD [44], FADD [48] and RIP [85, 86] leads to cell death. Fourth, purified ICE or CPP32 causes nuclear DNA fragmentation in vitro [87]. Fifth, viral protease inhibitors such as the cowpox gene product *CrmA* [88, 89] and baculovirus p35 [73, 74] simultaneously block ICE–like protease activity and TNF/Fas–mediated apoptosis.

Enari and coworkers hypothesized that several ICE–like proteases are activated sequentially [87]. ICE has been shown to activate ProYAMA to its active form, whereas CPP32 is unable to activate ICE [90]. These studies suggest that

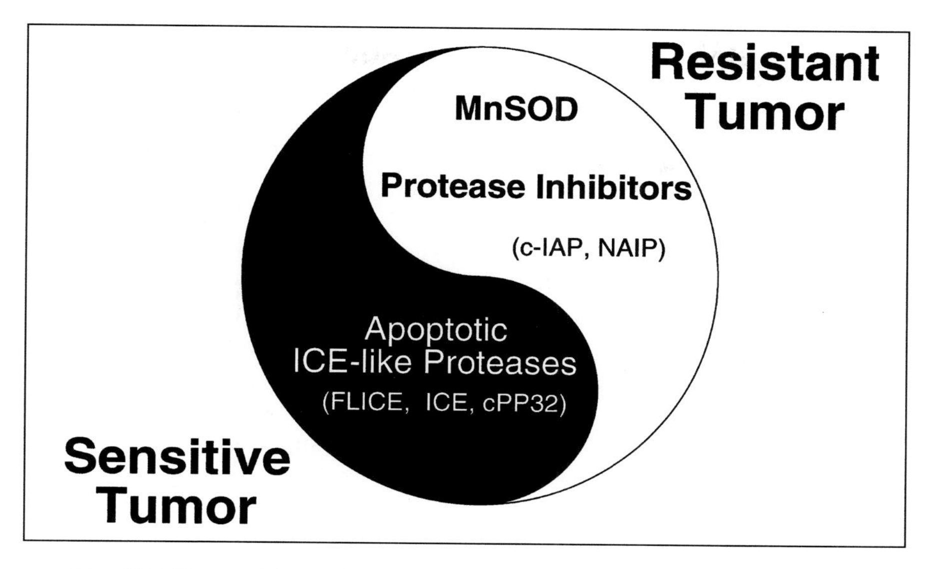

Fig. 13.2. The sensitivity of tumor cells to killing by irradiation or chemotherapy depends on a delicate balance between "killing" (FLICE, ICE, CPP32) and "protective" (e.g., MnSOD and/or protease inhibitors) proteins. Some cancer cells seem to have lost the ability to make protective proteins such as MnSOD and consequently are more susceptible to killing by irradiation or chemotherapy. Alternatively, protease–deficient tumor cells may be resistant due to an inability to activate the apoptosis pathway. Understanding how to modulate this balance may lead to new treatments for disease including cancer.

ICE is upstream of CPP32. Additional substrates of ICE–like proteases have been identified as poly(ADP–ribose) polymerase (PARP) [90, 91]. U1 ribonuclear protein [92–94], nuclear lamins [95], the catalytic subunit of the DNA–dependent protein kinase (DNA–PKcs) [94] and cytoskeletal proteins [96].

We propose that one of the reasons why TNF/LT can increase tumor sensitivity to killing by irradiation or chemotherapy is due to activation of these proteases (Fig. 13.2). We also propose that levels of these apoptosis–associated proteases (e.g., FLICE, ICE and CPP32) are high in TNF–sensitive tumor cells (Fig. 13.2). Opposite results are observed with MnSOD and protease inhibitors such as c–IAP and NAIP (Fig. 13.2). The balance of protective versus killing proteins may determine the outcome of cellular sensitivity to radiation or chemotherapy. On the basis of this hypothesis, small molecules specific for the activation of proteases or inhibition of MnSOD or protease inhibitors should be screened as potential therapeutic candidates for tumor sensitization.

Understanding how apoptosis is regulated may provide leads into several potential clinical applications. First, specific activation of the p55 TNF receptor can directly kill cancer cells [39]. Second, TNF/LT can be used as adjuncts to cancer therapy: chemo– or radioprotection of normal cells while sensitizing tumor cells. Third, TNF/LT have antiviral activity; they protect uninfected cells from viral infection while selectively killing cells infected with virus [20]. Some viruses have developed protease inhibitors, such as CrmA or p35, that inhibit apoptosis and allow viral replication [97–99] and agents that block viral inhibitors may act to decrease viral replication.

Cellular inhibitors of apoptosis (c–IAP) have first been identified by Goeddel and coworkers and are related to the p35 protein that interacts with the TRAF/p75 receptor complex [100]. We speculate that c–lAPs may play an essential role in inhibiting apoptosis. In support of this notion, we have found that TNF and IL–1 can induce the expression of c–IAP and the induction is further enhanced by cycloheximide (Wong et al, in preparation). In addition to c–IAP, a neuronal inhibitor of apoptosis (NAIP) which shares some homology with c–IAP has also been isolated [101]. Interestingly, mutations in NAIP appear to lead to spinal muscular atrophy [101]. Therefore, modulating the expression of c–IAP or NAIP may be also useful for disease intervention.

Understanding the genes involved in apoptosis may allow for screening of new diagnostic markers predicting cancer risk and/or identifying tumors that will respond to various cancer treatments. Since TNF/LT sensitive tumor cells have uninducible MnSOD protein levels, this and/or other protective proteins may serve as clinical surrogate markers to determine the likely usefulness of chemotherapy or radiation therapy for a particular tumor. Tumors that are unable to produce MnSOD following TNF/LT treatment might be suitable candidates for radiation or chemotherapy. If our hypothesis (that tumor sensitization by TNF/LT is due to activation of proteases) is correct, strategies could be designed to activate the proteases in order to enhance tumor sensitivity for cancer therapy.

Summary and Conclusions

TNF/LT are potent cytokines that induce complex and paradoxical responses. TNF/LT pretreatment can protect normal tissue from radiation– or chemotherapy–induced damage, while at the same time sensitizing tumor cells. The presence of protective proteins in normal cells following TNF/LT treatment seems to prevent damage from radiation or chemotherapy. Understanding the balance of "killing" and "protective" proteins, may be useful to modulate effective cancer treatments (Fig. 13.2). When this balance is disrupted, as with tumor cells which cannot produce protective proteins, the induction of killing proteins by radiation or chemotherapy may lead to the cell's demise. The ability to manipulate the balance of protective versus killing proteins may lead to novel approaches in cancer therapy.

Acknowledgments

We thank the manufacturing group at Genentech for providing pure recombinant human cytokines. We thank Cathy Carlson, Sharon Fong and Warren Young for technical assistance, and Nora Kolthoff for art work. R.L.K. is grateful for support from the Bireley Foundation and the BYU Cancer Center. G.H.W. thanks Alexander David Kamb, Jim Quinn and Boobam Oye for support.

References

1 Goeddel DV, Aggarwal BB, Gray PW et al. Tumor necrosis factors: gene structure and biological activities. Cold Spring Harb Symp Quant Biol 1986; 1:597–609.
2 Browning JL, Ngam–ek A, Lawton P et al. Lymphotoxin beta, a novel member of the TNF family that forms a heteromeric complex with lymphotoxin on the cell surface. Cell 1993; 72:847–56.
3 Beutler B, ed. Tumor Necrosis Factors: The Molecules and Their Emerging Role in Medicine. New York: Raven Press, 1992.
4 Moss M, Jin C, Becherer D et al. Structural features and biochemical properties of TNF–alpha convertase. European Cytokine Network 1996; 7:181.
5 Black R, Rauch C, Kozlosky C et al. Cloning of a membrane–bound TNF alpha converting enzyme (TACE). European Cytokine Network 1996; 7:180.
6 Beutler B, Milsark IW, Cerami AC. Passive immunization against cachectin/tumor necrosis factor protects mice from lethal effect of endotoxin. Science 1985; 229:869–871.
7 Brennan FM, Maini RN, Feldmann M. TNF alpha—a pivotal role in rheumatoid arthritis. Brit J Rheumatol 1992; 31:293–298.
8 Locksley RM, Heinzel FP, Shepard HM et al. Tumor necrosis factors alpha and beta differ in their capacities to generate interleukin 1 release from human endothelial cells. J Immunol 1987; 139:1891–1895.
9 Holt SJ, Grimble RF, York DA. Tumor necrosis factor–alpha and lymphotoxin have opposite effects on sympathetic efferent nerves to brown adipose tissue by direct action in the central nervous system. Brain Res 1989; 497:183–186.
10 Porter AG. Human tumor necrosis factors–alpha and –beta: differences in their structure, expression and biological properties. Fems Microbiol Immunol 1990; 2:193–199.
11 Qin Z, van Tits LJ, Buurman WA et al. Human lymphotoxin has at least equal antitumor activity in comparison to human tumor necrosis factor but is less toxic in mice. Blood 1995; 85:2779–2785.
12 Medvedev AE, Espevik T, Ranges G et al. Distinct roles of the two tumor necrosis factor (TNF) receptors in modulating TNF and lymphotoxin alpha effects. J Biol Chem 1996; 271:9778–9784.
13 Schuchmann M, Hess S, Bufler P et al. Functional discrepancies between tumor necrosis factor and lymphotoxin alpha explained by trimer stability and distinct receptor interactions. Eur J Immunol 1995; 25:2183–2189.
14 Crowe PD, VanArsdale TL, Walter BN et al. A lymphotoxin–b–specific receptor. Science 1994; 264:707–710.

15 Ware CF, VanArsdale TL, Crowe PD et al. The ligands and receptors of the lymphotoxin system. Curr Top Microbiol Immunol 1995; 198:175–218.
16 Grell M, Douni E, Wajant H et al. The transmembrane form of tumor necrosis factor is the prime activating ligand of the 80 kDa tumor necrosis factor receptor. Cell 1995; 83:793–802.
17 Fiers W. Tumor necrosis factor characterization at the molecular, cellular and in vivo level. *FEBS Lett* 1991; 285:199–212.
18 Schall TJ, Lewis M, Koller KJ et al. Molecular cloning and expression of a receptor for human tumor necrosis factor. Cell 1990; 61:361–370.
19 Wong GH, Goeddel DV. Tumor necrosis factors a and b inhibit virus replication and synergize with interferons. Nature 1986; 323:819–822.
20 Wong GH, Tartaglia LA, Lee MS et al. Antiviral activity of tumor necrosis factor is signaled through the 55 kDa type I TNF receptor J Immunol 1992; 149:3350–3353.
21 Englemann, H, Holtmann, H, Brakebush, C et al. Antibodies to a soluble form of a tumor necrosis factor (TNF) receptor have TNF–like activity. J Biol Chem 1990; 265:14497–14504.
22 Wong GH, Kamb A, Tartaglia LA et al. Possible protective mechanisms of tumor necrosis factors against oxidative stress. In: Molecular Biology of Free Radical Scavenging Systems. Cold Spring Harbor: Cold Spring Harbor Press, 1992:69–96.
23 Tartaglia LA, Weber RF, Figari IS et al. The two different receptors for tumor necrosis factor mediate distinct cellular responses. Proc Natl Acad Sci USA 1991; 88:9292–9296.
24 Wong GHW, Kamb A, Elwell JH et al. MnSOD induction by TNF and its protective role. In: Beutler B, ed. Tumor Necrosis Factors: The Molecules and Their Emerging Role in Medicine. New York: Raven Press, 1992:473–484.
25 Tartaglia LA, Goeddel DV, Reynolds C et al. Stimulation of human T cell proliferation by specific activation of the 75–kDa tumor necrosis factor receptor. J Immunol 1993; 151:4637–4641.
26 Rothe M, Sarma V, Dixit VM et al. TRAF2–mediated activation of NF–k B by TNF receptor 2 and CD40. Science 1995; 269:1424–1427.
27 Tartaglia LA, Pennica D, Goeddel DV. Ligand passing: the 75–kDa tumor necrosis factor (TNF) receptor recruits TNF for signaling by the 55–kDa TNF receptor. J Biol Chem 1993; 268:18542–18548.
28 Aderka D, Englemann H, Hornik V et al. Increased serum levels of soluble receptors for tumor necrosis factor in cancer patients. Cancer Res 1991; 51:5602–5607.
29 De Togni P, Goellner J, Ruddle NH et al. Abnormal development of peripheral lymphoid organs in mice deficient in lymphotoxin. Science 1994; 264:703–707.
30 Mariathasan S, Matsumoto M, Baranyay F et al. Absence of lymph nodes in lymphotoxin–a (LT alpha)–deficient mice is due to abnormal organ development, not defective lymphocyte migration. J Inflamm 1995; 45:72–78.
31 Pasparakis M, Alexopoulou L, Episkopou V et al. Immune and inflammatory responses in TNF a deficient mice: a critical requirement for TNF a in germinal center formation and in the maturation of the humoral immune response. European Cytokine Network 1996; 7:239.
32 Matsumoto M, Mariathasan S, Nahm MH et al. Role of lymphotoxin and the type I TNF receptor in the formation of germinal centers. Science 1996; 271:1289–1291.
33 Rothe J, Mackay F, Bluethmann H et al. Phenotypic analysis of TNFR1–deficient mice and characterization of TNFR1–deficient fibroblasts in vitro. Circ Shock 1994; 44:51–56.
34 Pfeffer K, Matsuyama T, Kundig TM et al. Mice deficient for the 55 kd tumor necrosis factor receptor are resistant to endotoxic shock, yet succumb to L. monocytogenes infection. Cell 1993; 73:457–467.
35 Erickson SL, de Sauvage FJ, Kikly K et al. Decreased sensitivity to tumor–necrosis factor but normal T cell development in TNF receptor–2–deficient mice. Nature 1994; 372:560–563.
36 Probert L, Keffer J, Corbella P et al. Wasting, ischemia, and lymphoid abnormalities in mice expressing T cell–targeted human tumor necrosis factor transgenes. J Immunol 1993; 151:1894–1906.
37 Alexopoulou L, Pasparakis M, Kollias G. Immunoregulatory activities of transmembrane TNF revealed in transgenic and mutant mice. Eur Cytokine Network 1996; 7:228.
38 Hale AJ, Smith CA, Sutherland LC et al. Apoptosis: molecular regulation of cell death. Eur J Biochem 1996; 236:1–26.

39 Greenblatt MS, Elias L. The type B receptor for tumor necrosis factor–alpha mediates DNA fragmentation in HL–60 and U937 cells and differentiation in HL–60 cells. Blood 1992; 80:1339–46.
40 Wong GH, Goeddel DV. Fas antigen and p55 TNF receptor signal apoptosis through distinct pathways. J Immunol 1994; 152:1751–1755.
41 Tartaglia LA, Ayres TM, Wong GH et al. A novel domain within the 55 kd TNF receptor signals cell death. Cell 1993; 74:845–853.
42 Baker SJ, Reddy EP. Transducers of life and death: TNF receptor superfamily and associated proteins. Oncogene 1996; 12:1–9.
43 Boldin MP, Mett IL, Varfolomeev EE et al. Self–association of the "death domains" of the p55 tumor necrosis factor (TNF) receptor and Fas/APO1 prompts signaling for TNF and Fas/APO1 effects. J Biol Chem 1995; 270:387–391.
44 Hsu H, Xiong J, Goeddel DV. The TNF receptor 1–associated protein TRADD signals cell death and NF–k B activation. Cell 1995; 81:495–504.
45 Boldin M, Goncharov T, Goltsev Y et al. Involvement of MACH, a novel MORT1/FADD–interacting protease in Fas/APO–1– and TNF receptor–induced cell death. Cell 1996; 85:803–815.
46 Muzio M, Chinnaiyan A, Kischkel F et al. FLICE, a novel FADD–homologous ICE/CED–3–like protease, is recruited to the CD95 (Fas/APO–1) death–inducing signaling complex. Cell 1996; 85:817–827.
47 Hsu H, Shu HB, Pan MG et al. TRADD–TRAF2 and TRADD–FADD interactions define two distinct TNF receptor 1 signal transduction pathways. Cell 1996; 84:299–308.
48 Chinnaiyan AM, O'Rourke K, Tewari M et al. FADD, a novel death domain–containing protein, interacts with the death domain of Fas and initiates apoptosis. Cell 1995; 81:505–512.
49 Varfolomeev EE, Boldin MP, Goncharov TM et al. A potential mechanism of "cross–talk" between the p55 tumor necrosis factor receptor and Fas/APO1: proteins binding to the death domains of the two receptors also bind to each other. J Exp Med 1996; 183:1271–1275.
50 Beg, AA, Baltimore, D. An essential role for NF–kB in preventing TNF–a–induced cell death. Science 1996; 274:782–784.
51 Wang C–Y, Mayo MM, Baldwin Jr AS. TNF– and cancer therapy–induced apoptosis: potentiation by inhibition of NF–kB. Science 1996; 274:784–787.
52 Van Antwerp DJ, Martin SJ, Kafri T et al. Suppression of TNF–a–induced apoptosis by NF–kB. Science 1996; 274:787–789.
53 Wong GH, Goeddel DV. Induction of manganous superoxide dismutase by tumor necrosis factor: possible protective mechanism. Science 1988; 242:941–944.
54 Wong GHW. Protective roles of cytokines against radiation: Induction of mitochondrial MnSOD. Biochim Biophys Acta 1995; 1271:205–209.
55 Wong GHW, Neta R, Goeddel DV. Protective roles of MnSOD, TNF–alpha, TNF–beta and D–Factor (LIF) in radiation injury. Nigam S, ed. In: Eicosanoids and other bioactive lipids in cancer, inflammation and radiation injury. 1992:353–357.
56 Hirose K, Longo DL, Oppenheim JJ et al. Overexpression of mitochondrial manganese superoxide dismutase promotes the survival of tumor cells exposed to interleukin–1, tumor necrosis factor, selected anticancer drugs, and ionizing radiation. Faseb J 1993; 7:361–368.
57 Neta R, Oppenheim JJ, Douches SD. Interdependence of the radioprotective effects of human recombinant interleukin 1 alpha, tumor necrosis factor alpha, granulocyte colony–stimulating factor, and murine recombinant granulocyte–macrophage colony–stimulating factor. J Immunol 1988; 140:108–111.
58 Neta R, Douches S, Oppenheim JJ. Interleukin 1 is a radioprotector. J Immunol 1986; 136:2483–2485.
59 Zsebo KM, Smith KA, Hartley CA et al. Radioprotection of mice by recombinant rat stem cell factor. Proc Natl Acad Sci USA 1992; 89:9464–9468.
60 Talley AK, Dewhurst S, Perry SW et al. Tumor necrosis factor a–induced apoptosis in human neuronal cells: protection by the antioxidant N–acetylcysteine and the genes *bcl–2* and *crmA*. Mol Cell Biol 1995; 15:2359–2366.
61 Opipari A, Jr., Hu HM, Yabkowitz R et al. The A20 zinc finger protein protects cells from tumor necrosis factor cytotoxicity. J Biol Chem 1992; 267:12424–12427.

62 Mehlen P, Preville X, Chareyron P et al. Constitutive expression of human hsp27, Drosophila hsp27, or human a B–crystallin confers resistance to TNF– and oxidative stress–induced cytotoxicity in stably transfected murine L929 fibroblasts. J Immunol 1995; 154:363–374.

63 Simon MM, Reikerstorfer A, Schwarz A et al. Heat shock protein 70 overexpression affects the response to ultraviolet light in murine fibroblasts. Evidence for increased cell viability and suppression of cytokine release. J Clin Invest 1995; 95:926–933.

64 Torti S, Kwak E, Miller S et al. The molecular cloning and characterization of murine ferritin heavy chain, a tumor necrosis factor inducible gene. J Biol Chem 1988; 263:12638–12644.

65 Lichtenstein A, Gera JF, Andrews J et al. Inhibitors of ADP–ribose polymerase decrease the resistance of HER2/neu–expressing cancer cells to the cytotoxic effects of tumor necrosis factor. J Immunol 1991; 146:2052–2058.

66 Sciavolino PJ, Lee TH, Vilcek J. Overexpression of metallothionein confers resistance to the cytotoxic effect of TNF with cadmium in MCF–7 breast carcinoma cells. Lymphokine and Cytokine Res 1992; 11:265–270.

67 Wong GHW, Goeddel DV. Tumor necrosis factors: modulation of synthesis and biological activities. In: Maini J, Dornand J, eds. Lymphocyte Activation and Differentiation. New York: Walter de Gruyter, 1988:217–226.

68 Kumar S, Baglioni C. Protection from tumor necrosis factor–mediated cytolysis by overexpression of plasminogen activator inhibitor type–2. J Biol Chem 1991; 266:20960–20964.

69 Dickinson JL, Bates EJ, Ferrante A et al. Plasminogen activator inhibitor type 2 inhibits tumor necrosis factor alpha–induced apoptosis: Evidence for an alternate biological function. J Biol Chem 1995; 270:27894–27904.

70 Yoon JB, Towle HC, Seelig S. Growth hormone induces two mRNA species of the serine protease inhibitor gene family in rat liver. J Biol Chem 1987; 262:4284–4289.

71 Medcalf RL, Kruithof EK, Schleuning WD. Plasminogen activator inhibitor 1 and 2 are tumor necrosis factor/cachectin–responsive genes. J Exp Med 1988; 168:751–759.

72 Tewari M, Telford WG, Miller RA et al. CrmA, a poxvirus–encoded serpin, inhibits cytotoxic T–lymphocyte–mediated apoptosis. J Biol Chem 1995; 270:22705–22708.

73 Beidler DR, Tewari M, Friesen PD et al. The baculovirus p35 protein inhibits Fas– and tumor necrosis factor–induced apoptosis. J Biol Chem 1995; 270:16526–16528.

74 Bump NJ, Hackett M, Hugunin M et al. Inhibition of ICE family proteases by baculovirus anti–apoptotic protein p35. Science 1995; 269:1885–1888.

75 Pitti RM, Marsters SA, Ruppert et al. Induction of apoptosis by Apo–2 ligand, a new member of the tumor necrosis factor cytokine family. J Biol Chem 1996; 271:12687–12690.

76 Chinnaiyan AM, O'Rourke K, Yu G–L et al. Signal transduction by DR3, a death domain–containing receptor related to TNFR–1 and CD95. Science 1996; 274:990–992.

77 Wiley, SR, Schooley, K, Smolak, PJ. Identification and characterization of a new member of the TNF family that induces apoptosis. Immunity 1995; 3:673–682.

78 Fraser A, Evan G. A license to kill. Cell 1996; 85:781–784.

79 Fernandes–Alnemri T, Litwack G, Alnemri ES. CPP32, a novel human apoptotic protein with homology to *Caenorhabditis elegans* cell death protein *ced–3* and mammalian interleukin–1 b–converting enzyme. J Biol Chem 1994; 269:30761–30764.

80 Miura M, Zhu H, Rotello R et al. Induction of apoptosis in fibroblasts by IL–1 b–converting enzyme, a mammalian homolog of the *C. elegans* cell death gene *ced–3*. Cell 1993; 75:653–660.

81 Los M, Van de Craen M, Penning LC et al. Requirement of an ICE/CED–3 protease for Fas/APO–1–mediated apoptosis. Nature 1995; 375:81–83.

82 Enari M, Hug H, Nagata S. Involvement of an ICE–like protease in Fas–mediated apoptosis. Nature 1995; 375:78–81.

83 Nicholson DW, Ali A, Thornberry NA et al. Identification and inhibition of the ICE/CED–3 protease necessary for mammalian apoptosis. Nature 1995; 376:37–43.

84 Schlegel J, Peters I, Orrenius S et al. CPP32/apopain is a key interleukin 1β converting enzyme–like protease involved in Fas–mediated apoptosis. J Biol Chem 1996; 271:1841–1844.

85 Stanger BZ, Leder P, Lee TH et al. RIP: a novel protein containing a death domain that interacts with Fas/APO–1 (CD95) in yeast and causes cell death. Cell 1995; 8145:513–523.

86 Hsu H, Huang J, Shu HB et al. TNF–dependent recruitment of the protein kinase RIP to the TNF receptor–1 signaling complex. Immunity 1996; 4:387–396.
87 Enari M, Talanian RV, Wong WW et al. Sequential activation of ICE–like and CPP32–like proteases during Fas–mediated apoptosis. Nature 1996; 380:723–726.
88 Miura M, Friedlander RM, Yuan J. Tumor necrosis factor–induced apoptosis is mediated by a CrmA–sensitive cell death pathway. Proc Natl Acad Sci USA 1995; 92:8318–8322.
89 Tewari M, Dixit VM. Fas– and tumor necrosis factor–induced apoptosis is inhibited by the poxvirus crmA gene product. J Biol Chem 1995; 270:3255–3260.
90 Tewari M, Quan LT, O'Rourke K et al. Yama/CPP32 beta, a mammalian homolog of CED–3, is a CrmA–inhibitable protease that cleaves the death substrate poly(ADP–ribose) polymerase. Cell 1995; 81:801–809.
91 Lazebnik YA, Kaufmann SH, Desnoyers S et al. Cleavage of poly(ADP–ribose) polymerase by a proteinase with properties like ICE. Nature 1994; 371:346–347.
92 Casciola–Rosen LA, Miller DK, Anhalt GJ et al. Specific cleavage of the 70–kDa protein component of the U1 small nuclear ribonucleoprotein is a characteristic biochemical feature of apoptotic cell death. J Biol Chem 1994; 269:30757–30760.
93 Tewari M, Beidler DR, Dixit VM. CrmA–inhibitable cleavage of the 70–kDa protein component of the U1 small nuclear ribonucleoprotein during Fas– and tumor necrosis factor– induced apoptosis. J Biol Chem 1995; 270:18738–18741.
94 Casciola–Rosen L, Nicholson DW, Chong T et al. Apopain/CPP32 cleaves proteins that are essential for cellular repair: a fundamental principle of apoptotic death. J Exp Med 1996; 183:1957–1964.
95 Lazebnik YA, Takahashi A, Moir RD et al. Studies of the lamin proteinase reveal multiple parallel biochemical pathways during apoptotic execution. Proc Natl Acad Sci USA 1995; 92:9042–9046.
96 Mashima T, Naito M, Fujita N et al. Identification of actin as a substrate of ICE and an ICE–like protease and involvement of an ICE–like protease but not ICE in VP–16–induced U937 apoptosis. Biochem Biophys Res Commun 1995; 217:1185–1192.
97 Duke RC. Apoptosis in cell–mediated immunity. In: Tomei LD, Cope FO, eds. Apoptosis. The Molecular Basis of Cell Death. Cold Spring Harbor: Cold Spring Harbor Laboratory Press, 1991:209–226.
98 Thompson CB. Apoptosis in the pathogenesis and treatment of disease. Science 1995; 267:1456–1462.
99 Ray CA, Black RA, Kronheim SR et al. Viral inhibition of inflammation: cowpox virus encodes an inhibitor of the interleukin–1 beta converting enzyme. Cell 1992; 69:597–604.
100 Rothe M, Pan MG, Henzel WJ et al. The TNFR2–TRAF signaling complex contains two novel proteins related to baculoviral inhibitor of apoptosis proteins. Cell 1995; 83:1243–1252.
101 Liston P, Roy N, Tamai K et al. Suppression of apoptosis in mammalian cells by NAIP and a related family of IAP genes. Nature 1996; 379:349–353.

Subject Index